ERRATA

STATISTICAL DYNAMICS
A STOCHASTIC APPROACH TO NONEQUILIBRIUM THERMODYNAMICS

R. F. STREATER

p. 4, line 10:	replace "$\log(\omega)$" by "$\log p(\omega)$".
p. 185, (i) line 6:	should be: "$\ldots = (n!)^{-1/2} a^{*n} \psi_0 = (n!)^{-1/2} \frac{\partial^n}{\partial \overline{z}^n} \exp\{a^* \overline{z}\} \psi_0 \big|_{\overline{z}=0}$".
(ii) line 8:	insert "$\exp(az)\exp(a^*\overline{z})$" at the end of the R.H.S.
(iii) line 9:	"$n \exp\{-\beta\varepsilon\}$" should be "$\exp\{-n\beta\varepsilon\}$".
p. 249, line −3:	in Eq. (13.7), "$-2\lambda J$" should be "$+2\lambda J$".
p. 250, line −13:	"Exercise (13.22)" should be "Exercise (13.25)".
p. 251, line −9:	"Exercise (13.24)" should be "Exercise (13.27)".
p. 252, (i) line 1:	"Exercise (13.25)" should be "Exercise (13.28)".
(ii) line 2:	insert "if $\beta > 0$".

STATISTICAL DYNAMICS

A Stochastic Approach to Nonequilibrium Thermodynamics

STATISTICAL DYNAMICS

A Stochastic Approach to Nonequilibrium Thermodynamics

R. F. Streater

Department of Mathematics
King's College
London

ICP Imperial College Press

Distributed by

World Scientific Publishing Co Pte Ltd
P O Box 128, Farrer Road, Singapore 9128
USA office: Suite 1B, 1060 Main Street, River Edge, NJ 07661
UK office: 57 Shelton Street, Covent Garden, London WC2H 9HE

British Library Cataloguing-in-Publication Data
A catalogue record for this book is available from the British Library.

Library of Congress Cataloging-in-Publication Data
Streater, R. F.
Statistical dynamics : a stochastic approach to nonequilibrium thermodynamics / R. F. Streater.
p. cm.
ISBN 1-86094-002-1 ISBN 1-86094-004-8 (pbk)
1. Nonequilibrium thermodynamics -- Statistical methods.
2. Stochastic processes. I. Title.
QC318.I7S768 1995
530.1'3--dc20 95-33011
CIP

This book is printed on acid-free paper.

Printed in Singapore by Uto-Print

Preface

This book grew out of lectures given at Virginia Tech. in 1989-1991, and continued at King's College, London. The first half is suitable as a text for a course of one semester, being part of a programme for students of mathematics or physics pursuing a degree at the level of the M. Sc., M. Sci., M. Phys. or M. Math. The second half, which is undoubtedly harder, might be used as a course-book for a more challenging course or as a source for project work at the M. Sc. level, or as a guide to research work leading to the M. Phil.

The author is indebted to many students and friends for help: E. B. Davies, A. Uhlmann, B. Crell, B. Luffman, L. Rondoni, S. Koseki, F. Behmardi and C. Connaughton have been particularly generous. He owes the idea of (12.3) to V. Jaksic. The author was introduced to cotransport by Professor R. J. Naftalin, of the Department of Physiology, King's College, who also kindly drew the pictures using the Stella II 2.2.1 programme for differential equations, produced by High Performance Systems.

R. F. S. May 1995.

Contents

Part I

Classical Statistical Dynamics

Chapter 1

Introduction

Probability theory has been used in physics since the work of Maxwell, Boltzmann and Gibbs. However, many-body theory was not very successful until Planck introduced the quantum of action. Then many puzzles about black-body radiation, specific heats, and the like, were resolved. Looking today at Planck's argument we see that he used classical probability theory, even though his energy was quantised, to get the famous Planck black-body law. The same can be said of Einstein's seminal paper on stimulated emission. Even Bose's paper on the derivation of Planck's law from radiation theory is a correct use of classical probability, if we admit that we are discussing the statistics of field configurations, rather than configurations of particles. It is not clear that Bose realised the difference. It was Einstein who noticed that what Bose had done was to use a new counting rule for identical particles (assuming that photons were particles). At first, the new quantum mechanics was regarded as a modification of *mechanics*, not probability theory. Indeed, it was not until 1933 that Kolmogorov formally defined what a probability theory is. The upshot of Bell's notable work is that the Copenhagen interpretation of quantum theory, formulated between 1925 and 1930, is not actually a consistent set of rules within classical probability, but needs a generalisation, which we call *quantum probability*. This was formalised by von Neumann in his book [1] before Kolmogorov set up the foundations of the classical theory. There were two reasons for the lack of success of the statistical mechanics of classical particles; the first is the continuous nature of the phase space, with the result that the statistical entropy of a distribution is infinite. In statistical mechanics, we hope to identify the statistical entropy with the experimental entropy, which is finite. We may introduce a coarse-graining by replacing phase-space by a discrete set of points, such as the set of labels for phase-cells; but then the entropy depends on the choice of coarse-graining, and increases as we make the divisions finer and finer. The second trouble with the statistical mechanics of Newtonian particles comes from their distinguishability. Thus, if there are two particles in three-dimensional space, phase-space is $\Gamma = \mathbf{R}^{12}$; the configuration $(\mathbf{x_1}, \mathbf{p_1}; \mathbf{x_2}, \mathbf{p_2})$ is taken to be a different point in Γ from $(\mathbf{x_2}, \mathbf{p_2}; \mathbf{x_1}, \mathbf{p_1})$ (unless the points 1 and 2 coincide). A consequence of this was noticed by Gibbs in the early days, and is known as the Gibbs paradox. The problem arises when we try to relate the physical entropy of a gas (say) to the statistical entropy. To define the latter, we need a sample space Ω with sample points $\omega \in \Omega$, and a distribution $p(\omega)$, that is, a probability measure on Ω. The Shannon entropy, for a discrete space, is

$$S(p) = -\sum_{\omega} p(\omega) \log(\omega).$$

The entropy is measured in bits if the logarithm is taken to base two, and could be infinite; it is always non-negative. If the sample space has a finite number of elements, N, then the probability giving rise to the largest entropy is the uniform distribution

$$p(\omega) = 1/N.$$

Then the entropy is $\log N$, which should be compared with Boltzmann's formula for the physical entropy of a gas in equilibrium,

$$S = k_B \log N$$

where here N is the number of configurations of a given number of the molecules with a given total energy, and k_B is Boltzmann's constant. For a system in the usual continuum $\mathbf{R}^3$ the number of configurations is infinite, but if we divide space into cells (and do the same in momentum space), and restrict to a finite volume, then the entropy can be calculated. Gibbs' paradox arises when we compare the number of configurations of n particles in m cells in phase space with the number of configurations of $2n$ particles in $2m$ cells, corresponding to twice the volume in x-space and the same volume in momentum space. According to thermodynamics, which is confirmed by experiment, the physical entropy of the second system is twice that of the first; we say that entropy is an *extensive* variable. Looking at Boltzmann's formula, we see that for this to be true, the number of configurations in the second system must be N^2, the product of the numbers in the two halves. This is just not true for Newtonian particles. Not only can each of the N configurations in the first half be accompanied by each of the N configurations in the second half (already giving us the N^2 configurations), but a particle in the first half of space can be in any cell in the second half, and vice versa; there are far more than N^2 configurations, and the entropy of twice the volume is more than twice that of its parts (other things being equal). Gibbs called this extra entropy *the entropy of mixing.* He noted that its presence is not observed experimentally when two gases of identical molecules are mixed. Gibbs' paradox has been clearly discussed by [2]. It is not so much a paradox within probability theory as a demonstration that molecules do not obey classical mechanical laws, given that we have faith in the use of probability theory. Gibbs' proposed solution to his paradox is to divide the number of configurations by suitable factorials. In modern parlance, this amounts to altering the configuration space of n particles, by collecting together the $n!$ configurations that differ only by a permutation of the particles into an equivalence class; one then takes the whole equivalence class as a single point in a new sample space. The Gibbsian sample space has $N/n!$ points; it is the quotient of the Newtonian phase space on n particles by the permutation group $S(n)$, which acts on it. It was not appreciated at the time, but this is tantamount to introducing Bose statistics for the particles, and using the field description of the configuration. For gases at room temperature and pressure, n is about 10^{23} for each cubic centimetre, so the over-counting of states in Newtonian physics is by a factor of $(10^{23})!$, one of the biggest errors in any theory still taught to students. No wonder that classical statistical mechanics was not able to explain even the main features of microscopic many-body systems. Statistical methods applied to Newtonian particles do give good answers for the bahaviour of large numbers of ball bearings or tossed coins, which are consequently said to obey Boltzmann statistics. This is true even if the samples are so accurately similar that no difference between them can be discerned. The failure of molecules to obey Boltzmann statistics is not related to the difficulty, in practice, of following such a small individual along its path, but is due to the need for a new formulation, which we call the "field" point of view. We describe this in the next section, and contrast it with the "particle" point of view of Newton. This will resolve Gibbs' paradox; but the first difficulty, the lack of a natural unit of size for a cell in phase space, will not be resolved without quantum theory.

The first half of the book is devoted to giving the theory of classical statistical dynamics, based on a discrete sample space and using the field point of view. The objective is to provide a theory of nonequilibrium thermodynamics which obeys the first and second laws. The formalism is constructed in a way that makes clear that the classical theory is a special case, rather than a limiting case, of the quantum version, which is the subject of Part II.

We start with Chapter 2, with some remarks on probability theory. It is emphasised that the sample space must be clearly defined; one's choice of sample space, denoted Ω, is just as much part of the physics as is the choice of dynamics, and indeed must come first. The number of points in Ω, for example, determines the entropy of the uniform state. We also mention the concept of *local structure*. There should be a set Λ, usually finite, a point of which represents a point or small region in physical space; to each $x \in \Lambda$ is assigned the local sample space Ω_x. A configuration of the system, a point $\omega \in \Omega$, is then a *field*, that is, a map ω from Λ such that $\omega(x) \in \Omega_x$. It is explained how this formalism avoids

Gibbs' paradox. Indeed, we prove that entropy is extensive. To make the models tractable, we choose a discrete, but possibly infinite, sample space at each $x \in \Lambda$. In this context, marginal probabilities and conditional expectations are entirely elementary, and can be introduced gently. The idea of a random variable, often disguised as a mystery in books on Physics, is simply a real function on Ω. To get a model, we must specify a random variable, $\mathcal{E}$, the energy; this divides Ω into energy-shells Ω_E; this is the set of all sample points with energy E. We shall always assume that Ω_E is finite; this is needed for thermodynamic stability, in that the microcanonical state, known to physicists as an "ensemble", then exists. The assumption of finiteness also enables us to avoid measure theory and the many niceties of functional analysis.

The bounded random variables form an algebra, always denoted by $\mathcal{A}$; its dual, the set of linear functionals on this algebra, contains the positive cone. The normalised part of this is the set of possible states of the system. We shall limit our attention to the set of *normal* states, each given by a (countably additive) probability measure. In the finite case we can forget this refinement, and the duality between states and algebras is mutual. The characteristic of Kolmogorov's probability is that the algebra of observables is *abelian*, that is, commutative. For this reason we call this part the *classical* theory, even though there are no Newtonian particles present. The discrete sample space leads to a discrete energy-spectrum, so we are able to incorporate this aspect of quantum theory without which a classical theory would give very poor results.

The main mathematical tool is convexity, which plays a large role in the entropy-estimates, the paradigm of which is Kullback's inequality.

In Chapter 3 we define reversible dynamics as a group $\{\tau_t\}$ of invertible maps on Ω, labelled by discrete time $t \in \mathbf{Z}$. Here $\mathbf{Z}$ denotes the group of all integers. We prefer the term reversible to "conservative dynamics", since all our dynamics will obey the first law, conservation of energy. This is called "linear" not because the equations of motion are given by linear equations, as in a noninteracting system, but because τ_t defines a linear operator on $\mathcal{A}$; more, it gives rise to a group of automorphisms of $\mathcal{A}$. The dynamics will be limited to energy-conserving maps. Thus τ_t maps Ω_E bijectively onto itself. This allows us to extend the duality theory (valid if $|\Omega| < \infty$) to all cases of interest to us, even if $|\Omega| = \infty$. Here we get our first success: we can construct models in which the mean energy is independent of time (the first law). We introduce random dynamics by forming the convex sum of reversible dynamics. This is a generalisation of the construction of ergodic means and leads to the class of bistochastic Markov chains with discrete time. These have the Markov property: the state at time $t+1$ depends on the state at time t, but not on the history, that is, the state at all times before t. Not only does the discrete time lead to a readily computable theory and avoids some hard questions on the existence and uniqueness of solutions to non-linear equations, but the time-step mitigates the Markov assumption. It may be a poor approximation to real systems to assume the Markov property if time is continuous, but if the time-step is chosen to be large on the time-scale of the fast variables, it should give a good account of the dynamics of the slow variables. This division into fast and slow variables is well explained in [3]. The dynamics given by a mixture of reversible maps is entropy-increasing, and so we get models obeying the second law. Here we meet the first "ensemble", the microcanonical state, and show that it is the limit of any ergodic bistochastic process if the initial state has a sharp energy. This uses Lyapunov's direct method, which is explained simply. We illustrate with models of discrete diffusion and of reacting particles.

In Chapter 4 we introduce the Boltzmann map. This is an abstract version of the Stosszahlansatz of Boltzmann. Given any algebra $\mathcal{A} = \mathcal{A}_1 \otimes \mathcal{A}_2$ (a tensor product), we can replace any state on $\mathcal{A}$ by the product of its marginal states. This is the state of greatest entropy having the same marginal distributions as the given state. It follows that the map increases the entropy (or leaves it the same if it was already a product). This was the step in Boltzmann's derivation of his equation that caused most controversy. We do not attempt to prove that validity, or approximate validity, of the step; rather, the step is added to the dynamics to model randomness which increases entropy but does not alter the mean energy, and which is not accounted for by the random reversible dynamics. It models the increase in entropy caused by the fast variables which break the correlations between the factors involved in the map. The resulting dynamics is non-linear on the space of states, which causes worry to some people. In particular, the powerful and popular theory of Markov chains, briefly described in Chapter 3, does not apply. But, as with Boltzmann's original equation, the non-linearity is necessary if the isolated system is to converge to equilibrium at its own temperature. It is the Boltzmann map that mixes up the different energy-shells, and leads to the canonical state at large times. A further simplifying map, called the LTE-map, standing for *local thermodynamic equilibrium* is introduced. This increases the entropy still further, by replacing the state after one time-step by the product of states which, on each local algebra, are in a canonical state relative to the local energy. After the LTE-map the state is independent at each site x. This map conserves energy only if there is no energy of interaction between sites. In that case it leads to very tractable equations of motion, in which the state is parametrised by a field of intensive variables, such as the beta and chemical potential, far simpler than the detail needed to specify a general state. The highly non-linear equations then converge to equilibrium. This is illustrated with simple models of chemical kinetics, and a free gas in a gravitational field. This last model allows us to give a natural definition of the pressure.

Then we introduce an idea possibly even more controversial than nonlinearity: the heat-particle. This is a convenient way to incorporate the first law of thermodynamics, which requires that the total energy, including heat, must be conserved by the dynamics. In statistical mechanics, heat is most often identified with random kinetic energy. More precisely, it is any form of energy that has been randomised. By introducing a new degree of freedom explicitly to carry the heat, we mitigate the omission of all velocity variables, such as would occur in the particle picture. Then we can formulate a stochastic law that conserves total energy. It is one of the tenets of statistical dynamics that in irreversible dynamics it is *information* that is dissipated as time progresses, not energy; energy simply moves between the different forms in which it can appear. In the simplest models there is just one heat-particle, which represents the shared, thermalised kinetic energy of the fields in the system. In more detailed models, the heat-particle is located on the bond between two molecules, and represents the energy of vibration of the bond, and the shared electromagnetic energy contained in the photons moving between them, and the thermalised phonons. The chapter ends with a general theory of chemical reactions. Chemical kinetics has been an active field since 1850, when the law of mass action was formulated in the form we learn at school: the rate of a reaction is proportional to the product of the concentrations of the reactants. Later, it was recognised that this could not be completely correct; the problem is that if the forward reaction rate is in balance with the backward rate, so that we are in equilibrium, then we get an equation between products of concentrations, and this equation is not satisfied by the canonical or grand canonical state. So the law of mass action was rewritten, so that the rate was proportional to the product of the *activities* of the reactants, rather than their concentrations. This fitted the data even better than the original form of the law, and, in addition, the grand canonical states are among its fixed points. The original law became known as the "density-led" form, and the new form was known as the "activity-led" version. A large part of the Chapter is used to show that the activity-led theory can be fitted into statistical dynamics by a suitable choice of sample space, energy-function, and bistochastic map. The good properties of the dynamics (energy-conservation, increaseing entropy, convergence to equilibrium) then follow from our general theory.

In Chapter 5 we introduce the concept of isothermal dynamics. This arises from the isolated dynamics by adjusting the state after each time-step so that the heat-particle is in thermal equilibrium at a specified temperature. The motion then reduces to a linear dynamics through the states of the system, and so the theory can again be described by a linear stochastic process. The adjustment to the state after each time-step models the heat-flows that must occur to keep the temperature constant. We show that this is related to the Legendre transform. The theory is applied to chemical reactions, and we quickly obtain the famous results of chemical kinetics: the principle of detailed balance, which says that for closed systems of chemical reactions at equilibrium each chemical reaction is separately in equilibrium; moreover the *law of Arrhenius*, which says that the forward and backward rates of a reaction are related by a Boltzmann factor $e^{-\beta E}$, where E is the energy of reaction, is a simple consequence of the dynamics. The main theorem is that the free energy decreases along the orbit, and so is a Lyapunov function. This contrasts with the isolated dynamics, in which the entropy increases. The difference between these is the heat that must be put in or taken out to maintain constant beta, and indeed we derive the axiom of thermodynamics, that entropy changes by $\geq \beta dQ$ when heat dQ is added at beta β. We use the free-energy theorem to prove the "fundamental theorem of chemical kinetics", which states that isothermal chemical reactions converge to thermal and chemical equilibrium at large times.

We end Chapter 5 with a convergence theorem and a structure theorem. The convergence of systems to equilibrium is shown to occur in the very strong sense defined by a large norm. We also state the law of "equivalence of ensembles". This says that the state of a subsystem of fixed size is close to a canonical state if the full system is large and is in a microcanonical state. In our version, the "closeness" is measured by the norm. This should not be construed as saying that the isothermal and isolated dynamics are nearly the same; as every chemist knows, isothermal dynamics needs carefully controlled heat-flows through the boundary. The distinction between the microcanonical and the canonical state disappears in the thermodynamic limit: it is one of the casualties of the idealisation implicit in the infinite-volume theory. This makes the boundary conditions difficult to specify, and some ambiguity in the dynamics seems inevitable. Ingarden [4] says "At present, many physicists think that this assumption [that the volume is infinite] is the only possible one for thermodynamical and statistical theories. Such an extreme point of view seems to be, however, absurd."

The structure theorem says that every Markov chain with positive fixed point, and small time step, can be regarded as the isothermal dynamics of the system coupled to its heat-particle. This shows that the theory, and in particular the introduction of the heat-particle, leads by passing to its isothermal version to the most general Markov chain with the canonical state as its fixed point.

In Chapter 6 we consider driven systems; these are open in the sense of chemistry, in that particles can flow in and out, in order to keep the chemical potentials of certain chemicals fixed. These are called the driving chemicals. In driven systems the fundamental theorem of chemical kinetics does not apply. This is illustrated by heat conduction through a lattice gas. There is a brief introduction to linear stability theory and the Lyapunov indices. It is applied to a simple version of the Brussellator of Prigogine, both with and without noise.

There is also a numerical study of this model. The noise in the model is *thermal* and we show how to include it in the equations. It is shown by example that noise can destroy a limit cycle. This ends the classical part; we return to some harder problems in Chapter 13.

Chapter 2

Probability Theory

2.1 Sample Spaces and States

The sample space of the system will be taken to be a countable space Ω, a point of which will represent a configuration. A point in Ω will not represent a point in phase space of a Newtonian particle, because of the difficulty with Gibbs' paradox, as explained in the Introduction. The velocity or momentum of the particle will not be explicitly mentioned; the kinetic energy will later on be represented in an average way by the energy of the heat-particle. Space will be approximated by a finite space Λ, and might have a regular structure such as the set of vertices of a periodic crystal. For example, the cubic lattice in three dimensions would use $\Lambda = \mathbf{Z}_N^3$, where $\mathbf{Z}_N$ is the group of integers $\{0, 1, 2, \ldots, N-1\}$, with addition modulo N. We want the theory to have a local structure, which means that at each point of space, $x \in \Lambda$, we are given the sample space Ω_x. This is the set of values that the field or fields at x can take. Then we require that

$$\Omega(\Lambda) = \prod_{x \in \Lambda} \Omega_x. \tag{2.1}$$

This notation means that an element ω of Ω, that is, a configuration of the field, is given by the labelled set $\{\omega_x\}$ as x runs over Λ. If all the sample spaces Ω_x are the same, say $\mathbf{N} = \{0, 1, 2, \ldots\}$, then the configuration ω is a map $\omega : \Lambda \to \mathbf{N}$. For example, if $\Lambda = \mathbf{Z}_N$, then ω is a finite sequence of natural numbers $\{\omega_1, \omega_2, \ldots, \omega_N\}$. The non-negative integer $\omega_x \in \Omega_x$ represents the number of molecules of a chemical that sit at x, or, lie in the volume labelled by x. We see that ω is indeed a map from Λ to $\mathbf{N}$. In this case ω is a configuration of the density field. This specification of the way particles are distributed is called the *field point of view*. It should be contrasted with the *particle point of view*; in the same example, with $\Lambda = \mathbf{Z}_N$, a configuration of n particles is given by, in turn, considering the first particle, and then specifying its location, at $x_1 \in \Lambda$, then considering the second particle, and specifying its location, at $x_2 \in \Lambda, \ldots$. Thus, from the particle point of view, a configuration is an ordered sequence of n elements of $\mathbf{Z}_N$. This is a different space from that given by the field point of view, Eqn. (2.1). In particular, the space $\Omega(\Lambda)$ of Eqn. (2.1) obeys the product law: if Λ_1 and Λ_2 are disjoint sets, then

$$\Omega(\Lambda_1 \cup \Lambda_2) = \Omega(\Lambda_1) \times \Omega(\Lambda_2). \tag{2.2}$$

Exercise (2.49) asks you to prove this when Ω is finite, and to show that it is not true in the particle point of view.

The product structure of the sample space is a feature of several very successful models of equilibrium statistical mechanics, such as the Ising model, interpreted as a gas of indistinguishable particles on a lattice [5]. The question then arises as to how to describe the dynamics of such a model, so that it approaches equilibrium at large times. This question was first considered by Glauber [6]. His solution, now called the Glauber dynamics, does not involve any momentum for the particles, and seems to omit the kinetic energy as well. Our sample space is constructed in the same spirit, and we shall be able to describe diffusion of the particles, but not streamline flow; the latter involves the hydrodynamic limit, and will be possible only in the quantum version.

In a probabilistic theory, in general we do not know the configuration of the system precisely at a given time; the best we can do is to describe the possible configurations by a probability. Thus, we describe the *state* of the system at a given time by a probability measure on Ω, typically denoted p:

Definition 2.3 *A probability measure on Ω is a function $p(\omega)$ obeying*

1. $p(\omega) \geq 0$
2. $\sum_\omega p(\omega) = 1$.

The set of all probability measures on Ω forms the set $\Sigma(\Omega)$, called the *state space* of the model. We shall use the word state to mean a probability measure. If we have complete information about the system, for example that the configuration is ω_0, then the state is the point measure $p(\omega) = \delta(\omega, \omega_0)$. At the other extreme, if we know nothing about the configuration, and Ω is finite, then it is sometimes reasonable to postulate that the state is the uniform measure $p(\omega) = 1/|\Omega|$; here, $|\Omega|$ denotes the number of points in Ω.

The use of probability theory in physics can lead to long and heated debates about philosophy. Historically, many people were worried by Boltzmann's work. This debate should not be surprising, since the theory of probability was not formalised until much later. To explain probability, Gibbs asked his readers to imagine all possible configurations of the system laid out on a table; together, these form the "ensemble"; the actual system will be one of these, picked out blindly at random. He had in mind that each configuration was equally likely. We recognise the ensemble as the sample space, provided with the uniform probability. Other states, with non-uniform probability on the sample space, are not so easy to explain using the ensemble. Gibbs was able to formulate mixing (ensembles of ensembles) and conditioning (subensembles), but there remained some confusion between the ensemble of atoms in a gas (an actual large collection of atoms) and the ensemble of configurations of the gas as a whole. In statistical dynamics, it is imperative to choose the sample space firmly at the start, as part of the specification of the model. We shall see that there is no sample space in quantum probability, and so the Gibbsian point of view can become even more confusing. The "many-worlds" interpretation of quantum mechanics is a modern relic, a brave (or foolhardy) attempt to rescue the notion of an ensemble in quantum mechanics.

There are three philosophies concerning what choice of probability, $p \in \Sigma(\Omega)$, we should make to describe the state at hand. The first, used in some books on statistics, is the "frequency" definition. The second uses the notion of a "prior" distribution, and has gained ground with the popularity of information theory. The third considers the state p to be part of the model, just like the sample space or the choice of dynamics.

In the frequency definition, we do not commit ourselves as to p until we have made several tests on the system. We take it that the system is reproducible in the same state as often as necessary. We find $p(\omega_1)$ for example by counting the number of times the configuration ω_1 occurs in a long run, and dividing it by the total number of trials. The method is rather deficient if we want to use it as the *definition* of p. It is difficult to think of one theorem that can be proved using this idea, which depends on an infinite sequence of experiments on a physical system.

In the definition using prior knowledge, we take it that in the absence of any knowledge as to what the configuration is, the "correct" probability distribution is the uniform one. Bernoulli and then Laplace used this in their work. It only makes sense when the sample space is finite; for example if we try to use the uniform distribution when Ω is continuous, the paradoxes of Bertrand are very puzzling. Suppose we are in the finite case, and we start with a known sample space but do not know the state. We therefore postulate the uniform distribution as the best (or rather, least bad) guess. Suppose we now get some information about the system. We must distinguish between two ways in which this can happen. In the first type, the sample point has already been chosen, but do not know which one it was. Imagine a punter with the problem of laying a bet on a throw of a die, which has already occurred, but in another room. The chance is $1/6$ that a particular number ω has occurred. If he finds out that the number was in fact even, then taking into account this information, the new improved probabilty is zero for odd ω, and 1/3 for even ω. More generally, the rules for conditioning allow us to modify the uniform distribution, or any other prior distribution, in a systematic manner, to take account of new information. This leads to the school of statistics known as Bayesian, whose methods are used in operations research, decision theory and planning. This type of information, in which some hints are given as to which sample point has occurred, should not be confused with a second type of knowledge, in which we are not dealing with one sample, but the whole sample space. Suppose that we do not know what the probability is, but that we have measured the mean of one or more random variables by examining a large number of samples. Information theory then tells us how to best choose the prior distribution, given these mean values. If no variables have been measured, then the method gives the uniform distribution as its best prediction. If the mean of at least one random variable is known, the method leads to analogues of the canonical states occurring in statistical physics. These states are not conditioned in the true sense. They can be modified by conditioning if information about the sample point is acquired, as in Bayes' theorem, and nowadays the whole theory is known as the Bayesian method. It has been introduced into physics and chemistry by Jaynes [7]. One can counter the criticism, that the hypothesis of equal probabilities (in the absence of information) is purely arbitrary, with the question "can you suggest anything better?". All we can say is that people with no information should not be taking decisions. We shall see that the uniform distribution is the one with the largest entropy (and least information, which is negative entropy); this is why information theory starts with this as the prior distribution. Some people find this a completely convincing justification; but this is somewhat deceptive. In fact, the finite sample space Ω is often obtained by setting up cells in a continuous space; this coarse graining is then quite arbitrary, and the corresponding uniform distribution is no more special than another based on a different division into cells. The Bayesian method can be quite misleading for the application of probability in physics. Consider for example the problem of setting up a theory of the interior of a distant star, about which we have no information except that it is composed of protons and electrons. Suppose that we set up a large but finite sample space Ω where a configuration expressed, for each point x in the star (where x runs over a finite set Λ) whether the site were occupied or not. Suppose that we formulate some dynamics for the system. We are then faced with the problem, what is the initial state of the system? In the absence of information, a Bayesian would say: take p to be the uniform distribution on Ω. Now, such a theory might well be able to predict that the star will collapse, or will explode, depending on the initial state. Say, that if the initial state is uniform, it will collapse. Thus a Bayesian with no further information would predict that it will collapse: this would be his best estimate. But since the initial state could have been anything, this conclusion is not justified. We could seek more information about the star, such as its brightness. This would allow the Bayesian to condition his probability, and he might now predict that it will not collapse, but will explode. We cannot believe that the puny human's knowledge, obtained by measuring the brightness, makes any change whatsoever to the state of the star, since it is large and far away. So we conclude that the uniform state was not a good description of the situation. A physical theory cannot do better than allow us to work out which p lead to a collapse, and which p lead to explosion. Physics is not in the gambling business, and is not obliged to hazard a guess as to what the initial state was. Another example of the odd conclusions that can be got from the uniform distribution is that all systems must be in equilibrium. Fowler wants to explain why equilibrium states are so common [8, 9], and he calculates the probability that a cubic centimetre of gas is not at equilibrium to within measurable accuracy; he find it astronomically small, and concludes that it will be driven to equilibrium not just because this is more probable than not, but because it is *overwhelmingly* more probable. Grandy [10] uses the same phrase. If it is so overwhelming, how come systems still manage to exist out of equilibrium for minutes, hours, years at a time? The idea that equilibrium states are the only possible ones has become part of the folk-lore of statistical physics, but it is a non-starter in a theory such as statistical dynamics, which is trying to describe non-equilibrium phenomena.

The third philosophy about probability is more recent than the Bayesian, but does use ideas from it, as well as from the frequency definition. It is to regard the choice of p as part of the model, along with the choice of Ω and the dynamics. We are led to this point of view by comparison with quantum theory, where the state of the system is not regarded as something one can predict *a priori*, by pure reasoning. Thus any $p \in \Sigma(\Omega)$ is a possible state of the system (at the time in question), and we do not attempt to give preference to some over others without recourse to experiment. In particular, we do not use the idea that there is an *a priori* probability that the probability is p. The procedure is as follows: we decide on p as the *hypothesis* H, and then test the theory by the method of significance tests; we reject the model if it fails the test. We might use the first method, frequency, to help us with the hypothesis. This comes under the legitimate subject of estimation theory. Bayes' rule can also be used if more information is available; it is the correct way to incorporate information. It should be kept in mind that the conditioned probability is still part of the model under the hypothesis H. This is the philosophy we shall adopt in this book. At the very end, we shall be drawn towards some of the ideas of information theory.

We are led to define a probabilistic model as a pair (Ω, p), where Ω is a countable space and p is a probability on Ω. This seems less elaborate than Kolmogorov's definition of a probability space, which involves a σ-ring of measurable sets and a countably additive measure μ [11]. However, because Ω is countable, these concepts are equivalent. Suffice it to say that each subset Ω_0 of Ω is measurable, and represents an *event*. We say that the event Ω_0 has occurred if the sample point ω chosen lies in Ω_0. The probability that the event Ω_0 occur is given by the measure of the set $\Omega_0 \subseteq \Omega$, which is given by

$$\mathrm{Prob}\{\Omega_0\} = \sum_{\omega \in \Omega_0} p(\omega).$$

You are asked to prove that Prob is countably additive in the exercises. Conversely, every measure Prob obeying $\mathrm{Prob}\{\Omega\} = 1$ defines a probability by

$$p(\omega) = \mathrm{Prob}\{\omega\}$$

where $\{\omega\}$ is the set containing the single point ω.

We now give the most important examples of states:

- Point measure at ω_0: $p(\omega) = \delta(\omega, \omega_0)$.
- Uniform state: here $|\Omega| = n < \infty$, and for all ω, $p(\omega) = 1/n$.
- Microcanonical state: here $|\Omega|$ can be infinite; we are given an energy function $\mathcal{E}$ on Ω such that the *level sets* Ω_E of $\mathcal{E}$ are finite. Ω_E is the "energy-shell", that is the set of points ω where $\mathcal{E}(\omega)$ takes the value E. Let $|\Omega_E| = n(E)$. Then the microcanonical state is

$$p(\omega) = \begin{cases} 0 & \text{if } \omega \notin \Omega_E \\ 1/n(E) & \text{if } \omega \in \Omega_E \end{cases} . \tag{2.4}$$

- The canonical states: here the energy function $\mathcal{E}$ is taken to be bounded below. Let β be a real number, and suppose that $Z_\beta = \sum_\omega \exp(-\beta\mathcal{E}(\omega))$ is finite. In this case, Z_β is called the *canonical partition function*. The canonical state is then

$$p_\beta(\omega) = Z_\beta^{-1} \exp\left(-\beta\mathcal{E}(\omega)\right). \tag{2.5}$$

  We note that if Ω is infinite, then β is necessarily positive.

- The grand canonical states: here we are given an energy function $\mathcal{E}$, and another function $\mathcal{N}(\omega)$ which takes the values $0, 1, \ldots$. Let β and μ be real numbers such that

$$Z_{\beta,\mu} = \sum_\omega \exp\left(-\beta\{\mathcal{E}(\omega) - \mu\mathcal{N}(\omega)\}\right) < \infty.$$

In this case, $Z_{\beta,\mu}$ is called the *grand partition function* and μ is the *chemical potential.* The grand canonical state is then

$$p(\omega) = Z_{\beta,\mu}^{-1} \exp\left(-\beta\{\mathcal{E}(\omega) - \mu\mathcal{N}(\omega)\}\right). \tag{2.6}$$

Please do Exercise (2.48) at this point. The parameter β is related to the inverse temperature: $\beta = 1/(k_B T)$, where k_B is Boltzmann's constant. The parameter μ is called the chemical potential.

There are other interesting states, the great grand canonical states with several numbers $\mathcal{N}$, that we shall meet later. One of our problems is to show why these arise as equilibrium states of systems, and why many systems settle down to one of these at large times, and what determines which one.

The set of states, Σ, is a convex set. It can be regarded as a subset of the set of all real functions on Ω; this set is a real vector space. The concepts of linear sums, and of linear independence are well defined for functions, and so have a meaning for states. If p and q are states and $\lambda \in [0,1]$, then $\lambda p + (1-\lambda)q$ is a function of ω which obeys Definition (2.3), and so lies in Σ. Geometrically, this means that if $p \in \Sigma$ and $q \in \Sigma$ then the whole segment of the line joining p and q lies in Σ. This is the definition of a convex set. Please prove that Σ is convex, and do Exercise (2.53). For any n, if $p_1, p_2, \ldots, p_n$ are linearly independent states, then the set of linear combinations with positive weights

$$p = \sum_j \lambda_j p_j, \qquad \text{where } \lambda_j \in [0,1] \text{ and } \sum_j \lambda_j = 1, \tag{2.7}$$

form a closed convex set, which is a simplex. The set $\Sigma(\Omega)$ is also a simplex. The meaning of simplex is the following. Given any convex subset K of a vector space, we say that a point $p \in K$ is an *extreme point* of K if it cannot be written as $p = \lambda p_1 + (1-\lambda)p_2$ for any pair of different elements p_1, p_2 of K, where $0 < \lambda < 1$. We say that a convex set K is generated by its extreme points if K is the smallest closed convex set containing the extreme points of K. When $K = \Sigma(\Omega)$ with Ω a countable set, we can prove this, which is a special case of the Krein-Milman theorem. For, a probability p can obviously be written as

$$p(\omega) = \sum_{j\in\Omega} p_j \delta(\omega, j).$$

It follows that p is extremal if and only if it is a point measure; also, this decomposition of p into point measures is unique (the coefficients p_j are unique). Non-extremal points of Σ are called *mixed states.* We say that a convex set K is a simplex if the decomposition of any point in it into extreme points is unique. Thus, we have proved that $\Sigma(\Omega)$ is a simplex. Note that when K is not the whole of Σ, the extreme points of K might not be point measures, since extreme points are limited to belong to K. An extreme point of K might not be an extreme point of a larger convex set containing K, as can be seen by choosing K to be a line, and extending one end to form a larger convex set. You are asked to do Exercise (2.54).

The interpretation of the act of forming the mixture of the states p and q with weights $\lambda, 1-\lambda$ is given by considering the states of the system got by physically mixing the states p and q with probabilities λ and $1-\lambda$. The statistics of the mixture can be got as follows: we start with two systems, one in the state p and the other in the state q. We put each in a black box, and juggle them up, and select p with probability λ and q with probability $1-\lambda$. Whereas this does give the same statistics as the state $\lambda p+(1-\lambda)q$ on the space Ω, there is no real mixing of the physical systems, which remain separated. The physical development in time of the two black boxes will be different from that of the state $\lambda p+(1-\lambda)q$, regarded as the state of a single system in one container. For example, if the two black boxes are Thermos flasks, each system will approach equilibrium in time at its own temperature, but if they are in a single container, the limit state will have a single, common temperature. We are led to conclude that not all physical information is contained in the state; we need to specify the dynamics, just as in a reversible theory the Hamiltonian must be given. The local dynamics is determined physically by what type of molecule we have, and this should be the same whether we put the molecules in one container or two; but the walls of the container should be specified (whether permeable to particles, or heat conducting, or not). These properties are usually modelled by imposing boundary conditions on the dynamics, and this will be how it is done in statistical dynamics. The boundary effects are casualties of the idealisation involved in the thermodynamic limit; by taking Λ to become infinitely extended, we take it that there is only one container. The mixing obtained by considering two unrelated black boxes is an every-day sort of mixing, coped with by ordinary probability theory. There is also the more intimate mixing of two states p and q, obtained by considering $\lambda p+(1-\lambda)q$ as a state of the system regarded as being in the one and the same container as the original system. This is the interpretation we shall put on this mixture, but how it is achieved physically is not going to be discussed.

The real vector space, called $\ell^1(\Omega)$, spanned by the states on Ω can be easily shown to consist of functions of the form $f=\alpha p-\beta q$, where p and q are states and α and β are non-negative. It is exactly the set of summable functions on Ω; (Exercise (2.55)). We can furnish this space with a norm by

$$\|f\|_1=\sum_{\omega\in\Omega}|f(\omega)|. \tag{2.8}$$

It can be shown that this is a norm and that $\ell^1(\Omega)$ is complete in the topology defined by this norm. This is Exercise (2.56). As usual, we can use the norm to define convergence: we say that a sequence $\{p_j\}$ of elements of $\ell^1(\Omega)$ converges to $p \in \ell^1(\Omega)$ if

$$\lim_{j\to\infty} \|p_j - p\|_1 = 0.$$

Thus the vector space spanned by the set of states Σ becomes a real Banach space.

The following device is sometimes useful. Let $\operatorname{Span}\Omega$ denote the formal vector space spanned by Ω. How can we add elements of Ω, which is a set without linear structure? Easy: we just do it. This is how. We number the elements of Ω, say $\omega_1, \omega_2, \ldots$, and define $\operatorname{Span}\Omega$ to be the set of terminating sequences of real numbers $\{\lambda_1, \lambda_2, \ldots\}$ with the addition and multiplication by reals given elementwise. This is clearly a vector space. Then we write

$$\{\lambda_1, \lambda_2, \ldots\} = \lambda_1\omega_1 + \lambda_2\omega_2 + \ldots.$$

We identify ω with the point measure at ω, and $\ell^1(\Omega)$ with the completion of $\operatorname{Span}\Omega$ in the norm $\|\bullet\|_1$. In this notation, Σ is the closure of the convex hull of Ω.

To understand the concept of *marginal probability*, consider two spaces $\Omega_1 = \{1, 2, \ldots, m\}$ and $\Omega_2 = \{1, 2, \ldots, n\}$, and take $\Omega = \Omega_1 \times \Omega_2$. A sample point is a pair of points, ω_1, ω_2, one in Ω_1 and the other in Ω_2. A probability on Ω is determined by an $m \times n$ matrix $p = [p_{ij}]$ of non-negative numbers, such that the sum of all the components is unity:

$$\sum_{i=1}^{m}\sum_{j=1}^{n} p_{ij} = 1. \tag{2.9}$$

This is called a bivariate distribution in statistics. The *marginal distribution on* Ω_1 gives the probability of the event $\{\omega \in \Omega : \omega_1 = i\}$, and this is the sum of the entries in the ith row of p, $\sum_j p_{ij}$. This is a probability on Ω_1, since its components are non-negative and, by Eqn. (2.9), they sum up to unity. We denote by $\mathcal{M}_1$ the map taking p to its marginal on Ω_1:

$$(\mathcal{M}_1 p)(\omega_1) = \sum_{\omega_2} p(\omega_1, \omega_2). \tag{2.10}$$

Similarly, we define the marginal map on Ω_2. In statistics, we say that i and j are independent if p is the product of its marginals, thus:

$$p_{ij} = (\mathcal{M}_1 p)(i) \times (\mathcal{M}_2 p)(j). \tag{2.11}$$

In statistical dynamics, the sample space is the product as in Eqn. (2.1), from which we can show that given that $\Lambda = \Lambda_1 \cup \Lambda_2$, with Λ_1 and Λ_2 disjoint, then putting $\Omega_1 = \Omega(\Lambda_1)$, and similarly for Ω_2, we have the product structure

$\Omega(\Lambda) = \Omega_1 \times \Omega_2$. A sample point ω can be written as $\omega = (\omega_1, \omega_2)$ and a probability $p \in \Sigma(\Omega)$ is determined by the function $p(\omega_1, \omega_2)$. The marginal map $\mathcal{M}_1$ takes p to its marginal on Ω_1, and is thus a map from $\Sigma(\Omega)$ to $\Sigma(\Omega_1)$. Explicitly,

$$(\mathcal{M}_1 p)(\omega_1) = \sum_{\omega_2 \in \Omega_2} p(\omega_1, \omega_2). \tag{2.12}$$

The map $\mathcal{M}_1$ can be extended to the whole of $\ell^1(\Omega)$ by linearity; it follows from the definition that $\mathcal{M}_1$ is linear on $\Sigma(\Omega)$. Similarly, $\mathcal{M}_2$ is defined. The sums involved obviously converge, and the map $\mathcal{M}_1$ is continuous; see Exercise (2.57).

Given two probability spaces (Ω_1, p_1) and (Ω_2, p_2), we can form their independent product as follows. Put $\Omega = \Omega_1 \times \Omega_2$ and $p(\omega_1, \omega_2) = p_1(\omega_1) \times p_2(\omega_2)$. We sometimes write $p = p_1 \otimes p_2$ for the independent product, since it agrees with the tensor product of elements of the Banach spaces $\ell^1(\Omega_1)$ and $\ell^1(\Omega_2)$. That is,

$$\ell^1(\Omega_1) \otimes \ell^1(\Omega_2) \equiv \ell^1(\Omega_1 \times \Omega_2). \tag{2.13}$$

If p is the independent product of p_1 and p_2 as above, then the marginals of p on Ω_1 and Ω_2 are exactly the p_1 and p_2 we started with. This, then, is the characteristic feature of independence: we say that Ω_1 and Ω_2 making up a probability space $(\Omega = \Omega_1 \times \Omega_2, p)$ are independent if p is the product of its marginals:

$$p = \mathcal{M}_1 p \otimes \mathcal{M}_2 p. \tag{2.14}$$

In statistical dynamics, we say that a state $p \in \Sigma(\Omega)$ is independent over Λ if

$$p = \bigotimes_{x \in \Lambda} \mathcal{M}_x p \tag{2.15}$$

which means, if we label the points $x \in \Lambda$ by $1, 2, \ldots$:

$$p(\omega) = (\mathcal{M}_1 p)(\omega_1) \times (\mathcal{M}_2 p)(\omega_2) \times \ldots. \tag{2.16}$$

The concept of independence is best formulated in terms of random variables, to which we turn in the next section.

The marginal map $\mathcal{M}$ will be used to restrict the state to a subsystem; this is a true subsystem, located in part of the space Λ, when Ω has the local structure described, and Ω_1 is the sample space of a subset of Λ. We can also define marginal maps onto subsystems that are not localised in a subset of space, but are described by some other type of subset of the observables; say, the subsystem described by the macroscopic observables. The marginal state then provides a reduced description. A full account of this idea needs the concept of random variable, coming next. Note that the reduced description does not involve putting the probability to zero on a region of sample space; rather, the probability p is replaced by $\mathcal{M}p$ which does not depend on some of the variables needed for a full description of the state p.

2.2 Random Variables, Algebras

There is no other simple mathematical idea that is taught in such a muddled way as random variables are taught to physics students. We must distinguish the concept of random variable, and its probability distribution, from the concept of sample space and the probability p on it. In our case, where Ω is countable and every subset is measurable, a random variable is a real-valued function on Ω, denoted by F or another capital letter. Thus, $F(\omega)$ is the value that F takes when the sample point takes the value ω, which it does with probability p. The state p itself does not enter the definition of F but it does enter into the probability distribution of F; this is denoted p_F, and is defined as follows. Since Ω is countable, the set of (real) values $\{f\}$ that $F(\omega)$ takes is discrete. The probability that $F = f$ is

$$p_F(f) = \sum_{\omega:F(\omega)=f} p(\omega). \tag{2.17}$$

For different p we obviously get different p_F.

In gambling with dice, Ω is the set of outcomes and $F(\omega)$ might denote the reward that a particular player gets if a particular outcome arises. A random variable F divides Ω into disjoint sets, on each of which F takes different values. Thus, define the events,

$$\Omega_f = \{\omega \in \Omega : F(\omega) = f\}$$

for all possible values f of F. Then Ω is the disjoint union of all possible Ω_f, and $p_F(f)$ is the probability of the event Ω_f. This idea, the division of Ω into disjoint sets, will be important when we come to discuss the random variable chosen to represent the energy. In this case, the slices of Ω defined by the level surfaces of the energy are called *energy-shells*. When Ω has a local structure, as in Eqn. (2.1), then a random variable is a function of all the points in $\Lambda = \{1, 2, \ldots\}$: $F(\omega) = F(\omega_1, \omega_2, \ldots)$.

The mean and variance of F, in the state p, are defined as

$$\text{mean}: \qquad \langle F \rangle = \langle p, F \rangle = \sum_{\omega} p(\omega) F(\omega) \tag{2.18}$$

$$\text{variance}: \qquad (\Delta F)^2 = \sum_{\omega} p(\omega)(F(\omega) - \langle F \rangle)^2. \tag{2.19}$$

These definitions assume that the sums are absolutely convergent. They can be expressed in terms of the probability distribution of F; see Exercise (2.58). The standard deviation Δ is often called the *uncertainty* in books on quantum mechanics. The correlation between two random variables F and G is defined to be

$$\mathcal{C}(F, G) = \langle FG \rangle - \langle F \rangle \langle G \rangle \tag{2.20}$$

where FG is the product function of F and G. The correlation cannot be expressed in terms of the distribution functions of F and G, but needs the joint probability distribution (also called the bivariate distribution) defined as

$$p_{F,G}(f,g) = \sum_{\omega:F(\omega)=f,G(\omega)=g} p(\omega). \tag{2.21}$$

At this point we would get into trouble if we tried to define probability theory in terms of the various p_F, without introducing the sample space and its state p, since the joint distribution would need a separate definition, which would need to satisfy some subtle relations in order to be derivable from a p. Now try Exercise (2.59).

In elementary books on probability, two random variables F and G are said to be independent if their joint probability distribution is the product of the two distribution functions: $p_{F,G}(f,g) = p_F(f)p_G(g)$ for all f and g. The correlation between F and G is a measure of the dependence of these random variables; for example, it is zero if they are independent, but this is not a sufficient condition. In addition, to ensure independence, all the higher correlation functions involving F and G must be zero too. The higher moments and correlation functions are best defined using the *characteristic function.* Let us first treat the case of one random variable. So let p be a probability on Ω and let F be a random variable. Then the function

$$\mathcal{F}_F(x) = E(\exp ixF) = \sum_\omega p(\omega)e^{ixF(\omega)} \tag{2.22}$$

is called the *characteristic function* of the random variable F (if p is given) or of the state p (if F is given). It always has the properties

1. $\mathcal{F}(0) = 1$;
2. $\mathcal{F}(x)$ is a continuous function on the whole real line;
3. for any N complex real numbers $\lambda_1, \ldots, \lambda_N$ and any real numbers $x_1, \ldots, x_N$ we have
$$\sum_{ij} \overline{\lambda}_i \lambda_j \mathcal{F}(x_j - x_i) \geq 0. \tag{2.23}$$

A function with these properties characterises the random variable; the famous Bochner theorem [12] says that given a function $\mathcal{F}$ with the properties (1), (2) and (3), there exists a sample space Ω, a probability $p \in \Sigma(\Omega)$ and a random variable F such that Eqn. (2.22) holds. A function satisfying (3) is said to be of *positive type.* If $\mathcal{F}$ has a power series expansion about $x = 0$ then it can be written

$$\mathcal{F}(x) = 1 + (ix)\sum_\omega p(\omega)F(\omega) + (ix)^2/2!\sum_\omega p(\omega)F^2(\omega) + \ldots \tag{2.24}$$

from which the n-th moment can be read:

$$i^n E(F^n) = \frac{\partial^n}{\partial x^n} F(x)|_{x=0}. \tag{2.25}$$

For this reason the function $\mathcal{G}(s) = \mathcal{F}(-is)$ is called the *moment-generating function*. The n-th cumulant is then defined by

$$K_n(F) = \frac{\partial^n}{\partial s^n} \log \mathcal{G}(s)|_{s=0}. \tag{2.26}$$

We say that $\log \mathcal{G}$ is the cumulant generating function. There is a generalisation of all this to any number of random variables. Let $F_1, \ldots, F_m$ be random variables; define

$$\begin{aligned} \mathcal{F}(x_1, x_2, \ldots, x_m) &= E\left(\exp(i \sum_j x_j F_j)\right) \\ &= \sum_\omega p(\omega) \exp\{i(x_1 F_1(\omega) + \ldots + x_m F_m(\omega))\}. \end{aligned} \tag{2.27}$$

This always has the properties

1. $\mathcal{F}(0, \ldots, 0) = 1$;
2. $\mathcal{F}$ is a continuous function of the m real variables $X = (x_1, \ldots, x_m)$;
3. $\sum_{i,j=1}^N \overline{\lambda}_i \lambda_j \mathcal{F}(X_j - X_i) \geq 0$ for all choices of N complex numbers λ_i and N real m-vectors X_i.

Parts (1) and (2) are not hard to prove; (3) follows simply from the positivity of the expectation of the random variable $|\sum \lambda_i \exp\{\sum_{j=1}^m x_{ij} F_j\}|^2$. If $\mathcal{F}$ is a bit more than continuous, say analytic in a neighbourhood of $X = 0$, then the multiple power series expansion converges and the coefficient of $x_1^{r_1} \ldots x_m^{r_m}$ is related by combinatorial factors to the moment

$$E\left(F_1^{r_1} F_2^{r_2} \ldots F_m^{r_m}\right).$$

Again, the multiple cumulants are similarly generated by $\log \mathcal{F}$. There is also a Bochner theorem, which says that any function obeying (1), (2) and (3) is the characteristic function of m random variables F_j on some space. If $\log \mathcal{F}$ is a quadratic form in x_j, then we say that the random variables are jointly Gaussian. There is the interesting result [13], which says that if all the cumulants vanish beyond some order, then they vanish beyond the second order, and the system of random variables is Gaussian.

The characteristic function is the Fourier transform of the joint distribution of the variables; see Exercise (2.61). There is no analogue of joint probability distribution in quantum mechanics, between non-compatible observables, but there are analogues of correlation functions. These are defined in terms of the state, and the algebraic properties of the observables; to keep the analogy and to allow a quantum generalisation, we develop the classical theory along these lines.

A random variable represents an *observable* in the theory. This is just another way of saying that an observable in a theory with randomness will in general be random. We shall use the terms random variable and observable interchangeably. The constant function $F(\omega) = 1$, or any other constant, is not actually random, as it has zero variance. Conversely, any random variable with zero variance is constant. See Exercise (2.60). Such a random variable is called a *sure* function, and is said to take its value with probability 1.

The set of random variables forms an *algebra*; an algebra is a vector space with a product, so we must specify the product in question. We choose the *point-wise* product (ordinary product of the values of the functions); the addition of functions, and real multiples of functions, is also taken point-wise. Of great technical usefulness is the set of bounded observables, denoted $\mathcal{C}(\Omega)$, which is a subalgebra of the set of all observables. A bounded observable is a map $F : \Omega \to \mathbf{R}$ such that the range of F, that is, its set of values, makes up a bounded set in $\mathbf{R}$. In symbols,

$$\|F\|_\infty = \sup\{|F(\omega)| : \omega \in \Omega\} < \infty. \tag{2.28}$$

Now try Exercise (2.64). Clearly any sure observable is bounded, as its range is a single point. The advantage of the bounded observables is that all means, variances and correlations are finite (the sums (2.18) etc. all converge). Moreover, the value of the supremum in Eqn. (2.28) is a norm for the algebra $\mathcal{C}(\Omega)$, which is complete in the metric given by this norm. $\mathcal{C}(\Omega)$ is a Banach algebra; in particular, it is a real Banach space. In Exercise (2.65), you are asked to show that $\mathcal{C}(\Omega)$ is the topological dual of $\ell^1(\Omega)$, the span of the set of states, introduced in Sec. (2.1) and with norm given by Eqn. (2.28). It is clear that each $F \in \mathcal{C}(\Omega)$ can be used to define a continuous linear functional on $\ell^1(\Omega)$ by the formula

$$p \mapsto \langle p \rangle = \langle p, F \rangle = \sum_\omega p(\omega) F(\omega). \tag{2.29}$$

This is clearly linear in the variable p, and it is part of the exercise to show that it is continuous in the sense that if

$$\|p_n - p\|_1 \to 0, \text{ then } \langle p_n, F \rangle \to \langle p, F \rangle. \tag{2.30}$$

The converse also holds: if $p \mapsto \langle p \rangle$ is a continuous linear functional on $\ell^1(\Omega)$ then there exists an element, say F of $\mathcal{C}(\Omega)$, such that Eqn. (2.29) holds (see

Exercise (2.65)). Thus, $\mathcal{C}(\Omega)$ is exactly the set of continuous linear functions from $\ell^1(\Omega)$ to $\mathbf{R}$, which is known as the topological dual. We say that ℓ^1 is the predual of $\mathcal{C}(\Omega)$. A Banach space which is equal to the dual of its dual is said to be *reflexive.* In fact, ℓ^1 is not reflexive, being only a tiny subspace of the set $\mathcal{C}^d$ of continuous linear functionals on $\mathcal{C}(\Omega)$. The missing states are not countably additive. We shall manage to avoid them. Imitating Dirac we write $\langle p, F\rangle$ for both the action of F on $p \in \Sigma$, and for the action of p on F. It obeys the basic inequality

$$|\langle p, F\rangle| \leq \|p\|_1 \quad \|F\|_\infty. \tag{2.31}$$

The Banach space $\mathcal{C}(\Omega)$ contains the *positive cone*, $\mathcal{C}_+$ which consists of bounded functions F obeying $F(\omega) \geq 0$ for all ω. The space $\ell^1(\Omega)$ contains the positive cone, ℓ^1_+ of summable functions with non-negative values. These cones are dual to each other, in that given $F \in \mathcal{C}(\Omega)$, then $\langle p, F\rangle \geq 0$ for all $p \in \ell^1_+$ if and only if $F \in \mathcal{C}_+$; conversely, given $p \in \ell^1$, then $\langle p, F\rangle \geq 0$ for all $F \in \mathcal{C}_+$ if and only if $p \in \ell^1_+$ (see Exercise (2.66)). We shall frequently use the separating property of a dual space: $p \in \ell^1$ is determined by its values $\langle p, F\rangle$ for all $F \in \mathcal{C}$ and $F \in \mathcal{C}$ is determined by its values $\langle p, F\rangle$ for all $p \in \ell^1$.

We have seen that the pure states (the point measures) form a natural basis $\{\omega_1, \omega_2, \ldots\}$ in the completion of Span Ω, the space $\ell^1(\Omega)$. The dual set in $\mathcal{C}(\Omega)$ is defined as the sequence of functions F_n such that $\langle \omega_m, F_n\rangle = \delta_{n,m}$; clearly F_n is zero except at ω_n, where it is 1. The algebraic span of these consist of functions that are zero except at a finite number of points; we say such a function has *finite support.* However the completion in the norm $\| \bullet \|_\infty$ of the algebraic span of all the F_n is not the whole of $\mathcal{C}$ but consists of its compact elements, $\mathcal{K}(\Omega)$. These are so called since they are compact operators when regarded as multiplication operators on the Hilbert space of square-summable sequences $\ell^2(\Omega) = \{F \in \mathcal{C} : \sum |F(\omega)|^2 < \infty\}$. For example, if $|\Omega| = \infty$, a non-zero sure function cannot be uniformly approximated by functions of finite support (Exercise (2.67)). This natural basis in Span Ω allows us to identify the $\ell^1(\Omega)$ with a subset of $\mathcal{K}(\Omega)$, where the point measure at ω_n is identified with F_n. We get only a subset, (the elements of trace-class). This identification of a space with part of its dual, although natural, can lead to confusion since the action of a linear operator on a space should be distinguished from the dual action on the dual. This is illustrated in quantum mechanics, where we shall see a similar identification of density-matrices as operators; nevertheless, the time-evolution of the density matrices is the conjugate of the time-evolution of the observables [1]. We shall see in the next Chapter that this is true in classical statistical dynamics too.

If $|\Omega| < \infty$, any linear transformation T on $\mathcal{C}$ can be "thrown" onto the dual space, which in the finite case is ℓ^1, to provide the dual linear transformation there. We shall denote the dual of T by T^d; its defining property is

$$\langle T^d p, F\rangle = \langle p, TF\rangle, \qquad p \in \ell^1, \;\; F \in \mathcal{C}. \tag{2.32}$$

This defines the linear map $T^d : p \mapsto T^d p$ uniquely, since there is only one functional that, applied to F, gives the right-hand side. Since $\mathcal{C}$ is a reflexive Banach space if $|\Omega| < \infty$, we have $T^{dd} = T$. More generally, the dual of a linear map from $\mathcal{V}_1$ to $\mathcal{V}_2$, where these are any vector spaces, is a map from $\mathcal{V}_2^d$ to $\mathcal{V}_1^d$.

Similarly, any continuous linear map on $\ell^1(\Omega)$ can be thrown onto $\mathcal{C}(\Omega)$, where now the condition of finiteness can be dropped. Consider the case where $\Omega = \Omega_1 \times \Omega_2$, and put $\mathcal{C} = \mathcal{C}(\Omega)$, $\mathcal{C}_1 = \mathcal{C}(\Omega_1)$, and ℓ^1, ℓ^1_1 their predual spaces. The marginal distribution Eqn. (2.12) gives us a linear map from ℓ^1 to ℓ^1_1. The dual of this must be a map from $\mathcal{C}_1$ to $\mathcal{C}$. We claim that this is the injection of $\mathcal{C}_1$ into $\mathcal{C}$. Any function of the variable ω_1 can be regarded as a function of $\omega = (\omega_1, \omega_2)$ which happens not to depend on ω_2. In this sense, $\mathcal{C}_1$ is a subset of $\mathcal{C}$. Let Υ denote this injection. We prove our claim by verifying Eqn. (2.32) for all $p \in \ell^1$ and all $F \in \mathcal{C}_1$. For then

$$\langle \mathcal{M}_1 p, F\rangle = \sum_{\omega_1} F(\omega_1) \sum_{\omega_2} p(\omega_1, \omega_2) = \sum_{\omega_1, \omega_2} p(\omega_1, \omega_2)(\Upsilon F)(\omega_1, \omega_2)$$

which shows that $\Upsilon^d = \mathcal{M}$ as claimed. The map Υ is called an *ampliation*. Now again suppose that $|\Omega| < \infty$, and that Ω as the product of Ω_1 and Ω_2; then corresponding algebras are related by the tensor product

$$\mathcal{C}(\Omega) = \mathcal{C}(\Omega_1) \bigotimes \mathcal{C}(\Omega_2). \tag{2.33}$$

This means that any function of two variables, ω_1 and ω_2, which run over a finite set can be written as a sum of product functions $F(\omega_1)G(\omega_2)$. We write $F \otimes G$ for the product function. In this notation, the ampliation is given by $F \mapsto F \otimes 1$. We can extend Eqn. (2.33) to any finite number of factors. In particular, when Ω has a local structure, as in Eqn. (2.1) we get

$$\mathcal{C}(\Omega) = \bigotimes_{x \in \Lambda} \mathcal{C}(\Omega_x). \tag{2.34}$$

In this way, the algebra of observables acquires a local structure, since a function depending only on the ω_x for $x \in \Lambda_0$ can be regarded by ampliation as an element of $\mathcal{C}(\Omega)$ that is localised in Λ_0. Note that this is not related to being zero outside Λ_0 in any sense. An important example is the local number, defined in the case where $\Omega_x = \{0, 1, 2, \dots\}$ as the function $\mathcal{N}_x(\omega) = \omega_x$, which is an integer, the value of the sample point at x.

The formulation of independence is best done using events. We specify a state in $\Sigma(\Omega)$.

Definition 2.35 *We say that two events Ω_1, Ω_2 (i.e., two subsets of Ω) are independent if*

$$\text{Prob}\{\Omega_1 \cap \Omega_2\} = \text{Prob}\{\Omega_1\}\text{Prob}\{\Omega_2\}.$$

Then two observables F and G are independent if every level set $\Omega_f = \{\omega : F(\omega) = f\}$ is independent of every level set $\Omega_g = \{\omega : G(\omega) = g\}$. Note that there is no connection between this concept, which is *statistical* independence, and depends on the choice of p, and the linear independence of F and G, although if F and G are functionally dependent they can be statistically independent only if they are constant (Exercise (2.68)). Similarly we say that three random variables, F, G, H are mutually independent if the joint probability distribution $p_{F,G,H}(f, g, h)$ is the product of its marginals. It can be shown that three random variables can be pairwise independent but not mutually independent. Mutual independence can be formulated for events, extending Def. (2.35) to more than two sets.

Another way to express independence involves the concept of a *ring of subsets* (to avoid confusion, I follow [14] in not calling it an algebra).

Definition 2.36 (Boolean Ring) *A collection $\mathcal{B}$ of subsets of Ω is said to be a* Boolean ring *if Ω itself belongs to $\mathcal{B}$, the complement of any member of $\mathcal{B}$ is also a member, and if the union of two elements of $\mathcal{B}$ is also in $\mathcal{B}$.*

By working with the axioms, we can show that the empty set is a member of $\mathcal{B}$, and that the union of any finite number of members of $\mathcal{B}$ is also a member. It is a ring, with product given by intersection, and sum given by symmetric difference [14]. In measure theory use is made of the concept of Boolean σ-ring, which is a Boolean-ring in which the union of countably many members is always a member. A σ-ring defines what is known as a *measurable structure* on Ω, and the elements are called measurable sets. The necessity for the introduction of this idea is seen in the Banach-Tarski "paradox", in which it is shown that a sphere can be cut up into parts, and reassembled to form a larger sphere. The sets involved are very pathological, and measure theory outlaws such sets (they are not measurable). Such troubles cannot arise if Ω is countable, but nevertheless Boolean σ-rings provide us with a useful heuristic. Given a measurable space $(\Omega, \mathcal{B})$, we say that a function $F : \Omega \to \mathbf{R}$ is $\mathcal{B}$-measurable if the inverse image of every interval (in $\mathbf{R}$) lies in $\mathcal{B}$. Here, the inverse image of the interval $[a, b]$ is the set $\{\omega \in \Omega : a \leq F(\omega) \leq b\}$. For example the indicator function of a measurable set is measurable. There is some similarity, in the idea of a measurable function, with the idea of a continuous function, defined in terms of a collection of sets $\mathcal{S}$, called the open sets: a function $F : \omega \to \mathbf{R}$ is continuous if the inverse image of every open interval is open. The open sets provide Ω with a topology (a topological structure), just as the measurable sets provide it with a measurable structure. The following manœuvres are very useful. Given a collection of sets, there is a unique smallest Boolean σ-ring containing every set in the collection; it is called the σ-ring generated by the sets. Similarly, given a collection of functions $\{F, G \ldots\}$ on Ω, there is a unique smallest Boolean σ-ring with respect to which they are all measurable; it is the σ-ring generated by all the level sets of the functions (when Ω is countable). We denote it by $\mathcal{B}(F, G, \ldots)$. We shall use this idea when we have a local structure for Ω, so that a sample point ω is written as $\omega = (\omega_1, \omega_2, \ldots)$. Here, either there are a finite number of points ω_x, one for each point in Λ or we have an infinite sequence, one for each time $t = 0, 1, \ldots$. The latter case arises in processes. In either case, we define a *cylinder set* to be any subset of Ω defined by a condition on some finite number of components ω_j, with no condition on the remaining components. For example, let $\Omega_{x_1,0}$ be a subset of Ω_{x_1} for one $x_1 \in \Lambda$. Then the condition $\omega_{x_1} \in \Omega_{x_1,0}$ defines the cylinder set of all configurations whose component at the point x_1 takes one of the values in $\Omega_{x_1,0}$. Where the label x refers to the time, as in a process, we shall denote it by t; then consider the cylinder set which contains all paths which at time t_1 sit in $\Lambda_{t_1,0}$, which can be pictured as all paths passing through the "gate" $\Lambda_{t_1,0}$ at time $t = t_1$. In this case, $\Lambda_{x_1,0}$ or $\Lambda_{t_1,0}$ is the base of the cylinder set. Similarly, cylinder sets with bases given by conditions on two, three or more variables ω_j can be pictured. To each region Λ_0 we may consider all cylinder sets based on variables $\omega_x : x \in \Lambda_0$. These generate a σ-ring associated to Λ_0, which we call $\mathcal{B}(\Lambda_0)$. Then (Exercise (2.69)) a function on Ω depends only on the variables in Λ_0 if and only if it is $\mathcal{B}(\Lambda_0)$–measurable.

We now formulate the concept of independence using σ-rings. Given a countable probability space (Ω, p) we say that two σ-rings $\mathcal{B}_1$ and $\mathcal{B}_2$ are independent if all the events in one are independent of all the events in the other. Recall the definition (2.35) for events. Then it follows that all random variables measurable with respect to $\mathcal{B}_1$ are independent of all random variables measurable with respect to the other. Conversely, given two independent random variables, the σ-rings they generate are independent. This can be generalised to a family of mutually independent σ-rings. Thus, our definition of a state's being independent over Λ can be restated in dual terms, by saying that the σ-rings $\mathcal{B}_x$ generated by the cylinder sets are mutually independent; we might also say that the algebras $\mathcal{C}_x$ are mutually independent.

We can use Boolean rings to give a neat formulation of conditional probability. If $\Omega_1 \subseteq \Omega$ is an event in the probability space (Ω, p), such that $\text{Prob}(\Omega_1) \neq 0$, then the conditional probability of an event Ω_2, given Ω_1, is defined to be

Definition 2.37 (Conditional Probability)

$$\text{Prob}(\Omega_2|\Omega_1) = \frac{\text{Prob}(\Omega_1 \cap \Omega_2)}{\text{Prob}(\Omega_1)}.$$

This is the correct probability to use if, after a measurement on (Ω, p), we found that the sample point was certainly in Ω_1, but that no more information was available. If Ω_1 and Ω_2 are independent, then conditioning relative to Ω_1 makes no difference to the probability of Ω_2. This is obvious, as $\text{Prob}(\Omega_1 \cap \Omega_2)$ is the product of the probabilities of Ω_1 and Ω_2. We shall denote by $p(\omega|\Omega_1)$ the function of ω got by choosing Ω_2 to be the set containing the single point ω. It is to be shown in Exercise (2.70) that this is indeed a probability on Ω. A slight variation of this concept is the conditional probability, given the value of the random variable F. If we are given that $F = f$, then we can say that the event, the level set Ω_f, has happened. Then the conditional probability, given F, is the collection of states $p(\omega|\Omega_f)$, as f runs over the possible values of F. We might be given more information, such as the values of F and G, or any collection of observables. Now, giving G when F is known might entail giving some redundant information, especially when G is functionally related to F. Thus if $G = F^3$, conditioning given F and G is the same as giving just F. What is relevant is the Boolean ring $\mathcal{B}(F, G, \ldots)$ of joint level sets of the functions. In our case, where Ω is countable, we can write Ω as the union of disjoint, minimal elements of $\mathcal{B}(F, G, \ldots)$, and the probability is conditioned by the information telling us which of these sets the sample point is in. This leads to the concept of $\text{Prob}(\Omega_1|\mathcal{B})$, the conditional probability given the Boolean ring $\mathcal{B}$: it is the collection of conditional probabilities of Ω_1, conditioned by each of the sets in $\mathcal{B}$.

The *conditional expectation* of a random variable G, given F, naturally uses the conditional probability, and forms the mean in the usual way. It is denoted $E(G|F)$. Thus

$$E(G|F)(f) = E(G|\mathcal{B}(F)) = \sum_{\omega} G(\omega) p(\omega|\Omega_f). \tag{2.38}$$

Again, if G is independent of F, then these conditional expectations coincide with the usual expectation of G. The conditional expectation can be regarded as a new random variable, as it is a function on Ω, constant on the level-sets of F. Thus, define $E(G|F)(\omega) = E(G|F)(f)$ where $f = F(\omega)$. This may be said to be the part of the function G that is a function of F. This can be more exactly expressed as follows. Although we have avoided problems of measurability by using a discrete sample space, we find it useful to consider random variables that are measurable relative to the Boolean σ-ring $\mathcal{B}$. Then we denote by $\ell^\infty(\Omega, \mathcal{B})$ the algebra of bounded $\mathcal{B}$-measurable functions; if $\mathcal{B} = \mathcal{B}(F)$, then this is the set of bounded functions of F. Similarly we define the space $\ell^2(\Omega, p, \mathcal{B})$ as the set of $\mathcal{B}$-measurable functions $F(\omega)$ such that $\sum p(\omega)|F(\omega)|^2 < \infty$. This is a Hilbert space. You are asked in Exercise (2.71) to show that if $\mathcal{B} = \mathcal{B}(F)$ for some function, then the map $G \mapsto E(G|F)$ is the orthogonal projection from

$\ell^2(\Omega, p)$ to $\ell^2(\Omega, p, \mathcal{B}(F))$. It is then natural to define, for any $\mathcal{B}$, the conditional expectation $E(G|\mathcal{B})$ to be the orthogonal projection of G onto $\ell^2(\Omega, p, \mathcal{B})$.

2.3 Entropy

The entropy of a probability p on a countable space Ω is given by

Definition 2.39 $S(p) = -\sum_{\omega\in\Omega} p\log p$.

In this definition, we adopt the convention that $0\log 0 = 0$, and that a divergent sum means that the entropy is infinite (all the terms are non-negative). This formula is a special case of the quantum entropy defined by von Neumann [1], but it is usually attributed to Shannon, who took it to represent missing information, and proved some theorems about this expression. Thus, if $\Omega = \Omega_1 \times \Omega_2$ and $p = p_1 \otimes p_2$, then we have:

$$\begin{aligned} S(p) &= -\sum_{\omega_1,\omega_2} p(\omega)\log\left(p_1(\omega_1)p_2(\omega_2)\right) \\ &= -\sum_{\omega_1,\omega_2} p_1(\omega_1)p_2(\omega_2)[\log p_1(\omega_1) + \log p_2(\omega_2)] \\ &= -\sum_{\omega_2} p_2(\omega_2)\sum_{\omega_1} p_1(\omega_1)\log p_1(\omega_1) - \sum_{\omega_1} p_1(\omega_1)\sum_{\omega_2} p_2(\omega_2)\log p_2(\omega_2) \\ &= S(p_1) + S(p_2) \end{aligned}$$

since $\sum_{\omega_1} p_1(\omega_1) = \sum_{\omega_2} p_2(\omega_2) = 1$. For the converse, for conditions under which S is the only function with the additive property, see Renyi [15]. We can now see that entropy is an extensive quantity in a model with local structure, whenever the state is independent over Λ. Exercise (2.52). Another easy result is that the entropy is zero if and only if the state is a point measure (recall that such is an extremal point of the set Σ). We measure S in bits of information, if the logarithm is to base 2. We shall choose to measure in units such that the logarithm is to base e. The physical entropy is obtained by multiplying S by k_B, where k_B is Boltzmann's constant. The entropy is an unbounded function on Σ, but can be shown to be finite on a dense set of states. (Exercise (2.51).) It cannot be defined on the whole dual space ℓ^1, since it relies on the fact that the values of $p(\omega)$ are non-negative. But if needed it can be extended to positive but not normalised elements.

We now prove the intuitively obvious idea that the uniform distribution contains no information about the system.

Theorem 2.40 *Suppose that* $|\Omega| = n < \infty$. *Then the state* $p \in \Sigma(\Omega)$ *that maximises* $S(p)$ *is unique, and is given by the uniform distribution* $p(\omega) = 1/n$.

Proof. We must maximise $S(p)$ over the variables $p_1, p_2, \ldots, p_n$ subject to the constraint $\sum_j p_j = 1$. We use the method of Lagrange multipliers; so, maximise $L = S(p) + \lambda \sum_j p_j$ subject to no constraint, and then find λ by fitting the solution to the constraint. We find the turning points in the region $0 < p_j < 1, \ 1 \leq p_j \leq n$ by

$$\frac{\partial L}{\partial p_j} = -\log p_j - p_j(1/p_j) + \lambda = 0.$$

The solution is $p_j = \exp(\lambda - 1)$ for all j; since they sum to 1, the unique answer is $p_j = 1/n$, and $S = \log n$. The second derivative $\partial^2 L/\partial p_j \partial p_k$ without constraint is $-\delta_{j,k}/p_j$ and this is negative. This shows that we are at a maximum, and *a fortiori* the constrained function is a maximum. The function $S(p)$ is smaller than $\log n$ on the boundary of Σ: if we are on a face of the simplex Σ, such that $m > 0$ components of p are zero, the same method shows that the maximum of S on the face is $\log(n - m)$, which is less. □

Inasmuch as information is negative entropy, it is therefore minimised by the uniform distribution.

In Exercise (2.50) it is shown by a similar method that the canonical state is the state of maximum entropy among all states with a given value of the mean energy.

The entropy is a concave function: this means that the area under the graph of S is a convex set. The exact definition is

Definition 2.41 *Let Σ be a convex set. A function $F : \Sigma \to \mathbf{R}$ is said to be* concave *if for all $\lambda \in (0, 1)$, we have*

$$F(\lambda p + (1 - \lambda)q) \geq \lambda F(p) + (1 - \lambda)F(q)$$

for all p and q in Σ. It is said to be strictly concave *if $\geq$ is replaced by $>$ whenever $p \neq q$.*

Note that linear functionals are concave but never strictly concave. The space Σ must be convex (as, indeed, it is if it is $\Sigma(\Omega)$) in order for the left-hand side to make sense. We shall prove a much stronger result, which estimates the amount of concavity of S; in the following theorem, $\|p\|_2^2 = \sum |p_j|^2$ is the ℓ^2-norm.

Theorem 2.42 (Concavity of the Entropy) *Let $p, q \in \Sigma(\Omega)$ and $0 \leq \lambda \leq 1$. Then*

$$S(\lambda p + (1 - \lambda)q) - \lambda S(p) - (1 - \lambda)S(q) \geq \frac{1}{2}\lambda(1 - \lambda)\|p - q\|_2^2. \tag{2.43}$$

Proof. We use Taylor's theorem with Lagrange remainder up to second order, and shall use ξ and η to denote intermediate abscissae. Let $f(x) = -x\log x$; then $f' = -\log x - 1$ and $f'' = -1/x$. First let p and q denote numbers in $[0,\ 1]$. Let $x = \lambda p + (1-\lambda)q$; then $p = x + (1-\lambda)(p-q)$ and $q = x + \lambda(q-p)$. Then the difference in entropy is

$$\begin{aligned}
\delta S &= f(x) - \lambda f(p) - (1-\lambda)f(q) \\
&= f(x) - \lambda\{f(x) + (1-\lambda)(p-q)[-\log x - 1] - (1-\lambda)^2(p-q)^2/(2\xi)\} \\
&- (1-\lambda)\{f(x) + \lambda(q-p)[-\log x - 1] - \lambda^2(p-q)^2/(2\eta)\} \\
&= \lambda(1-\lambda)^2(p-q)^2/(2\xi) + (1-\lambda)\lambda^2(p-q)^2/(2\eta) \\
&> \left(\lambda(1-\lambda)^2 + (1-\lambda)\lambda^2\right)(p-q)^2/2 \\
&= \frac{1}{2}\lambda(1-\lambda)|p-q|^2.
\end{aligned}$$

Now do the same for each component p_j and q_j, and sum, to get the theorem. □

This proof is typical; the estimate involves the second derivative $-f''$ which is positive and its intermediate values are all larger than 1. We interpret the theorem as saying that if we mix two different states of a gas or liquid, then there is a gain in entropy overall. The theorem applies to states p, q and $\lambda p + (1-\lambda)q$ of a solid, but it is hard to see how to do the mixing experimentally.

Suppose that Ω has a product structure, so that $\Omega = \Omega_1 \times \Omega_2$, and write $\mathcal{B}_1$ for the Boolean ring consisting of cylinder sets based on subsets of Ω_1; similarly, write $\mathcal{B}_2$ for the cylinder sets based on Ω_2. Then a state in which $\mathcal{B}_1$ and $\mathcal{B}_2$ are independent has higher entropy than a state with the same marginals in which they are not independent. Before proving this, we need an important result. It concerns the *relative entropy* $S(p|q)$ of p given q, defined to be

Definition 2.44 (Relative Entropy) *Let p and q be states in $\Sigma(\Omega)$. Then*

$$S(p|q) = \sum_\omega p(\omega)(\log p(\omega) - \log q(\omega)).$$

Lemma 2.45 (Kullback) $S(p|q) \geq \frac{1}{2}\|p-q\|_2^2$.

Proof. If p and q lie in the interval $(0,\ 1)$, then

$$p\log p = q\log q + (p-q)(1+\log q) + (p-q)^2/(2\xi) \geq p - q + p\log q + (p-q)^2/2.$$

Do this for each p_j and q_j, and sum, giving

$$\sum_j p_j(\log p_j - \log q_j) \geq \sum_j (p_j - q_j)^2/2$$

since $\sum p_j - \sum q_j = 0$. □

Now let $p \in \Sigma(\Omega_1 \times \Omega_2)$, and let p_1, p_2 be the marginals onto Ω_1 and Ω_2. That is

$$p_1(\omega_1) = (\mathcal{M}_1 p)(\omega_1) = \sum_{\omega_2} p(\omega_1, \omega_2),$$

and similarly for p_2. Then we have

Theorem 2.46

$$S(p_1 \otimes p_2) - S(p) \geq \|p - p_1 \otimes p_2\|_2^2/2.$$

Proof. Let $q = p_1 \otimes p_2$ in Kullback's lemma; then

$$\sum p(\omega) \log p(\omega) - \sum p(\omega) \log (p_1(\omega_1) p_2(\omega_2)) \geq \frac{1}{2}\|p - q\|_2^2.$$

The second sum on the left-hand side is

$$-\sum_{\omega_1,\omega_2} p(\omega_1,\omega_2) \log p_1(\omega_1) - \sum_{\omega_1,\omega_2} p(\omega_1,\omega_2) \log p_2(\omega_2) = S(p_1) + S(p_2). \quad (2.47)$$

Collecting terms gives the result. □

We shall use this result a lot when we come to discuss the stoss map. It can be used to give a proof of Theorem (2.42).

2.4 Exercises

Exercise 2.48 Let $\mathcal{E}$ be a random variable such that the canonical partition function is finite. Show that any level set Ω_E of $\mathcal{E}$ obeys $|\Omega_E| < \infty$.

Exercise 2.49 Suppose that $\Lambda_1 \cup \Lambda_2 = \emptyset$. Show that

$$\Omega(\Lambda_1 \cup \Lambda_2) \equiv \Omega(\Lambda_1) \times \Omega(\Lambda_2)$$

and that this is not true in the particle point of view.

Exercise 2.50 If $\Omega = \{0, 1, 2, \ldots\}$ and $\mathcal{E}(\omega) = \kappa\omega$, show that the canonical distribution is

$$p(\omega) = (1 - \exp\{-\beta\kappa\}) \exp\{-\beta\kappa\omega\}.$$

Hence show that the mean energy is given by Planck's distribution.

Show that in general the state with the greatest entropy among all states with the same given mean energy is the canonical state. Show that the state having the greatest entropy among all states with the same given mean energy and the same given mean particle number is the grand canonical state.

Exercise 2.51 Show that if $|\Omega| < \infty$ then $S(p) < \infty$ on a dense set.

Exercise 2.52 Show that if p is independent over Λ then S is extensive.

Exercise 2.53 Show that if $p_1, p_2, \ldots, p_n$ lie in a convex set Σ, then the smallest convex set containing all these points also lies in Σ.

Exercise 2.54 Show that a closed triangle in the plane and a closed tetrahedron in $\mathbf{R}^3$ are simplices. Show that the state spaces of the sample spaces $\{1,2,3\}$ and $\{1,2,3,4\}$ are simplices.

Exercise 2.55 Show that any element of $\ell^1(\Omega)$ can be written as $\alpha p - \beta q$, where $\alpha, \beta \in \mathbf{R}_+$, $p, q \in \Sigma(\Omega)$ even if $|\Omega| = \infty$.

Exercise 2.56 Show that $\| \bullet \|_1$ is a norm, and that $\ell^1(\Omega)$ is complete in this norm.

Exercise 2.57 Let $\Omega = \Omega_1 \times \Omega_2$; show that the map $\mathcal{M}_1 : \Sigma(\Omega) \to \Sigma(\Omega_1)$ is continuous in the norm $\| \bullet \|_1$.

Exercise 2.58 Show that the mean and variance of a random variable F on a sample space Ω furnished with a probability p are given by

$$\begin{aligned} \langle F \rangle &= \sum_f f p_F(f) \\ \mathcal{V}F &= \sum_f (f - \langle F \rangle)^2 p_F(f). \end{aligned}$$

Exercise 2.59 Show that the joint probability distribution $p_{F,G}$ must satisfy the inequality

$$\left| \sum_{f.g} fg\, p_{F,G}(f,g) \right|^2 \leq \mathcal{V}F.\mathcal{V}G.$$

Exercise 2.60 Show that any random variable with zero variance is sure.

Exercise 2.61 Show that the variance is given by

$$E(F^2) - (E(F))^2 = \frac{\partial^2}{\partial s^2} \log \mathcal{G}(s); \tag{2.62}$$

show that in two variables, the covariance is given by

$$E(FG) - E(F)E(G) = \frac{\partial^2}{\partial s \partial t} \log \mathcal{G}(s,t). \tag{2.63}$$

Exercise 2.64 Show that $\|F\|$ is a norm obeying $\|\,|F|^2\,\| = \|F\|^2$.

Exercise 2.65 Show that $\mathcal{C}(\Omega)$ is the topological dual of $\ell^1(\Omega)$. If $|\Omega| < \infty$, show that $\mathcal{C}(\Omega)$ and $\ell^1(\Omega)$ are mutual duals, and obtain the equations

$$\begin{aligned}\|p\|_1 &= \sup_F\{|\langle p, F\rangle| : \ F \in \mathcal{A}, \ \|F\|_\infty = 1\} \text{ for any } p \in \ell^1 \\ \|F\|_\infty &= \sup_p\{|\langle p, F\rangle| : \ p \in \ell^1, \ \|p\|_1 = 1\} \text{ for any } F \in \mathcal{A}.\end{aligned}$$

Exercise 2.66 Show that $\Sigma(\Omega)$ and the cone of positive bounded random variables on Ω are dual cones.

Exercise 2.67 Show that the function 1 on Ω cannot be a uniform limit of functions of finite support if $|\Omega| = \infty$.

Exercise 2.68 Show that two functionally dependent random variables can be statistically independent only if they are both sure.

Exercise 2.69 Let $\Omega = \prod_{x=1}^n \Omega_x$. Show that a function $F(\omega_1, \ldots, \omega_n)$ depends only on $\omega_1, \ldots, \omega_j$ if and only if F is measurable relative to the σ-ring generated by the cylinder sets based on all gates defined by sets in $\Omega_1, \ldots, \Omega_j$.

Exercise 2.70 Show that any conditional probability is a probability.

Exercise 2.71 Show that the map $G \mapsto E(G|F)$ is the orthogonal projection of $G \in \ell^2(\Omega, p)$ onto the subspace $\ell^2(\Omega, p, \mathcal{B}(F))$.

Chapter 3

Linear Dynamics

3.1 Reversible Dynamics

In classical mechanics, the dynamics means the motion of a point in phase space as time goes by; this is determined by Newton's laws. In a field theory, the field equations of motion are expressed as partial differential equations. In our much simplified models with a countable configuration space we cannot formulate differential equations. We choose the time t to be discrete, labelled by the group $\mathbf{Z} = \{\ldots, -1, 0, 1, 2, \ldots\}$. Then one time step in the dynamics is specified by giving a bijection of Ω, that is, a permutation, denoted by τ. The set of permutations of a set form a group, denoted $AUT\,\Omega$. A permutation is invertible, and τ is thus non-dissipative. We shall write the action as a right action, thus: $\omega \mapsto \omega\tau$. The effect of this is that the product $\tau_1\tau_2$ means the permutation in which τ_1 is done first, followed by τ_2. The invariance of the system under translations in time is expressed by choosing τ to be independent of time. Then the time-evolution through n steps is given by the composition of n single time steps, given by the permutation

$$\tau(n) = \underbrace{\tau \circ \tau \circ \ldots \circ \tau}_{n}.$$

We express this by saying that the dynamics is given by a group homomorphism $\tau : \mathbf{Z} \to AUT\Omega$. This means that $\tau(m) \circ \tau(n) = \tau(m+n)$ and $\tau(-n)$ is the inverse permutation $(\tau(n))^{-1}$. These relations are obvious from the definition of $\tau(n)$.

Given a dynamics τ, and a point ω_0, the set of points got by moving ω_0, backwards as well as forwards in time, is called the *orbit* of ω_0. It is denoted by $\omega_0\mathbf{Z}$, that is, we have

$$\omega_0\mathbf{Z} = \{\omega \in \Omega : \omega_0\tau(n) = \omega \text{ for some } n \in \mathbf{Z}\}.$$

Being on the same orbit is an equivalence relation (Exercise (3.28)) and so Ω is divided up as the union of disjoint orbits. We may also move subsets around by time evolution: if Ω_0 is an event, then the orbit of Ω_0 is the set

$$\Omega_0\mathbf{Z} = \{\omega \in \Omega : \omega = \omega_0 \tau(n) \text{ for some } n \in \mathbf{Z} \text{ and some } \omega_0 \in \Omega\}.$$

We have chosen the measurable structure so that every set is measurable, but we might be interested in a Boolean subring $\mathcal{B}_1$. We say that a map $\tau : \Omega \to \Omega$ is $\mathcal{B}_1-$measurable if the inverse image $\Omega_1\tau^{-1}$ of every element Ω_1 of $\mathcal{B}_1$ is also in $\mathcal{B}_1$. We say that a subset Ω_0 is invariant under time evolution if $\Omega_0\mathbf{Z} = \Omega_0$. Given a state $p \in \Sigma(\Omega)$, some people say that a dynamics $\{\tau(n)\}_{n=\ldots,-1,0,1\ldots}$ is ergodic if there are no invariant subsets except those of measure 0 or 1. Alternatively, if τ is given, and the same property holds, we say that the state p is ergodic. The word is constructed out of the erg, the unit of energy, and should really be related to an energy function in the theory. Thus let $\mathcal{E}$ be a random variable, chosen to represent the energy of the system. In order to get a theory obeying the first law of thermodynamics (the conservation of energy), we must choose the dynamics τ so that it leaves all the level sets of $\mathcal{E}$ invariant. This still leaves many choices of τ. This contrasts with Hamiltonian mechanics, where the dynamics uniquely determines the Hamiltonian up to an overall constant. Alternatively, if τ is given, then we must choose $\mathcal{E}$ so that the orbits are the level sets; but then the value of the energy on each orbit is not determined uniquely. We are left with rather a lot of freedom in constructing an energy-conserving dynamics. In order for the canonical state to exist, each energy-shell must be finite (Exercise (2.48)). So as to allow for energy-conserving dynamics, we don't use the above definition of ergodicity, but limit the notion to energy-conserving maps, and use the following

Definition 3.1 *Let $\mathcal{E}$ be a random variable on a countable set Ω, giving rise to energy-shells Ω_E. Let τ be one time-step of an energy-conserving dynamics. We say that τ is* ergodic *if for each E, there is no subset of Ω_E that is invariant except the empty set and Ω_E itself.*

This definition does not mention a measure; we have implicitly used the uniform measure. For any finite set (with n elements) it is easy to construct an ergodic permutation: we just choose any $n-$cycle such as

$$\begin{pmatrix} 1 & 2 & \ldots & n-1 & n \\ 2 & 3 & \ldots & n & 1 \end{pmatrix}$$

It follows that for any $\mathcal{E}$ with a finite partition function we can find many ergodic time-evolutions. Historically the idea of ergodicity on energy-shells was introduced in classical Hamiltonian mechanics in an attempt to explain irreversibility; unfortunately, Hamiltonian systems are generally not ergodic. This is the upshot of the deep work of Kolmogorov, Arnol'd and Moser, known as the KAM theorem. A very few cases of ergodic Hamiltonian motion have been established, but not anything interesting. In response to this, in the attempt to explain statistical mechanics, the ergodic *theorem* has given way to the ergodic *hypothesis*, which says that whatever the unknown Hamiltonian is, it leads to an ergodic action on phase space. This enables one to show that the time average of an observable is the same as the average over phase space using the microcanonical state. One can then argue that, since we take a large amount of time to measure an observable (large compared to atomic times), we really measure the time average, which is the same as the mean in the microcanonical state. This idea is not helpful to us, as it is entirely geared to explaining equilibrium statistical mechanics, and there it still leaves open the question of what happens in a particular model. The fact is that the ergodic programme for explaining irreversibility in classical Hamiltonian systems is moribund and has ground to a halt. For our countable space Ω, split into finite energy-shells, on the contrary, it is easy to obtain examples for which the ergodic theorem can be proved.

In many models of interest, such as occur in the theory of chemical reactions, the dynamics is not ergodic in exactly the sense given here, for the simple reason that there is one or more further conserved quantities, typically charge and atomic number. If we denote these random variables by $\mathcal{N}_i$, then the sample space is split into the disjoint union of energy-number-shells Ω_{E,N_i}. The time-evolution must conserve all these, and will be said to be ergodic (in the general sense) if no non-trivial subset of any energy-number shell is mapped to itself.

Whenever we have a permutation τ of a set Ω acting on the right, we can obtain from it a transformation τ_* of the algebra $\mathcal{C}(\Omega)$ of bounded functions on Ω, acting on the right:

Definition 3.2 $(F\tau_*)(\omega) = F(\omega\tau)$.

This defines an automorphism of the unital Banach algebra $\mathcal{C}$ (Exercise (3.29)). The properties of an automorphism are these; let $\sigma : \mathcal{C}(\Omega) \to \mathcal{C}(\Omega)$. Then we say that σ is an automorphism if σ is:

1. linear: $(\lambda_1 F_1 + \lambda_2 F_2)\sigma = \lambda_1 F_1\sigma + \lambda_2 F_2\sigma$.

2. unital: $1\sigma = 1$.

3. multiplicative: $(F_1 F_2)\sigma = (F_1\sigma)(F_2\sigma)$.

4. bijective: σ maps onto and is invertible.

An automorphism automatically preserves the norm: $\|F\sigma\|_\infty = \|F\|_\infty$. It follows from (3) that any automorphism is a positive map, since it takes any element of the positive cone into the positive cone: $F^2\sigma = (F\sigma)^2 \geq 0$. We shall see that in more general dynamics, including dissipation, that we replace (3) by positivity, and drop (4).

There is a converse: if $|\Omega| < \infty$, any automorphism of $\mathcal{C} = \mathcal{C}(\Omega)$ is of the form τ_*, where τ is some permutation of Ω. See Exercise (3.30).

If we specify an energy $\mathcal{E}$, and τ is energy-preserving, then $\mathcal{E}\tau_* = \mathcal{E}$, at least if $\mathcal{E}$ is bounded. For, the spectral theorem says

$$\mathcal{E} = \sum_E E\chi_{_E}(\omega)$$

where $\chi_{_E}$ is the indicator function for Ω_E, the energy-shell of energy E, and all these functions are invariant under τ_*. If $\mathcal{E}$ is not bounded, then all finite sums in the spectral theorem are invariant. Similarly, any function of $\mathcal{E}$ is invariant under τ_*. We shall need a partial converse to this: if the energy-shells are finite and in addition τ_* is ergodic, then any invariant element F of $\mathcal{C}$ is a function of $\mathcal{E}$. For, if it were not, it would not be measurable with respect to the Boolean ring $\mathcal{B}_1$ generated by $\mathcal{E}$, and so its level sets would not all lie in $\mathcal{B}_1$. A non-empty level set of F is invariant, and its intersection with the level sets of $\mathcal{E}$ give us a part of an energy-shell that is invariant, contrary to the ergodicity.

The time-evolution generated by τ_* is the analogue of the Heisenberg picture; to each $n \in \mathbf{Z}$, define $\tau_*(n)$ as the composition of n maps τ_*. Then the map $n \mapsto \tau_*(n)$ is a homomorphism of $\mathbf{Z}$ into the group $AUT\,\mathcal{C}$ of automorphisms. We can move to the analogue of the Schrödinger picture by throwing the action onto the states. The assumption that the automorphism preserves the energy-shells, which are finite, ensures that the dual action maps ℓ^1 to ℓ^1, and does not take us to non-normal states. Indeed the algebra of functions on an energy-shell $\Omega_{_E}$ is the dual of linear span of the set of probabilities on the same energy-shell; the question reduces to the finite-dimensional case. Thus, define the map $\tau^* : \Sigma(\Omega) \to \Sigma(\Omega)$, acting on the left, by the action of $\tau^* p$ on a general element of $\mathcal{C}$:

$$\langle \tau^* p, F\rangle = \langle p, F\tau_*\rangle \text{ for all } F \in \mathcal{C}. \tag{3.3}$$

This defines $\tau^* p$ as a linear functional on $\mathcal{C}$.

The components of the functional $\tau^* p$ are given by using the *inverse* automorphism τ^{-1}:

$$(\tau^* p)(\omega) = p(\omega\tau^{-1}). \tag{3.4}$$

To see this, write the definition Eqn. (3.3) as a sum over ω, and change the summation variable from ω to $\omega' = \omega\tau$:

$$\begin{aligned} \langle \tau^* p, F\rangle &= \langle p, F\tau_*\rangle \\ \sum_\omega (\tau^* p)(\omega)F(\omega) &= \sum_\omega p(\omega)F(\omega\tau) \\ &= \sum_{\omega'} p(\omega'\tau^{-1})F(\omega') \end{aligned}$$

which gives the result on changing the dummy label back to ω, as F is arbitrary. It is now clear that τ^* maps states to states; for the condition $p(\omega) \geq 0$ for all ω is identical with the condition $p(\omega\tau^{-1}) \geq 0$, and the condition $\sum_\omega p(\omega) = 1$ is the same if we sum over $\omega' = \omega\tau$.

The transformation law is compatible with the writing of p as a vector in the span of Ω: in a formal way, $p = \sum_\omega p(\omega)\omega$, and the permutation τ acts on the ω, but not on the $p(\omega)$, which are just numerical coefficients. Thus the action of τ is

$$\begin{aligned} p &= \sum_\omega p(\omega)\omega \mapsto \sum_\omega p(\omega)\omega\tau \\ &= \sum_{\omega'} p(\omega'\tau^{-1})\omega' \end{aligned}$$

from which we read off the result, because the elements of Ω are linearly independent. This convention, with F transforming with τ, and p with τ^{-1}, is the correct way round. When we come to consider dissipative dynamics, we shall allow endomorphisms, obeying (1), (2) and (3) of the axioms of automorphisms, but not necessarily being invertible. Then we could not define $F(\omega\tau^{-1})$ but we can give a meaning to $p(\omega\tau^{-1})$ as the probability of the inverse image of ω: $(\tau^*p)(\omega) = \text{Prob}\{\omega_1 : \omega_1\tau = \omega\}$.

The dynamics $\tau^*(n) = \tau^* \circ \ldots \circ \tau^*$ traces out an orbit in $\Sigma(\Omega)$, which is thus split into the union of disjoint orbits. All the points on the same orbit have the same entropy. For, S is a sum over all points of Ω, and τ just rearranges the order. Thus, reversible dynamics obeys the second law of thermodynamics, but in the trivial sense that while entropy does not decrease, it does not increase either. The partial order given by the concept of "more chaotic" [16] is also preserved along the orbit. The first law of thermodynamics also holds in the Schrödinger picture, for any τ that maps the energy-shells to themselves, for by duality

$$\langle \tau^* p, \mathcal{E} \rangle = \langle p, \mathcal{E}\tau_* \rangle = \langle p, \mathcal{E} \rangle.$$

If τ is ergodic, then the only invariant states are constant on the energy shells. Thus they are mixtures of microcanonical states of various energies; see Exercise (3.31). We now turn to some examples of dissipative dynamics.

3.2 Random Dynamics

Let Ω be a countable sample space, with state space Σ, and suppose that the dynamics depends on some parameters that are unknown. We can treat this using probability. Let us assign probabilities $\lambda_1, \lambda_2, \ldots, \lambda_n$ to the dynamics $\tau_1, \tau_2, \ldots, \tau_n$ respectively. We then compute a quantity of interest using each of the possible choices of dynamics, and then compute the mean over the λ's. This works well; its success is either due to the fact that a small system cannot be shielded from random influences, or to the chaotic nature of the actual dynamics over the time-interval represented by one time-step. The mean dynamics can be defined as

$$\tau_* = \sum_{i=1}^{n} \lambda_i \tau_{i*} \qquad \tau^* = \sum_{i=1}^{n} \lambda_i \tau_i^*, . \tag{3.5}$$

where τ_* acts on the algebra and τ^* acts on the states. With no more difficulty, we can also deal with the case where $n = \infty$ and $\{\lambda_i > 0\}$ is a sequence whose sum is 1. It is clear that τ_* maps $\mathcal{C}$ into itself, and also that τ^* maps Σ to Σ, since each τ_i^* does this, and Σ is convex.

Example 3.6 Take $\Lambda = \{1, 2, \ldots, n\}$, representing the points on a segment of a line, and take $\Omega_x = \{0, 1\}$ for all x, with 0 representing an empty site and 1 an occupied site. As always, we take $\Omega = \prod_x \Omega_x$. Let τ_x denote the flip operator, which exchanges the situation at the sites x and $x + 1$. Thus

$$(\omega_1, \ldots, \omega_x, \omega_{x+1}, \ldots, \omega_n)\tau = (\omega_1, \ldots, \omega_{x+1}, \omega_x, \ldots, \omega_n).$$

The part of the space involving x and $x + 1$ has four points,

$$(0,0), \ (0,1), \ (1,0), \ (1,1),$$

and in this basis T and $T^\dagger$ both are given by the matrix

$$\begin{pmatrix} 1 & 0 & 0 & 0 \\ 0 & 0 & 1 & 0 \\ 0 & 1 & 0 & 0 \\ 0 & 0 & 0 & 1 \end{pmatrix}.$$

The last row expresses that the two particles at the two sites are indistinguishable. The operator on the whole algebra $\mathcal{C}(\Omega)$ is the tensor product of this matrix with the unit operator on $\mathcal{C}(\prod' \Omega_x)$ where the $\prod'$ omits x and $x + 1$. The map τ_x conserves the random variable representing the number of particles, $\mathcal{N}(\omega) = \sum_x \mathcal{N}(x)$, where $\mathcal{N}(x)$ is the density. This is the random variable $\mathcal{N}(x)(\omega) = \omega_x$. $\mathcal{N}$ can also be interpreted as energy if the particles are molecules, each with a chemical energy of 1 unit. It is clear that the uniform distribution on an energy-shell (the microcanonical state) is a fixed point of each τ_x, and therefore of the uniform convex sum

$$\tau = (n-1)^{-1} \sum_{x=1}^{n-1} \tau_x.$$

This gives rise to a symmetric matrix in the natural basis. The action on the random field $\mathcal{N}(x)$ follows at once: only τ_x and τ_{x-1} affect $\mathcal{N}(x)$, and these shift it to the right and left respectively. So we get, if x is not an end-point:

$$\begin{aligned}\mathcal{N}\tau_*(x) &= (n-1)^{-1}\left(\mathcal{N}(x+1)+\mathcal{N}(x-1)\right)+(n-3)/(n-1)N(x)\\ &= \mathcal{N}(x)+(n-1)^{-1}\left(\mathcal{N}(x+1)-2\mathcal{N}(x)+\mathcal{N}(x-1)\right).\end{aligned} \tag{3.7}$$

We recognise the increment in the density in one time-step to be proportional to a finite difference approximation to minus the second derivative. When the elements of Λ are taken as a basis, we can express Eqn. (3.7) in matrix form: $N(t+1) = T\mathcal{N}(t)$ where T is the matrix

$$T = \begin{pmatrix} 1-\lambda & \lambda & 0 & \dots & 0 \\ \lambda & 1-2\lambda & \lambda & \dots & 0 \\ 0 & \dots & \dots & \dots & 0 \\ 0 & \dots & \dots & 1-2\lambda & \lambda \\ 0 & \dots & \dots & \lambda & 1-\lambda \end{pmatrix}. \tag{3.8}$$

Here, $\lambda = (n-1)^{-1}$.

In Exercise (3.32) you will show that at the end points the operator obeys Neumann boundary conditions, corresponding to no flow into or out of the region. One shows that if the initial state is a point measure with a certain value of total number, then the system converges to the corresponding microcanonical state. In fact, the dynamics is not only ergodic but has a spectral gap, a concept that will be met in the next section. This explains why it converges.

We get a variant of this model by considering the n points of Λ to lie on a ring. That is, we take $\Lambda = \mathbf{Z}_n$ the cyclic group on n elements. Then 1 and n become neighbours and if we add the permutation exchanging these, making n altogether, and add them with equal weights we get, instead of Eqn. (3.8) the following transition matrix, in which $\lambda = 1/n$:

$$T = \begin{pmatrix} 1-2\lambda & \lambda & 0 & \dots & \lambda \\ \lambda & 1-2\lambda & \lambda & \dots & 0 \\ \dots & \dots & \dots & \dots & \dots \\ 0 & 0 & \dots & 1-2\lambda & \lambda \\ \lambda & 0 & \dots & \lambda & 1-2\lambda \end{pmatrix} \tag{3.9}$$

which is the finite-difference approximation to the second derivative with periodic boundary conditions. □

In this example the transition matrix is symmetric; this reflects the fact that each of the permutations τ_i making up the dynamics is equal to its own inverse. More generally, we would get a symmetric matrix if, whenever the sum τ

contained a permutation τ_i it also contained its inverse, and with the same probability λ_i. This property is justified in classical mechanics by invoking invariance under time-reversal: to each possible dynamics, its time-reversed dynamics is possible, and equally likely. The symmetry of τ leads to a symmetry in the dynamics, connected with the "Onsager relations", and the theory becomes part of the subject of symmetric Markov chains. However, in biological systems such as neural nets a more general form of the transition matrix is sometimes needed; for example, when the transition matrix element T_{jk} represents the probability that the firing of neuron j causes the firing of neuron k, there is no reason why T should be symmetric. To allow for this, we prefer not to require that the symmetry of T be a fundamental part of the theory.

The next example has two conserved quantities and is ergodic in the generalised sense.

Example 3.10 The space $\Lambda = \{1, 2, \ldots, n\}$ is the same as in example (3.6), but now Ω_x has three points, called 0, A and B. We will interpret 0 as saying that x is an empty site, and the others as saying that it is occupied, by a molecule of type A or B. Then Ω is the space of "words" of length n, and alphabet $\{0, A, B\}$. It is split into the number-shells Ω_{N_1,N_2}, where N_1 is the number of A's in the word ω and N_2 is the number of B's. There are three types of exchange permutations of Ω that conserve the number of A-molecules and the number of B-molecules. We can exchange A at x with an empty site at y: call this $\tau(A, x, y)$; we can exchange B at x with an empty site at y: call this $\tau(B, x, y)$; or we can exchange an A at x with a B at y: call this $\tau(A, x; B, y)$. Any mixture of these permutations, as we vary x and y, gives rise to an induced map τ_* on $\mathcal{C}(\Omega)$ that leaves the indicator functions of the shells Ω_{N_1,N_2} fixed. It is easy to ensure that the action is ergodic in the generalised sense, if we take the energy to be $\mathcal{N}_1 + \mathcal{N}_2$. If we only use exchanges of neighbours, where $x = y \pm 1$, this model will possess only short-ranged forces. We could express that B is more mobile than A by choosing the weight of the permutation $\tau(B, x, y)$ to be larger than the weight of the permutation $\tau(A, x, y)$. The same effect could be achieved if we include some permutations $\tau(B, x, y)$ with a large distance between x and y. This model differs from two copies of the Example (3.6) since A and B interfer with each other; thus, if we do not use $\tau(A, x; B, y)$, and only exchange neighbours, then the action is not ergodic, even in the generalised sense using the two conserved numbers, since the initial order of the letters A and B is not changed by the dynamics. In fact, this order gives us more conserved quantities, labelled by a partition of $N_1 + N_2$. That the particles A and B cannot both sit at the same site effectively gives them an infinite repulsion, called a *hard core*. □

Our third model allows linear chemical reactions

$$A \rightleftharpoons B \tag{3.11}$$

as well as diffusion of each species.

Example 3.12 We take Λ to have n points as in the previous examples, and consider all permutations $\tau(A,x,y)$, $\tau(B,x,y)$ and $\tau(A,x;B,y)$, which cause diffusion, and include the permutation causing a transition from A to B and *vice versa*. It is said to be a local transition if it occurs at a site x. Thus define $\tau(A,B,x)$ to be the permutation

$$(\omega_1,\ldots,\omega_x=A,\omega_{x+1},\ldots,\omega_n)\tau(A,B,x)=(\omega_1,\ldots,\omega_x=B,\omega_{x+1},\ldots,\omega_n)$$

and the inverse, expressed as $\tau(A,B,x)=\tau(B,A,x)=\tau(A,B,x)^{-1}$. Let us mix the permutations giving diffusion with one weight, and those giving reactions with another. We get a linear reaction-diffusion system, in which $\mathcal{N}_1+\mathcal{N}_2$ is conserved, but not each separately. □

This example is easily generalised to m species of chemical.

Let us now return to the general theory. We take Ω to be countable, and consider a permutation τ that is the product of finite cycles. Since τ^* maps Σ to Σ, it follows that the entropy of τ^*p is well defined (with infinity as a possible value if Ω is infinite). In fact, entropy can increase under the action of τ^* if τ in Eqn. (3.5) is a proper mixture. Thus if two permutations τ_1 and τ_2, are involved, and these are different then there exists a state p such that $\|\tau_1^*p-\tau_2^*p\|_2=\delta>0$. Then

$$\begin{aligned} S(\tau^*p)-S(p) &= S\left(\lambda\tau_1^*p+(1-\lambda)\tau_2^*p\right)-S(p)\\ &\geq \lambda S\left(\tau_1^*p\right)+(1-\lambda)S\left(\tau_2^*p\right)+\frac{\lambda(1-\lambda)}{2}\|\tau_1^*p-\tau_2^*p\|_2^2-S(p)\\ &\quad \text{by theorem (2.42)}\\ &= \lambda(1-\lambda)\delta^2/2 \end{aligned}$$

since the permutations do not alter the entropy. We conclude that irreversibility generally holds if we postulate random dynamics. You might wonder how our not knowing the true dynamics can possibly lead to a physical law such as the increase of entropy with time. The answer is that we have not computed the entropy gain (zero) for each of the possible τ_i^*, and then found the mean of these (zero), but found the entropy gain of the mean dynamics. This is a clever model of the observed fact, the second law. You could ask, why do it this way? Well, I truly believe in the usefulness of probabilistic methods in science, and doing it this way gives the answer we need. Koseki [17] has remarked that probability is successful in thermodynamics just because the properties of the theory, such as convexity, are so similar to those of thermodynamics.

In order to get a model satisfying the first law as well, we recall the examples (3.6, 3.10, 3.12) and generalise: choose an energy $\mathcal{E}$, with level sets Ω_E, and some maps τ_i mapping each Ω_E to itself. We consider the case where $|\Omega_E| < \infty$ for each E. Then the mean energy is conserved by the mean dynamics. For, if the mean energy is finite in the state p, then

$$\begin{aligned}\langle \tau^* p, \mathcal{E}\rangle &= \sum_{i,E} \lambda_i E\langle \tau_i^* p, \chi_E\rangle = \sum_{i,E} \lambda_i E\langle p, \chi_E \tau_{*i}\rangle \\ &= \sum_{i,E} \lambda_i E\langle p, \chi_E\rangle = \langle p, \mathcal{E}\rangle.\end{aligned}$$

Thus the mean energy in the state $\tau^* p$ is the same as in the state p. We can also see that any microcanonical state is a fixed point of τ^*. To see this, note that τ^{-1} also leaves each Ω_E invariant. See (and do) Exercise (3.34). It follows that τ^* leave the state $p = \chi_E/|\Omega_E|$ invariant. This is true for all E. We thus have in place many elements of a theory of non-equilibrium statistical mechanics: the first law, the second law and the stationarity of the microcanonical states. But even if τ is ergodic, the system need not converge to equilibrium. It could move a point measure ω around and around. We need a stronger concept, that of a spectral gap. In the traditional approach to statistical mechanics via the ergodic theorem, it is argued that an actual observable is not a random variable at time t, since a measurement takes some time, indeed a very, very long time on the time-scale of the atomic variables. So we measure not F but

$$(1 + \tau_* + \tau_*^2 + \ldots, \tau_*^n)F/(n+1),$$

for some n; this is a special case of the mixing of permutations studied just now; a typical ergodic theorem (Exercise (3.35)) says that if τ is ergodic, then this converges, to the *ergodic mean* of F, as $n \to \infty$. It is this limit that is supposed to describe the macroscopic dynamics. We have seen how it leads to the increase of entropy. We shall study the question of the approach to equilibrium for a more general class of dynamics, the stochastic maps, and a special subclass, the bistochastic maps, which will turn out to be mixtures of permutations. This is Birkhoff's theorem, which was established for finite Ω in [18]. Indeed, more holds if the energy-shells are finite: if a bistochastic map leave $\chi_E \in \mathcal{C}$ and $\chi_E \in \mathcal{C}^d$ invariant for all E, then it is a mixture of energy-conserving permutations. Thus all linear dynamics obeying the first and second laws are in fact of the form of random dynamics. Similarly, if there are one or more conserved atomic numbers as well as energy, then any map increasing entropy is a mixture of permutations each of which conserves the same numbers. To explain this result, we develop the general theory a bit more.

Although the map τ_* might not be invertible, it is linear and positive, and maps 1 to 1. This is an easy exercise (3.36). These are all the properties of a stochastic map; we adopt the convention of writing T, acting on the left, for the general stochastic map, as in the

Definition 3.13 (Stochastic Maps) *Let $\mathcal{C}(\Omega)$ be the usual algebra of random variables on the countable space Ω. A map $T : \mathcal{C}(\Omega) \to \mathcal{C}(\Omega)$ is said to be* stochastic *if*

- *T is linear;*
- *T is positive; that is, T maps the positive cone to itself;*
- *$T\,1 = 1$.*

It is usual to add the condition of W^*-continuity to the definition, in order that its dual map leave ℓ^1 invariant. We shall soon impose a much stronger condition which will ensure this. We get an important class of stochastic maps as follows. Let $\tau : \omega \mapsto \omega\tau$ be a mapping of Ω into itself, not necessarily bijective. For $p \in \Sigma$, let $(\tau^* p)(\omega) = p(\omega\tau^{-1})$. This means that the measure of any set $S \subset \Omega$ in the state $\tau^* p$ is $p(S\tau^{-1})$. More, the map on $\mathcal{A}(\Omega)$ given by $(F\tau_*)(\omega) = F(\omega\tau)$ is an *endomorphism* of $\mathcal{A}$. This is explained and proved in Exercise (3.37).

If $|\Omega| < \infty$, a stochastic map induces by duality a linear map $T^\dagger$ on ℓ^1: to a given $p \in \ell^1$, $T^\dagger p$ is the functional which gives the value $\langle p, TF\rangle$ for the random variable F. The dual map has the following properties:

1. $T^\dagger$ is linear;
2. $T^\dagger$ is positive, that is, takes non-negative functionals to non-negative functionals;
3. $\|T^\dagger p\|_1 = 1$ for all $p \in \Sigma$.

Thus, T^* takes Σ to itself. The first one is obvious. To prove property (2), suppose p is in the positive cone. If F is in the positive cone of $\mathcal{A}$, then so is TF, and so $\langle p, TF\rangle \geq 0$. Hence $\langle T^\dagger p, F\rangle \geq 0$; since this is true for all F in the positive cone, we have that p is in the positive cone, since they are dual. Property (3) is also easy: since $T^* p$ is in the positive cone,

$$\|T^* p\| = \langle T^* p, 1\rangle = \langle p, T\,1\rangle = \langle p,\, 1\rangle = \|p\|_1. \quad \square$$

The same result follows immediately if $|\Omega| = \infty$, but is divided up as the union of finite sets Ω_i, and the indicator function of each Ω_i is invariant under the action of T.

For any Ω, it follows from positivity that a stochastic map is bounded, in the sense of an operator on the Banach space $\mathcal{C}$. For, F can be written in terms of its positive and negative parts: $F = F_+ - F_-$; where $F_\pm \geq 0$. Then $TF_\pm \geq 0$, and

$$|(TF)(\omega)| = |(TF_+)(\omega) - (TF_-)(\omega)| \leq (TF_+)(\omega) + (TF_-)(\omega) = (T|F|)(\omega).$$

This is true for all ω, so we get the same result for the sup-norm: $\|TF\|_\infty \leq \| T|F| \|_\infty$. Now, for each ω, $|F(\omega)| \leq \|F\|_\infty$, so $\|F\|1 - |F|$ is in the positive cone. Hence $T(\|F\|1 - |F|)$ is a non-negative function, so

$$(T|F|)(\omega) \leq (T\|F\|_\infty 1)(\omega) = \|F\|_\infty,$$

since $T1 = 1$. Take the sup over ω, to get $\| T|F| \|_\infty \leq \|F\|_\infty$, which combined with the result above, $\|TF\|_\infty \leq \| T|F| \|_\infty$, gives the following important

Theorem 3.14 *A stochastic map T on $\mathcal{C}$ is a contraction in the norm, with norm 1.*

The norm of T is not smaller than 1, since $T1 = 1$. The set of stochastic maps is a convex set (Exercise (3.38)). This raises the question as to what the extreme points are; the answer is nice: if Ω is countable, a stochastic map is extremal if and only if it is induced by an endomorphism (Exercise (3.39)). This result is implicit in [16] p.13. There is a general result (the Krein-Milman Theorem) giving conditions under which a general point in a convex set is a mixture of the extreme points of the set. Usually it involves the integral over the extreme points, rather than just a sum. Since the number of endomorphisms of Ω is uncountable, even when Ω is countable, this is outside the scope of this book. In Exercise (3.39) you are asked to prove these results when $|\Omega| < \infty$. Simon [5] gives an accessible proof of the Krein-Milman theorem.

Now suppose that $T^\dagger$ maps $\ell^1(\Omega)$ to itself. That $T^\dagger$ is bounded follows from the inequality

$$|\langle p, TF\rangle| \leq \|p\|_1 \|TF\|_\infty \leq \|p\|_1 \|F\|_\infty,$$

so $T^\dagger p$ is a continuous functional of norm $\|p\|_1$. Thus $T^\dagger$ is also a contraction (in the ℓ^1-norm).

On iterating a contraction map we generate a dissipative dynamics which quite often converges to a fixed point, which could be called equilibrium. For example, if $T^\dagger$ is a proper contraction, with norm less than 1, then $T^{\dagger n} p$ converges to zero as n becomes large. Such maps, however, are excluded from our theory by the condition $T1 = 1$, which leads to the conservation of probability. Many models of the reaction-diffusion equation have been assiduously studied which violate this condition, in that the dynamics does not preserve the ℓ^1-norm of the density. Such equations could not be realistic models of dissipative systems.

The dual $T^{\dagger}$ of a general stochastic map T does not necessarily increase the entropy; this has puzzled some writers, because stochastic maps are widely used by physicists and chemists to represent one time-step. This has been elevated to an axiom by Penrose [19], a book that, if not the father of the present one, is at least an uncle. The answer to the puzzle is that the dynamics generated by $T^{\dagger}$ (with a strictly positive fixed point p_0) is not the dynamics of an isolated system, but that of a system in contact with a heat bath. Entropy flows in and out with the heat-flows, and it is another thermodynamic function, sometimes called Ψ, that increases. Penrose shows that the relative entropy $S(p|p_0)$, given in Definition (2.44), decreases along the orbit. It is noted in [20] p. 68 that this is related to the free energy. We shall obtain a sharp version of this result on the decrease in free energy in Theorem (5.30). In thermodynamics a system must be isolated (also called adiabatic) before we can be certain that the entropy increases, and the most general subclass of stochastic maps which does this is the set of *bistochastic maps*:

Definition 3.15 (Bistochastic Map) *Let Ω be a countable sample space, and T a stochastic map on $\mathcal{C}(\Omega)$. Let T map the subspace ℓ^1 of $\mathcal{C}$ to itself, and denote by T_1 the restriction of T to ℓ^1. Then we say T is bistochastic if and only if T_1 is trace-preserving.*

The set of bistochastic maps themselves form a convex set (Exercise (3.38)). Since a bistochastic map is *a fortiori* a stochastic map, it can be written as a convex sum of stochastic endomorphisms. In finite dimensions, the celebrated theorem of Birkhoff states more: the extreme points of the set of bistochastic maps is exactly the set of permutations, and so by Minkowski's theorem, any bistochastic map is a convex sum of permutations. The hard part of the proof, as explained by Alberti and Uhlmann [16], p. 15, is not that permutations are extreme points (Exercise (3.40)), but to show that there are no extreme points of the set of bistochastic maps other than these. We shall not prove Birkhoff's theorem; a reasonable proof is to be found in Ando's paper [18].

Obviously a stochastic map that is symmetric (in the natural basis, Ω), is bistochastic. They generate the special class of *symmetric* Markov chains, in which entropy increases in time. It seems to me that the generalisation to bistochastic maps is both natural and necessary for applications.

In Exercise (3.41) it is shown that if a bistochastic map conserves energy, then the permutations making it up must also conserve energy. In (3.43) it is shown that a stochastic map that does not decrease the entropy of any state must be bistochastic. Together this shows that any linear map obeying both the first and second laws of thermodynamics must be a random dynamics. The next bit of theory shows that under a condition on the spectrum of $TT^{\dagger}$, such dynamics leads to the convergence of the system to a mixture of microcanonical states (but not to the canonical state) for any initial state.

3.3 Convergence to Equilibrium

In this section we find a criterion, called the spectral gap which ensures that a random conservative dynamics converges to equilibrium. We shall show that there are large classes of tractable examples. We shall also find that the limit state, at large times, is a microcanonical state, which explains why these occur. The occurrence of canonical and grand canonical states will be explained in the next Chapter.

We begin with a sharp estimate for the entropy gain under a bistochastic map.

Lemma 3.16 *Let $w_k \geq 0$, $\sum w_k = 1$ with $\{w_k\}$ a sequence in ℓ^1 and let $x = \sum w_k x_k$ with x_k a sequence with $0 \leq x_k \leq 1$. Let $s(y) = -y \log y$. Then*

$$\sum_k w_k s(x_k) \leq s(x) - \frac{1}{2} \sum_k w_k (x - x_k)^2. \tag{3.17}$$

Proof. By Taylor's theorem,

$$\begin{aligned} s(x_k) &= s(x) + (x_k - x)s'(x) - (x - x_k)^2/(2\xi_k) \\ &\leq s(x) + (x_k - x)s'(x) - (x - x_k)^2/2. \end{aligned}$$

Multiply by w_k and sum; the middle term drops out and we get the result. □

We now come to the main estimate. In this theorem, a *semiprobability* is an element of ℓ^1 with components in [0, 1].

Theorem 3.18 *Let T be a bistochastic map on $\mathcal{C}(\Omega)$, where Ω is countable. Let p be a semi-probability. Then*

$$S(T^\dagger p) - S(p) \geq \frac{1}{2} \left(\|p\|_2^2 - \|T^\dagger p\|_2^2 \right). \tag{3.19}$$

Proof. Choose the natural basis, the points in Ω, and write T as an infinite matrix of non-negative numbers with rows and columns in ℓ^1. The condition that T is stochastic says that the rows of T_{jk} sum to 1, and the condition that it is the dual of a stochastic map says that the columns of T_{jk} sum to 1. The action of $T^\dagger$ on the states is given by $p \mapsto q$, where $q_j = \sum_k T_{kj} p_k$ is a convergent sum. Fix j and put $x_k = p_k$ and $w_k = T_{kj}$ for all k in the lemma. Then $x = \sum_k T_{kj} p_k = q_j$. By lemma (3.17) we get for each j:

$$-\sum_k T_{kj} p_k \log p_k \leq -q_j \log q_j - \sum_k T_{kj} (q_j - p_k)^2/2.$$

Sum over j and use $\sum_j T_{kj} = 1$ and $\sum_k T_{kj} = 1$:

$$\begin{aligned} -\sum_j q_j \log q_j + \sum_k p_k \log p_k &\geq \sum_{k,j} T_{kj}(q_j^2 - 2q_j p_k + p_k^2)/2 \\ &= \left(\sum_k p_k^2 - \sum_j q_j^2\right)/2 \\ &= \left(\|p\|_2^2 - \|T^\dagger p\|_2^2\right)/2 \end{aligned}$$

since $\sum_k T_{kj} p_k = q_j$. □

The right-hand side is non-negative, since a bistochastic map is a contraction. This estimate gives us our first result for the long-time behaviour of random dynamics. Let us take the case of interest, when the bistochastic map T maps the energy-shells into themselves. Since these are finite, the restriction T_E of T to each shell is a finite-dimensional matrix. Then $T_E T_E^\dagger$ is a positive semi-definite matrix, with 1 as an eigenvalue and with the microcanonical state, denoted p_E, as eigenvector. Let us say that T has a *spectral gap* if 1 is a simple eigenvalue of every $T_E T_E^\dagger$. We also note that if p is any probability, then its part on the energy-shell is a semi-probability with a finite number of components. We can therefore apply Eqn. (3.19) for a finite-dimensional matrix with spectral gap, for which we have the following improvement:

Theorem 3.20 *Let $|\Omega| = n < \infty$ and let T be a bistochastic map with a spectral gap, such that $TT^\dagger$ has no eigenvalues between $1-\gamma$ and 1. Let e denote the uniform probability on Ω. Then for any semi-probability p we have*

$$S(T^\dagger p) - S(p) \geq \frac{1}{2}\gamma \, \|p - \|p\|_1 e\|_2^2 . \tag{3.21}$$

Proof. We note that

$$\|p\|_2^2 - \|T^\dagger p\|_2^2 = \langle (1 - TT^\dagger)p, p\rangle = \langle (1 - TT^\dagger)(p - \|p\|_1 e), (p - \|p\|_1 e)\rangle$$

since $(1 - TT^\dagger)$ annihilates e. Also, $p - \|p\|_1 e$ is orthogonal to e since

$$\langle p - \|p\|_1 e, e\rangle = \langle p, e\rangle - \|p\|_1 \langle e, e\rangle = \frac{1}{n}\|p\|_1 - \|p\|_1 \frac{1}{n} = 0.$$

Now $TT^\dagger$ is bounded by γ on the space orthogonal to e, so

$$\langle (1 - TT^\dagger)(p - \|p\|_1 e), (p - \|p\|_1 e)\rangle \geq \gamma \|p - \|p\|_1 e\|_2^2.$$

Combining this with Eqn. (3.19) gives the theorem. □

This shows that unless we are at the fixed point of the map the entropy is increased by a non-zero amount when we apply $T^\dagger$. This is a very useful property, in non-linear as well as linear dynamics: it allows us to prove a fundamental law of chemistry called (by Tolman [9]) "the law of detailed balance". It is the following. Suppose that a dynamics τ is a mixture of two maps, τ_1 and τ_2, and that S is increased by τ_1 unless we are at a fixed point of τ_1, and by τ_2 unless we are at a fixed point of τ_2. Then a fixed point of τ must be a simultaneous fixed point of both τ_1 and τ_2. The proof is easy, using the concavity of the entropy: if p is a fixed point of τ, then

$$S(p) = S(\tau p) \geq \lambda S(\tau_1 p) + (1-\lambda)S(\tau_2 p) > \lambda S(p) + (1-\lambda)S(p) = S(p),$$

a contradiction unless p is a simultaneous fixed point. Tolman interprets this as saying that in a complex chemical reaction (where several processes can occur together), chemical equilibrium is not reached until all the constituent processes are separately in balance, meaning at equilibrium. There is no perpetual motion, where $A \to B \to C \to A$. Our proof of this, above, needed the inequality rather than equality in the estimate (3.21), and depends on $\gamma > 0$.

Our inequality also allows us to apply Lyapunov's "direct method" to show that the system converges at large times. We now give a simple version of this theory.

Definition 3.22 (Lyapunov function) *Let τ^* be a continuous map of a separable compact space Σ to itself. Let S be a continuous function on Σ such that $S(\tau^* p) > S(p)$ unless p is a fixed point. Then we say that S is a* Lyapunov function *for the dynamics given by τ^*.*

We now give the main theorem, which applies to both linear and non-linear dynamics.

Theorem 3.23 (Lyapunov) *Let Σ be a compact separable space and let τ^* be a continuous map from Σ to Σ. Suppose that there is a unique fixed point p_0 under τ^*. Let S be a Lyapunov function for τ^*. Then for any initial point p, we have*

$$\tau^{*n} p \to p_0 \text{ as } n \to \infty.$$

Proof. The sequence $\tau^{*n}p$ has a convergent subsequence $\{\tau^{*n_i}p\}_{i\in\mathbf{N}}$, by the Bolzano-Weierstrass theorem. Let p_∞ be the limit. Since S is continuous, $S(\tau^{*n_i}p) \to S(p_\infty)$. Also, as τ^* is continuous,

$$\tau^*(\tau^{*n_i}p) \to \tau^* p_\infty. \tag{3.24}$$

Note that $\tau^*\tau^{*n_i} = \tau^{*(n_i+1)}$ might not be equal to $\tau^{*n_{i+1}}$ so we cannot immediately say that $\tau^*\tau^{*n_i}p \to p_\infty$. We must use the existence of the Lyapunov function, S, which is bounded as it is a continuous function on a compact set. Then $S(\tau^{*n}p)$ is bounded and non-decreasing, and so converges. This implies that each of its subsequences $S(\tau^{*n_i}p)$ and $S(\tau^{*(n_i+1)}p)$ converges to the same limit. The first converges to $S(p_\infty)$ as we have seen; so the second must too. But it is $S(\tau^*\tau^{*n_i}p)$ which converges to $S(\tau^*p_\infty)$ by Eqn. (3.24) since S is continuous. Hence $S(p_\infty) = S(\tau^*p_\infty)$. Since the entropy is strictly increasing, except at a fixed point, we have that p_∞ is a fixed point. Since there is only one fixed point, p_0 and the above argument applies to any convergent subsequence of the orbit, we have proved that *every* convergent subsequence of the orbit converges to p_0. This means that the orbit itself converges to the fixed point, by the following curious theorem: if $\{x_n\}$ is a sequence in a compact space such that every convergent subsequence has the same limit, then $\{x_n\}$ itself is convergent. This is Exercise (3.42). □

It is necessary for the function S to strictly increase along an orbit, to get the result. For example, if $\Omega = \{0, 1\}$ and Σ is the set of states on Ω, then the permutation τ which exchanges the two elements of Ω is ergodic, since there is no proper non-trivial invariant subset. The induced action on Σ, τ^*, is the permutation matrix

$$\tau^* = \begin{pmatrix} 0 & 1 \\ 1 & 0 \end{pmatrix}.$$

This conserves entropy but does not increase it. The sequence $\tau^{*n}p$ has two subsequences, those for odd and even values of n, which are constant and so convergent, but not to the same limit unless p is a fixed point. In this case, $\tau\tau^*$ is the unit matrix, of which 1 is a double eigenvalue. This means that the spectral gap γ of Eqn. (3.21) is zero. The proof breaks down since 1 is a repeated eigenvalue and $\tau\tau^*$ has an eigenvector orthogonal to the microcanonical state.

In our application of this theorem in this Chapter, the entropy is a Lyapunov function if γ is positive; in finite dimensions, this is the same as saying that 1 is a simple eigenvalue of $TT^\dagger$. To ensure this it is enough for $T^\dagger$ to be a matrix with positive entries. This is part of the Perron-Frobenius theorem. Let us say that a vector is *positive* if its components are non-negative, and *strictly positive* if its components are positive. Also a matrix $T^\dagger$ is *positively improving* if $T^\dagger p$ is strictly positive whenever p is positive. Clearly, a matrix with positive entries is positively improving.

Theorem 3.25 (Uniqueness part of Perron-Frobenius) *Let $T^\dagger$ be a positively improving $n \times n$ matrix, and having 1 as an eigenvalue with a positive eigenfunction. Then 1 is simple.*

Proof. Since $T^\dagger p = p$, p must be strictly positive. Suppose if possible that there are two linearly independent eigenvectors p_1 and p_2 where p_1 is strictly positive. If p_2 is complex, then the real and imaginary parts are real eigenfunctions, since $T^\dagger$ is real. So we may assume that p_2 is real with some negative entries (if it is positive, then take $-p_2$ instead). For real λ, consider the eigenvector

$p(\lambda) = p_1 + \lambda p_2$, which has eigenvalue 1. For λ sufficiently close to zero, this is strictly positive, but as we increase λ a value is reached when $p(\lambda)$ has some zero entries, and the rest positive. Then $T^\dagger p = p$ must be strictly positive, contradicting the presence of zero components. □

The theorem has a useful corollary; if the matrix $T^\dagger$ obeys the conditions of the theorem, except that only some power $T^{\dagger n}$ has positive entries, then 1 is simple.

We can get a bistochastic matrix some power of which is strictly positive by mixing, with non-zero weights, permutations enabling us to get from every point of Ω to any other. Thus, let us say that a set of permutations $\{\tau_i\}$ is *irreducible* if, given any two points ω_0, ω, there exists a subset of these permutations $\{\tau_1, \tau_2, \ldots, \tau_M\}$ such that

$$\omega_0 \tau_1 \tau_2 \ldots \tau_M = \omega.$$

The length M of the chain may depend on ω_0 and ω, and if $|\Omega| = \infty$ we require that

$$M_{\max} = \sup_{\omega_0, \omega} M(\omega_0, \omega) < \infty.$$

Mixing all the permutations from an irreducible set with non-zero weights gives rise to an irreducible matrix, which is one for which any two pure states are connected by a chain of non-zero transition probabilities. This has the property that its $M_{\max}$th power is strictly positive. Such a matrix, then, has 1 as a simple eigenvalue, and if Ω is finite, the spectral gap γ is positive. It is easy to see that the j-th power of the diffusion matrix with Neumann boundary conditions (3.8) has non-zero elements making a band of width $2j + 1$ down the diagonal, which causes spatial diffusion of a particle at the rate of one unit per unit time. Clearly, some power of it is strictly positive on each number-shell.

We are now in a position to prove the main theorem for random conservative dynamics:

Theorem 3.26 *Let Ω be a countable space and let $\mathcal{E}$ be a random variable (the energy) such that each energy-shell is finite. Let τ_* be a bistochastic map on $\mathcal{C}(\Omega)$, made up of permutations, each leaving every energy-shell invariant. Suppose that τ_* restricted to each energy-shell Ω_E is irreducible. Then for any initial state p, we have $\tau^{*n} p \to p_\infty$ as $n \to \infty$. The convergence is in ℓ^1, and the limit state is a mixture of microcanonical states*

$$p_\infty = \sum_E p(E) \chi_E, \text{ where } p(E) = \sum_{\omega \in \Omega_E} p(\omega).$$

The mean energy of the state is constant in time, and entropy properly increases unless the initial state is a mixture of microcanonical states.

Proof. Restrict the map τ to an energy-shell Ω_E. It is a finite-dimensional problem and the bistochastic matrix representing the map $\tau_{_E}^*$ is irreducible. Hence there is a non-zero spectral gap $\gamma_{_E}$, and by Eqn. (3.21), entropy is a strict Lyapunov function. Also, as the eigenvalue 1 is simple (for the restricted map), we can apply Lyapunov's theorem, which says that iterates of the map $\tau_{_E}^*$ applied to p converge, preserving the ℓ^1-norm. Since the fixed point is the uniform distribution, we get the result. □

In particular, if all the probability is initially in one energy-shell, then the system converges to the microcanonical state with that energy. This tendency to move to the uniform distribution is called the *equipartition of energy*.

This is a partially satisfactory result. We have not had to use the ergodic *hypothesis*, since we have a good criterion allowing us to prove the stronger property of having a spectral gap. For example, the diffusion model satisfies this, and so converges to the uniform distribution (Exercise (3.33)). But we are not able to set up an energy-conserving model which redistributes probability between energy-shells, and so are still unable to understand why systems move to the canonical state. This is achieved in the next Chapter.

3.4 Markov Chains

Let $\Omega_0 = \{\omega_1, \ldots, \omega_{_N}\}$ and let $\mathcal{A} = \mathcal{C}(\Omega_0)$. Let T be a stochastic map on $\mathcal{A}$ and $T^\dagger$ its dual, acting on the set Σ of probabilities on Ω_0. We can regard Σ as a subset of $\operatorname{Span}\Omega_0$, the vector space that has as its basis the N points of Ω_0. The dual basis in $\mathcal{A}$ consists of Kronecker δ-functions δ_j at the N points ω_j. Then we define the *stochastic matrix* of $T^\dagger$ to be

$$T^\dagger_{ij} = \langle T^\dagger \omega_j, \delta_i \rangle.$$

It has the property that $\sum_i T^\dagger_{ij} = 1$ for all j. For

$$\begin{aligned}\sum_{i=1}^N T^\dagger_{ij} &= \langle T^\dagger \omega_j, \sum_i \delta_i \rangle = \langle T^\dagger \omega_j, 1 \rangle \\ &= \langle \omega_j, T1 \rangle = \langle \omega_j, 1 \rangle = 1.\end{aligned}$$

We can interpret $T^\dagger_{ij}$ as the probability that the system is in state ω_i at time $t+1$ given that it is in state ω_j at time t. It is then a conditional probability. Suppose that the initial probability is $\overset{\circ}{p} \in \Sigma$. We can then use the definition (2.37) of conditional probability to define the probability that at times $t = 0$ and $t = 1$ the sample points are $(\omega(0), \omega(1))$:

$$\operatorname{Prob}\{\omega(0), \omega(1)\} = \overset{\circ}{p}(\omega_0) P(\omega(1)|\omega(0)) = \overset{\circ}{p}(\omega_0) T^\dagger_{10}.$$

Proceeding at successive times, we can define the probability that the sample point $\omega(t)$ trace out a given walk $(\omega(0),\omega(1),\ldots,\omega(n))$ in n time-steps to be

$$\text{Prob}\{\omega(0),\omega(1),\ldots,\omega(n)\} = \overset{\circ}{p}(\omega(0))T^{\dagger}_{10}T^{\dagger}_{21}\ldots T^{\dagger}_{n,n-1}.$$

If we are not interested in the intermediate states, but just want the probability that at time $t=n$ the state is $\omega(n)$, we sum over the positions at all previous times, to get

$$\text{Prob}\{\omega(n)\} = \sum_{\omega(n-1),\ldots,\omega(0)} T^{\dagger}_{n,n-1}\ldots T^{\dagger}_{10}\,\overset{\circ}{p}(\omega_0) = T^{\dagger n}\,\overset{\circ}{p}(\omega(n))$$

since this is just matrix multiplication; so our rule for calculating probabilities agrees with the given dynamics on Σ when limited to the arrival point.

In the theory of the random walk we are interested in the path by which $\omega(n)$ was arrived at, as well as the final point of arrival. This extra detail can be incorporated at the cost of a much larger sample space. Suppose for simplicity that we know that the sequence started at $\omega(0)$. Then for a sequence of indefinite length we need the space

$$\Omega = \{\omega(0)\}\times\Omega_0\times\Omega_0\ldots$$

(one copy of Ω_0 for each time $n\geq 1$). If the sequence is infinite, care is needed in the definition of measurable sets, measures and so on. We cannot assign a non-zero probability to a sample point, a sequence of points in Ω_0, since in general the product of infinitely many numbers less than one diverges to zero (as we say). This is how we proceed; let $\mathcal{B}_n$ be the Boolean ring of all cylinder sets having as base any subset of Ω_0 located at time n; that is, any set of the form

$$\Omega_0\times\Omega_0\times\ldots\times\Omega_n\times\Omega_0\ldots \qquad \text{with } \Omega_n\subseteq\Omega_0 \text{ in the } n\text{-th place.}$$

These represent events that can be measured at time n. Physically, we can measure the present, and might have a record of past measurements, but we cannot know the future. It makes sense to define a Boolean ring $\mathcal{B}_{\leq n}$ as that generated by $\{\mathcal{B}_j : j\leq n\}$. Then a measurable set (relative to $\mathcal{B}_{\leq n}$) is a cylinder set defined by gates $\Omega_1, \Omega_2, \Omega_n$, in which event Ω_j happened at time j, with $1\leq j\leq n$. Clearly, $\mathcal{B}_{\leq n+1}\supseteq\mathcal{B}_{\leq n}$. The increasing family of Boolean rings $\{\mathcal{B}_{\leq n}\}$ is called a *filtration.* We can define the probability of any cylinder set, as the sum of the probabilities of all random walks passing through the (finite number of) gates. This is called a cylinder measure. We shall denote it by μ. Books on probability theory [11] prove the Kolmogorov extension theorem, which says that this measure on the cylinder sets can be uniquely extended to a countably additive measure on the σ-ring $\mathcal{B}$ generated by the filtration. The semi-group of time-evolution acts naturally on the sample-space, by left-shift: $\omega\tau(j) = \omega(j+1)$. This is $\mathcal{B}$-measurable, and induces as usual an action τ_* on the random variables and τ^* on the states. The process we have obtained has the Markov property: the future, conditioned by the present, is independent of the past. This can be expressed in every-day terms by saying that the random walk starts afresh at each time, from where it's at. In terms of conditional expectations, the Markov property can be expressed neatly:

Theorem 3.27 *Let $F : \Omega \to \mathbf{R}$ be $\mathcal{B}$-measurable. Then*

$$E(F|\mathcal{B}_{\leq n}) = E(F|\mathcal{B}_n) \qquad \textit{for each } n.$$

We shall not prove this. For further details on Markov chains, see [21].

One of the problems of the non-linear dynamics, which we meet in the next Chapter, is that the time-evolution of p in n steps is not obtained by summing the probabilities of all walks leading to the final point; so the traditional theory of stochastic processes needs modifying. Work along these lines was initiated by McKean [22]. If the dynamics $p \mapsto T^\dagger p$ is linear we can extend it to act on all functions on Ω. We can equally well consider one step in the dynamics to act on the algebra of random variables, $F \mapsto TF$. Let $t_1 < t_2 \ldots < t_n$ be n integers denoting times, and let the initial state be p. The many-time correlation functions of random variables $F_1, \ldots, F_n$ can be expressed in terms of the algebraic dynamics or in terms of the Markov chain and its measure μ:

$$\langle p, T^{t_1}F_1 \ldots T^{t_n}F_n\rangle = \langle \mu, \prod_j F_j(\omega_j)\rangle.$$

In the next chapter we shall consider dynamics where the map on the states is non-linear, and cannot easily be used to define a "Heisenberg" picture $F \mapsto TF$. In this case the definition of the many-time correlation functions is not obvious. I offer a suggestion in Chapter 13.

3.5 Exercises

Exercise 3.28 Show that the property of being on the same orbit under a group is an equivalence relation.

Exercise 3.29 Let τ be a permutation of Ω. Show that the map τ_* is an automorphism of $\mathcal{C}(\Omega)$.

Exercise 3.30 Suppose that $|\Omega| < \infty$. Show that any automorphism of $\mathcal{C}(\Omega)$ is of the form τ_*, where τ is a permutation of Ω. Show that the group of automorphisms of $\mathcal{C}(\Omega)$ is isomorphic to the permutation group of Ω. Hint: two groups G_1 and G_2 are said to be isomorphic if there exists a bijective map $\phi : G_1 \to G_2$ such that ϕ is a homomorphism. Show that the map $\phi(\tau) = \tau_*$ is such.

Exercise 3.31 Let τ be a permutation of Ω leaving invariant the energy-shells defined by a random variable $\mathcal{E}$. Suppose that τ is ergodic relative to energy. Prove that the only states invariant under τ^* are mixtures of microcanonical states.

Exercise 3.32 Show that the finite difference equation for the density $N(x)$, coming from the matrix in Eqn. (3.8) corresponds to the Neumann boundary condition $\partial\mathcal{N}/\partial x = 0$. Write the marginal distribution of $p \in \Sigma(\Omega)$ onto the factor Ω_x as (q_x, p_x). Show that the time evolution of p_x along the orbit of τ^* obeys the same equation. This is the Fokker-Planck equation for this dynamics.

Exercise 3.33 Show that the dynamics given by the matrix in Eqn. (3.8) takes any initial state to the uniform state at large times.

Exercise 3.34 Let Ω be a countable set, and τ a permutation of Ω. Let Ω be the disjoint union of Ω_1 and Ω_2, and suppose that Ω_1 and Ω_2 are invariant under τ. Show that each is invariant under τ^{-1}.

Exercise 3.35 A sample space Ω is the union of energy-shells Ω_E, with $|\Omega_E| < \infty$, and τ is an ergodic permutation of Ω. Let p be a point measure with support in Ω_E. Show that

$$(1 + \tau^* + \tau^{*2} + \ldots + \tau^{*n})p/(n+1) \to \chi_E \text{ as } n \to \infty$$

in ℓ^1–norm.

Exercise 3.36 Show that the mapping $\tau_* = \sum \lambda_i \tau_{i*}$, where τ_i are permutations of Ω and the λ_i are probabilities on $\{1, 2, \ldots, n\}$, is a stochastic map.

Exercise 3.37 Let $\tau : \Omega \to \Omega$ be any map, and let $\tau_* : \mathcal{C}(\Omega)$ be the induced map $(\tau_* F)(\omega) = F(\omega\tau)$. Prove that τ_* is a stochastic endomorphism, that is, obeys the two properties of an endomorphism

1. For $F_i \in \mathcal{C}$ and real λ_i, we have linearity:

$$\tau_*(\lambda_1 F_1 + \lambda_2 F_2) = \lambda_1 \tau_* F_1 + \lambda_2 F_2.$$

2. $\tau_*(F_1 F_2) = \tau_* F_1 \, \tau_* F_2$,

as well as the "stochastic" condition $\tau_* 1 = 1$. Also prove the converse, that any stochastic endomorphism of $\mathcal{C}(\Omega)$ is induced from a unique map τ. Hint: construct τ from the action of the endomorphism on functions with support at a single point.

Exercise 3.38 Show that the set of stochastic maps on $\mathcal{C}(\Omega)$ is convex. Show that the subset of bistochastic maps is also convex.

Exercise 3.39 Suppose that $|\Omega| < \infty$; show that a stochastic map T is an extreme point of the convex set of stochastic maps if and only if it is induced by an endomorphism. Show that any stochastic map is a convex mixture of endomorphisms.

Exercise 3.40 Show that a permutation is an extreme point of the set of bistochastic maps.

Exercise 3.41 Let T be a bistochastic map on $\mathcal{C}(\Omega)$ that conserves energy: $T\chi_E = \chi_E$ and $T^\dagger\chi_E = \chi_E$ for all E, where $\Omega = \bigcup \Omega_E$ is the division of Ω into energy-shells. If each Ω_E is finite, show that T is a mixture of energy-conserving permutations.

Exercise 3.42 Let $\{x_n\}$ be a sequence of points in a compact space such that every convergent subsequence has the same limit. Show that $\{x_n\}$ is convergent.

Exercise 3.43 Show that a stochastic matrix that increases the entropy of every state must be bistochastic.

Exercise 3.44 Let $\Omega_x = \{0, 1\}$ and $\Lambda = \{1, 2, \ldots, n\}$, $\Omega = \prod_x \Omega_x$, and $\mathcal{E}(\omega) = \sum_x \omega_x$. Let τ_x denote the permutation which exchanges site x if occupied, with the site $x+1$, if unoccupied, and makes no other changes to ω; here τ_n exchanges 1 and n. Let τ^* be any mixture of all these τ_x^*, for $x = 1$ to $x = n$, with non-zero weights. Show that the system converges to a distribution with uniform energy-density in space from any initial state.

Chapter 4

Isolated Dynamics

4.1 The Boltzmann Map

We have seen that the random mixture of reversible maps can lead to thermodynamics, including the convergence to equilibrium; however in this way we get only the microcanonical states and not the canonical states. Within the linear theory there is no mechanism for mixing up the various energy-shells, or the number-energy-shells if there is more than one conserved quantity, if we wish to retain the conservation laws. The extra randomness needed to mix the shells can be achieved by a construction hidden in Boltzmann's equation, and called the Stosszahlansatz by him. This is called the hypothesis of molecular chaos in early English text-books [8, 9]. We shall express this in a general but very simple form. Suppose that $\Omega = \Omega_1 \times \Omega_2$, and also that the energy of the whole system is the sum of the energies of the two subsystems, thus:

$$\mathcal{E}(\omega) = \mathcal{E}_1(\omega_1) + \mathcal{E}_2(\omega_2) \text{ where } \omega = (\omega_1, \omega_2).$$

This expresses that there is no interaction between the two systems. Then for a general state $p \in \Sigma(\Omega)$, the mean energy is

$$\begin{aligned} \langle p, \mathcal{E} \rangle &= \sum_{\omega_1, \omega_2} p(\omega_1, \omega_2)\,(\mathcal{E}_1(\omega_1) + \mathcal{E}_2(\omega_2)) \\ &= \sum_{\omega_1} \mathcal{E}_1(\omega_1) \sum_{\omega_2} p(\omega_1, \omega_2) + \sum_{\omega_2} \mathcal{E}_2(\omega_2) \sum_{\omega_1} p(\omega_1, \omega_2) \\ &= \langle p_1, \mathcal{E}_1 \rangle + \langle p_2, \mathcal{E}_2 \rangle \end{aligned}$$

p_1 and p_2 are the marginal probabilities. From this it follows that if we replace p by $p_1 \otimes p_2$, which is the stoss map, then the mean energy is unchanged. We saw that the entropy of $p_1 \otimes p_2$ is not less than that of p (Theorem (2.46)). We can then define the dynamics as follows; let $T^{\dagger}$ be an energy-conserving bistochastic map on $\Sigma(\Omega)$, and define one time-step by

$$p \mapsto T^{\dagger} p \mapsto \mathcal{M}_1 T^{\dagger} p \otimes \mathcal{M}_2 T^{\dagger} p = \tau^* p. \tag{4.1}$$

We call this the Boltzmann map. Since each step conserves the average energy and does not decrease the entropy, our map τ^*, which is nonlinear, obeys both the first and the second laws of thermodynamics. The stoss map introduces a bit of mixing between energy-shells, while preserving the mean energy, and quite often is enough to drive the system to the canonical or grand canonical state. The discrete Boltzmann equation ([23]) is actually of this form: after a scattering, the marginal distributions of the two scattered particles are calculated, and the particles then re-enter the population as independent particles. This was the "hypothesis". Since in general $T^{\dagger} p$ is actually not equal to $\mathcal{M}_1 T^{\dagger} p \otimes \mathcal{M}_2 T^{\dagger} p$, it is fruitless to try to prove the hypothesis. We shall use it as a part of the model; it expresses random effects such as exterior noise that is not included in the mapping T. This noise does not inject any energy, but destroys the correlations set up between the two systems by the scattering T. The system for which the Boltzmann map is ideal is a pair of "well stirred" chemical reactors, joined by a semi-permeable membrane. Chemicals in pots (i) and (ii) diffuse into the membrane, where they react or diffuse to one side or the other. After each time step, each reactor is stirred by a paddle, which does not change the amount of each chemical, or add a significant quantity of heat. All it does is to destroy the correlation between the two reactors. The Boltzmann map then should not be regarded as a possibly bad approximation to the true Hamiltonian dynamics, but as theory in its own right. Thus, the next example is a model of two well-stirred reactors containing three chemicals, and both pots are densely occupied, as in a liquid rather than a gas.

Example 4.2 Let $\Omega_1 = \{A, B, C\} = \Omega_2$, and $\Omega = \Omega_1 \times \Omega_2$. This falls into the field picture if Λ has two points and $\Omega_x = \Omega_1 = \Omega_2$. We wish to describe the chemical process

$$A + C \rightleftharpoons 2B.$$

For this to obey the first law, our energy function must satisfy $\mathcal{E}(A) + \mathcal{E}(C) = 2\mathcal{E}(B)$. Assume this, and that there are no accidental relations, such as all being zero. Then Ω, which has nine points, splits into the energy-shells

$$\{AA\}, \quad \{AB, BA\}, \quad \{AC, CA, BB\}, \quad \{BC, CB\}, \quad \{CC\}.$$

An energy-conserving $T^{\dagger}$ must be the identity on the one-point shells $\{AA\}$ and $\{CC\}$. Let us define $T^{\dagger}$ on the sets $\{AB, BA\}$ and $\{BC, CB\}$ to be the symmetric matrix

$$T^{\dagger} = \begin{pmatrix} 1-\kappa & \kappa \\ \kappa & 1-\kappa \end{pmatrix}. \tag{4.3}$$

The number κ is the probability that the particles A and B exchange places in one time-step, and so $0 < \kappa < 1$. We have chosen the same rate for the exchange of B and C. The action of $T^\dagger$ on the remaining energy-shell is chosen to be the 3×3 matrix

$$\begin{array}{c} \\ AC \\ BB \\ CA \end{array} \begin{array}{c} \begin{array}{ccc} AC & BB & CA \end{array} \\ \left(\begin{array}{ccc} 1-\lambda-\kappa & \lambda & \kappa \\ \lambda & 1-2\lambda & \lambda \\ \kappa & \lambda & 1-\lambda-\kappa \end{array} \right) \end{array} . \tag{4.4}$$

So κ is the rate that A and C exchange places, and λ is the rate of reaction. We get a strictly positive bistochastic matrix on the energy-shell if $\kappa > 0$ $\lambda > 0$ and $\kappa + \lambda < 1$. The action of $T^\dagger$ on Ω is the product of the actions on the various energy-shells; these commute, so the order of the product makes no difference. The restriction of $T^\dagger$ to each energy-shell is positively improving. Thus the map $T^\dagger$ increases the entropy unless p is constant on each of these energy-shells. The stoss map increases entropy unless the probability p is a product, so any fixed point has the form $p_1 \otimes p_2$, and whether at equilibrium or not, this is the form of the state just after the stoss map. One step in the dynamics starts with $p = p_1 \otimes p_2$, then transforms this to $T^\dagger p = p'$ which then leads to new marginals p'_1 and p'_2. For our example, the first step is

$$\begin{aligned}
p'(AA) &= p(AA) \\
p'(AB) &= (1-\kappa)p(AB) + \kappa p(BA) \\
p'(BA) &= (1-\kappa)p(BA) + \kappa p(AB) \\
p'(CC) &= p(CC) \\
p'(BC) &= (1-\kappa)p(BC) + \kappa p(CB) \\
p'(CB) &= (1-\kappa)p(CB) + \kappa p(BC) \\
p'(AC) &= (1-\lambda-\kappa)p(AC) + \lambda p(BB) + \kappa p(CA) \\
p'(BB) &= (1-2\lambda)p(BB) + \lambda p(AC) + \lambda p(CA) \\
p'(CA) &= (1-\lambda-\kappa)p(CA) + \lambda p(BB) + \kappa p(AC).
\end{aligned}$$

From this the marginals p'_1 and p'_2 can be computed. Thus

$$\begin{aligned}
p'_1(A) &= p'(AA) + p'(AB) + p'(AC) \\
&= p(AA) + (1-\kappa)p(AB) + \kappa p(BA) \\
&\quad + (1-\lambda-\kappa)p(AC) + \lambda p(BB) + \kappa p(CA) \\
&= p_1(A)p_2(A) + (1-\kappa)p_1(A)p_2(B) + \kappa p_1(B)p_2(A) \\
&\quad + (1-\lambda-\kappa)p_1(A)p_2(C) + \lambda p_1(B)p_2(B) + \kappa p_1(C)p_2(A) \\
&= p_1(A) + \kappa\,(p_2(A) - p_1(A)) - \lambda\,(p_1(A)p_2(C) - p_1(B)p_2(B))
\end{aligned}$$

where we have simplified using $p_i(A) + p_i(B) + p_i(C) = 1$, $i = 1, 2$. Similarly the other five marginals $p_1'(B) \ldots p_2'(C)$ can be computed; see Ex. (4.73). □

This example shows that the resulting dynamics is non-linear; the state after one time step is a quadratic function of the components of the state before the step. Because of this, the time evolution of a mixed state p is not the mean of the evolution of the pure states into which p decomposes. We have replaced the random path through the sample space with a non-linear, non-random path through the state space $\Sigma(\Omega)$. There is a kind of coherence to the mixture p, which we contrast with the incoherence of the every-day mixture; in the every-day mixture the pure states represent systems in different boxes.

The example describes a model with a diffusion term, $\kappa(p_1 - p_2)$, giving a flow proportional to the gradient of the density. If we extend the model to the case where $\Lambda = \{1, 2, \ldots, n\}$ then the gradient term builds up to the second derivative, and we get a discrete approximation to the heat equation. The non-linear term $\lambda\left(p_1(A)p_2(C) - p_1(B)p_2(B)\right)$ causes the reaction. This is of the form used by chemists under the name "the law of mass action". The limit to continuous time is called the reaction-diffusion equation. If κ is much bigger than λ then p_1 and p_2 rapidly become equal, and the system behaves as two particles in a single container.

From the equations of motion, one can see that the following are conserved quantities:

$$
\begin{aligned}
p_1(A) + p_1(B) + p_1(C) &= 1 \\
p_2(A) + p_2(B) + p_2(C) &= 1 \\
2\left(p_1(A) + p_2(A)\right) + p_1(B) + p_2(B) &= 2N_A + N_B \\
2\left(p_1(C) + p_2(C)\right) + p_1(B) + p_2(B) &= 2N_C + N_B \; .
\end{aligned}
$$

The first two say that there is always exactly one particle of some sort at each point of Λ. The third says that the removal of one A from the total number of A is accompanied by the increase by two in the total number of B. The fourth is not independent of the others, since $N_A + N_B + N_C = 2$. The value of $2N_A + N_B$ on the orbit is fixed by the initial state. Since each of the restrictions of T^1 to the energy-shells is positively improving, the law of detailed balance proves that the only fixed points are fixed points of all the maps simultaneously. The diffusion terms require $p_1 = p_2$ at equilibrium, and, putting this in the reaction term, we get $p(B)^2 = p(A)p(C)$ at equilibrium. If both $2N_A + N_B$ and $2N_C + N_B$ are initially non-zero, then either some B is present, or if there is no B, then there is a non-zero amount of both A and C. The chemical interpretation of this situation is that the reaction can proceed. If one of these, $2N_A + N_B$ or $2N_C + N_B$ is zero, then there is no B present, and one of A and C is absent. Then the reaction cannot proceed in either the forward or backward direction. In the first case, there is a unique fixed point, obtained by solving the equilibrium equations and the conservation laws. It follows by Lyapunov's theorem that the system converges to equilibrium for large times. The fixed point can be interpreted as the canonical state for some beta. Let $\mathcal{N}_A(\omega)$ denote the number of A-particles in the configuration ω; it is the random variable taking values 0, 1, or 2, whose mean is N_A. Similarly $\mathcal{N}_B$ and $\mathcal{N}_C$ are defined. Then, the state $p_\beta \in \Sigma(\Omega)$ given by

$$p(\omega) = Z^{-1} \exp -\beta(E_A\mathcal{N}_A + E_B\mathcal{N}_B + E_C\mathcal{N}_C)$$

obeys all the equilibrium conditions, where β is determined by the initial conditions, and $E_A + E_C = 2E_B$. The conservation law of $2\mathcal{N}_A + \mathcal{N}_B$ is equivalent to the conservation of energy $\mathcal{E} = E_A\mathcal{N}_A + E_B\mathcal{N}_B + E_C\mathcal{N}_C$, provided that $E_A + E_C = 2E_B$. Thus the stoss map adds enough entropy to take the system to thermal equilibrium, provided that the initial point is not one of the unstable equilibrium points on the boundary of Σ.

Example 4.5 *COTRANSPORT THROUGH A MEMBRANE*
Suppose that we have two cells containing chemicals $\{A, B, C, D\}$ which can undergo the reaction

$$A + B \rightleftharpoons C + D. \tag{4.6}$$

Suppose that the cells are separated by a membrane, which is permeable to C and D but not to A and B. What are the possible equilibria? We model this by choosing Λ to have four points, two lying in one cell and two in the other. The sample space is

$$\Omega = \Omega_1 \times \Omega_2, \qquad \text{where} \qquad \Omega_1 = \Omega_2 = \{A, B, C, D\} \times \{A', B', C', D'\}.$$

We place the second factor of Ω_1 and the first factor of Ω_2 next to the membrane. Introduce a bistochastic map on Ω_1 which allows the reaction Eqn. (4.6) to occur in one cell with rate λ, in which the configuration (C, D') produces with rate $\lambda/2$ each of the configurations (A, B') and (B, A'). Similarly, we allow Eq. (4.6) to occur in the other cell with rate λ. We introduce diffusion terms allowing all the molecules to move inside each cell, but include diffusion between the cells for C and D only, since the membrane is not permeable to A and B. Let $p_1(A)$, $p_1(A')$, $p_2(A)$, $p_2(A')$ denote the marginal probabilities that the four sites in Λ are occupied at time t by an A, with $p_1'(A)$ and so on denoting these probabilities at time $t+1$. We use a similar notation for the others. This leads by a calculation similar to that in Ex. (4.2) to the following equations for the marginal probabilities:

$$p_1'(A) \;=\; p_1(A) + \kappa\left(p_1(A') - p_1(A)\right)$$

$$\begin{aligned} & +\lambda/2\,(p_1(C)p_1(D') + p_1(D)p_1(C') - 2p_1(A)p_1(B')) \\ p_1'(C') &= p_1(C') + \kappa\,(p_1(C) - 2p_1(C') + p_2(C)) \\ & +\lambda/2\,(p_1(A)p_1(B') + p_1(B)p_1(A') - 2p_1(D)p_1(C')) \end{aligned}$$

with similar equations for the other variables. The fixed points of this map must be fixed points of each of the reactions taking part, by the theorem on detailed balance. The existence of diffusion means that at equilibrium, A and B must be of uniform density in each cell, and C and D must be uniform throughout:

$$p_1(A) = p_1(A'); \quad p_2(A) = p_2(A'); \quad p_1(B) = p_1(B'); \quad p_2(B) = p_2(B').$$

$$p_1(C) = p_1(C') = p_2(C) = p_2(C'); \quad p_1(D) = p_1(D') = p_2(D) = p_2(D').$$

The chemical balance of the reaction then means that the law of mass action

$$p_1(A)p_1(B) = p_1(C)p_1(D) = p_2(C)p_2(D) = p_2(A)p_2(B)$$

holds at equilibrium. Eliminating C and D gives us the relation at equilibrium

$$\frac{p_1(A)}{p_2(A)} = \frac{p_2(B)}{p_1(B)}.$$

This ratio is called the "static head". The imbalance between the concentrations of A on the two sides of the membrane, $p_1(A) \neq p_2(A)$ is balanced at equilibrium by the opposite ratio for B. It is thought that living bodies make use of this to efficiently transport desired chemicals, such as sugar, through a membrane, such as the wall of the gut. To push the chemical B say from cell 1 to cell 2, we add chemical A, say sodium, to cell 1 in large quantities. In this way all but a few percent of the chemical B can be taken out of cell 1, a much better system than relying on diffusion, which would lead at equilibrium to equal concentrations on the two sides. It should be said that equilibrium may not be reached, as the products in cell 2 could be swept away by flowing blood. Nevertheless, the mechanism of cotransport is a useful addition to simple diffusion. More elaborate models are described in [24].

These two examples illustrate a quite reasonable theory of chemical kinetics, which can be extended to include any number of points in Λ and any number of chemicals. But it has some limitations; only balanced reactions, where the number of reactants is the same on each side, are included; there is no concept of heat, in that the energy is a linear sum of particle-numbers, and all the energy is "chemical". It is not possible to discuss the heat produced in an exothermic reaction. Finally, the rate of the forward reaction is the same as the reverse rate, contrary to experiment. These criticisms led to the following developments of the theory.

To allow unbalanced reactions we borrow the idea of Fock space from quantum mechanics. Instead of allowing only one particle at a site $x \in \Lambda$, the sample space Ω_x is $\{0, 1, 2, \ldots\}$, for each type of particle. This will be called "classical Fock space". Molecules take up some space, and we cannot have more than one at each position in space; thus if we make this choice of Ω_x the site x should represent a macroscopic volume, which can contain a large number of particles. In this case we are omitting all microscopic information, and will get only phenomenological laws of dynamics. Thus the entropy will not include that of the omitted microscopic degrees of freedom. When the particles have a hard core, and we want to make a mesoscopic description, we choose instead of classical Fock space the hard-core space $\Omega_x = \{\emptyset, A, B, \ldots, C\}$, which says that we can have either no particles at x, or one of the named particles. The hard-core is more severe than using Fermi Fock space, which would allow two different particles to sit at the site. If we use the hard-core space it is natural to allow diffusion only from an occupied site to a nearby unoccupied site. The exchange of particles, as in the example, might also be allowed, but might be given a much smaller rate. In either case we can set up dynamical laws for reactions that are not balanced. In the case of the hard-core space, it just amounts to calling one of the particles in a balanced reaction by the symbol $\emptyset$, and giving it no energy or atomic number. In Exercise 4.74 you will consider the reaction

$$A + C \rightleftharpoons B$$

using the hard-core model $\Omega = \Omega_1 \times \Omega_2$, where $\Omega_{1,2} = \{\emptyset, A, B, C\}$. The reaction goes forward only when one site is occupied by an A and the other is occupied by a C, in which case a B is produced with equal probability, λ, at each site. Omitting the possibility of exchange and diffusion, you will get the rate equation

$$p_1'(A) = p_1(A) + \lambda\left(p_1(\emptyset)p_2(B) + p_1(B)p_2(\emptyset) - 2p_1(A)p_2(C)\right).$$

This differs from the equations often seen in chemistry books, and called the law of mass action; our theory leads to the inclusion of the extra factors $p_i(\emptyset)$. These represent an inhibition of the reaction caused by the hard-core: unless one site is empty, we cannot produce two particles, A and C, from a B at the other site. This leads to the slowing down of the rate per molecule of reactions producing an increase in the number of molecules, as the density increases. This is because the pressure increases.

The second criticism, summarised in the question "where is the heat?", has been answered in studies of the classical Boltzmann equation; the sample space (of a single particle) is taken to be $\mathbf{R}^6$, phase space, a point of which gives us the position and velocity of the test particle. The distribution function $f(\mathbf{x}, \mathbf{v})$

allows us to define the density of kinetic energy per particle as

$$\mathcal{T}(\mathbf{x}) = \int f(\mathbf{x}, \mathbf{v}) m\mathbf{v}.\mathbf{v}/2\, d^3x. \tag{4.7}$$

If we postulate that the kinetic energy thermalises almost instantly, because the time-scale of particle exchange is very fast compared with most chemical rates, then this energy is identified with heat, and the velocity distribution is taken to be of the Maxwell-Boltzmann form

$$f(\mathbf{x}, \mathbf{v}) = \text{const. } \exp\{-\beta m\mathbf{v}.\mathbf{v}/2\}$$

where β is a function of $\mathbf{x}$. In this way we can track the production and flow of heat [23]. So far our model is not able to do this. As a start, we can enlarge the hard-core sample space to include a label telling us the kinetic energy of the particle as well as its type; this is less than telling us its velocity, but is enough to keep track of the heat. This avoids the trouble that arises if we use discrete velocity variables, and try to maintain the conservation of energy during reaction. In fact, even without reactions, the possible velocities in discrete velocity models is very limited [25].

Even with the simplification brought about by replacing the velocities by the kinetic energy, the dynamics $p \mapsto T^{\dagger}p$ carries at any time the detailed distribution of the energy among the energy-levels, and this is far more information than is usual in non-equilibrium theories in Physics and Chemistry. In fact, we might guess that very little difference will be made to any macroscopic property by a coarser picture. It has been remarked by Fowler [8] that the transfer of heat within a local region $x \in \Lambda$ happens 10^8 times faster than most chemical reactions. This is also true of the exchange of identical molecules between neighbouring small regions. So we might expect that a state at x, the marginal state, to be almost instantly thermalised. This is known as the *hypothesis of local thermodynamic equilibrium*, or *LTE*. This can be implemented mathematically as follows. Given the state p at time t, and the random dynamics $T^{\dagger}$, we first find the marginal map $\mathcal{M}_x T^{\dagger} p$, and then find the grand canonical state with the same mean energy and atomic numbers as this. This state will be specified by the beta and one chemical potential for each additional atomic number. We call this the local thermalised state. Then we form the product of the local thermalised states over the space Λ. This is the complete map for one time-step. Since each stage in the map can only increase the entropy, or leave it the same, the map obeys the second law of thermodynamics. By construction, it will conserve the energy and just those atomic numbers that are conserved by $T^{\dagger}$. Whether or not it gives nearly the same dynamics we would get from the same $T^{\dagger}$ without *LTE* does not really matter; we are not trying to prove the hypothesis, but to construct a model in which it is true. There are many situations, argues Fowler, when *LTE* replaces unspecified fast variables, and gives a better fit to experiment than the theory without *LTE* and without the fast variables.

The following model [26] shows these features.

Example 4.8 *FREE PARTICLES IN GRAVITY.*
Let us take $\Lambda = \{0, 1, \ldots, N\}$, which we interpret as points in a vertical direction. We introduce one type of molecule with a hard core, whose kinetic energy can be any non-negative multiple of a unit. Thus, the sample space at a site x is

$$\Omega_x = \{\emptyset, 0, 1, 2, \ldots\} \tag{4.9}$$

where $\omega_x = \emptyset$ means that no particle is present at x, and $\omega_x = j$ means that there is one particle present at x and its kinetic energy is j. We only allow one state with each kinetic energy; this is for simplicity. For a more realistic model, see Exercise (4.75). In taking this choice of sample space, we also assume in Eqn. (4.9) that the molecule has only one spin state. Let us write $p_x(\emptyset) = q_x$, and $p_x(j)$ for the marginal probabilities of the state p. The number of molecules at x is the random variable $\mathcal{N}_x(\omega) = 0$ if $\omega_x = \emptyset$, and $= 1$ otherwise. The energy of a molecule at x is the sum of its kinetic energy and its potential energy. Thus $\mathcal{E}(\omega) = \sum_x \mathcal{E}_x(\omega_x)$ where

$$\mathcal{E}_x(\omega_x) = \begin{cases} 0 & \text{if } \omega_x = \emptyset \\ 2\pi\hbar\nu j + mgdx & \text{otherwise} \end{cases} \tag{4.10}$$

where ν is the frequency of a photon with the same kinetic energy, g is the acceleration due to gravity, m is the mass of the molecule and d is the distance between the sites. The factor $2\pi\hbar\nu$ is just to give the energy the right dimensions. We assume that g is independent of x for simplicity. The grand canonical state at x with beta β_x and chemical potential μ_x is by definition

$$q_{\beta_x,\mu_x} = Z^{-1}(\beta_x, \mu_x) \tag{4.11}$$

$$p_{\beta_x,\mu_x}(j) = Z^{-1} \exp\{-\beta_x(2\pi\hbar\nu j + mgdx - \mu_x)\} \tag{4.12}$$

where Z is the grand partition function for the system at x:

$$Z(\beta, \mu) = 1 + e^{-\beta(mgdx-\mu)} \sum_j e^{-\beta 2\pi\hbar\nu j} = 1 + \frac{e^{-\beta(mgdx-\mu)}}{1 - e^{-2\pi\beta\hbar\nu}} . \tag{4.13}$$

Every state has density bounded by $1/d$, since there can be at most one particle at each point. There are, however, some states for which the mean kinetic energy at a point x, $\langle \mathcal{T}_x \rangle = \sum_{j=0}^{\infty} j p_x(j) 2\pi\nu\hbar$ is infinite. Let $\mathcal{D}$ be the affine subspace of $\Sigma(\Omega)$ consisting of states having finite mean kinetic energy at each

$x \in \Lambda$. Let us denote the grand canonical state at x, the pair $(q_{\beta_x,\mu_x}, p_{\beta_x,\mu_x}(j))$ for all j, by p_{β_x,μ_x}. We define the *LTE* map $Q : \mathcal{D} \to \mathcal{D}$ by

$$Qp = \otimes_x p_{\beta_x,\mu_x} \tag{4.14}$$

where μ_x and β_x are uniquely determined by the requirement that $(Qp)_x$ have the same mean density and kinetic energy as p_x. We know that $(Qp)_x$ is the state of greatest entropy among all states with this property, and that the state Qp, which is independent over Λ, is the state of greatest entropy among all states with the given marginals. It follows that Q is entropy nondecreasing. Note that the Shannon entropy is finite for any state p in $\mathcal{D}$, since it is not greater than

$$S(Qp) = \sum_x S(p_{\beta_x,\mu_x}).$$

The linear part of the dynamics is given by a convex sum of local terms, denoted $T_x^\dagger(j)$ or $T_x(j)$, as it is symmetric. This acts on $\Sigma(\Omega_x \times \Omega_{x+1})$, and transfers a particle at x with energy $2\pi\nu\hbar j$ into a hole at $x+1$, using up energy mgd to overcome gravity. It also transfers in the reverse direction, and with the same probability. In order to conserve energy, the ratio of the parameters $2\pi\nu\hbar$ and mgd must be an integer. We shall assume they are equal, for simplicity. Then in going up the ladder from x to $x+1$ a particle loses one unit of kinetic energy, and in going down the same step, it gains one unit. So $T_x(j)$ is the symmetric stochastic matrix

$$T = \begin{array}{c} \\ \omega_1 \\ \omega_2 \end{array} \begin{array}{c} \begin{array}{cc} \omega_1 & \omega_2 \end{array} \\ \left(\begin{array}{cc} 1-\lambda & \lambda \\ \lambda & 1-\lambda \end{array} \right) \end{array}. \tag{4.15}$$

In this matrix, the first row and column are labelled by a point $\omega_1 \in \Omega$ such that $\omega_1(x) = j+1$, $\omega_1(x+1) = \emptyset$ and the second row and column are labelled by a point $\omega_2 \in \Omega$ such that $\omega_2(x) = \emptyset$, $\omega_2(x+1) = j$. For simplicity we have chosen λ to be independent of j, although we might expect that the rate of hopping to increase for larger j, and this is easy to arrange. The sample point ω_1 is connected to only finitely many points by such matrices. So by forming a convex sum of all the matrices involving ω_1 we get a stochastic matrix taking care of all its possible transitions. Similarly for any other element of Ω. Thus we get an infinite stochastic matrix, any of whose rows or columns has only finitely-many nonzero elements. It therefore acts on $\operatorname{Span}\Omega$ (the algebraic sum) in a well-defined way, and preserves the mean total energy and total particle number. This follows from the fact that T connects only states with the same number of particles and the same energy. We shall see it explicitly in a while. It follows that $T^\dagger$ maps $\mathcal{D}$ into $\mathcal{D}$. Since the hopping particle carries energy with it, the model describes the *convection* of heat.

To find the fixed points of $\tau^* = Q \circ T^\dagger$ note that entropy is a strict Lyapunov function for each $T_x(j)$, and is strictly concave, so a fixed point of a convex sum of these stochastic maps is a fixed point of each of them (Tolman's principle of detailed balance is true here). So a fixed point of τ is a grand canonical state that is left unchanged by all the $T_x(j)$.

Take $p \in \Sigma(\Omega)$ to be independent over Λ, and consider the action of $T_x(j)$ on the relevant part of p, namely $p_x \otimes p_{x+1} \in \Sigma(\Omega_x \times \Omega_{x+1})$. As before, put

$$p_x = (q_x, p_x(j)) \qquad p_{x+1} = (q_{x+1}, p_{x+1}(j)) \ .$$

Let $p' \in \Sigma(\Omega_x \times \Omega_{x+1})$ be the relevant part of $T_x(j)p$. Then p' differs from p at only two points, as we see from Eqn. (4.15):

$$\begin{aligned} p'(j+1, \emptyset) &= (1-\lambda)p_x(j+1)q_{x+1} + \lambda q_x p_{x+1}(j) \\ p'(\emptyset, j) &= \lambda p_x(j+1)q_{x+1} + (1-\lambda)q_x p_{x+1}(j) \ . \end{aligned}$$

Now

$$p'_x(j+1) = p'(j+1,\emptyset) + \sum_{l=0}^{\infty} p'(j+1,l) = p'(j+1,\emptyset) + \sum_{l=0}^{\infty} p_x(j+1)p_{x+1}(l) \quad (4.16)$$

since $T_x(j)$ does not alter $p(j+1,l) = p_x(j+1)p_{x+1}(l)$. Thus

$$\begin{aligned} p'_x(j+1) &= p'(j+1,\emptyset) + p_x(j+1)\sum_{l=0}^{\infty} p_{x+1}(l) \\ &= (1-\lambda)p_x(j+1)q_{x+1} + \lambda q_x p_{x+1}(j) + p_x(j+1)(1-q_{x+1}) \\ &= p_x(j+1) - \lambda\left(p_x(j+1)q_{x+1} - p_{x+1}(j)q_x\right) . \end{aligned} \quad (4.17)$$

This could have been guessed, since the change in $p_x(j+1)$ can come about from the absence of a particle at x (probability: q_x) and the presence of one at $x+1$ of energy $j2\pi\hbar\nu$ (probabilty: $p_{x+1}(j)$), or from a hole at $x+1$ and a particle of energy $(j+1)2\pi\hbar\nu$ at x (probability: $p_x(j+1)q_{x+1}$). Our calculation shows that this intuitively natural dynamical law comes from a symmetric stochastic matrix followed by Q, and is therefore entropy-increasing. Similarly we get

$$p'_{x+1}(j) = p_{x+1}(j) + \lambda\left(p_x(j+1)q_{x+1} - p_{x+1}(j)q_x\right) . \quad (4.18)$$

We remark that $p_0(0)$, the ground-state occupation density, does not appear in either equation of motion, since $j \geq 0$ and $x \geq 0$, and so is a constant of the motion under the Markov chain generated by T. It is Q that causes the time-dependence of this variable, and leads us to equilibrium. We can now see explicitly that τ conserves the mean total energy; it is enough to show this for T_x acting on the two sites x and $x+1$. For, the energy is equal to

$$\begin{aligned} ((j+1)2\pi\hbar\nu + dmgx)p_x(j+1) \ &+ \ (j2\pi\hbar\nu + dmg(x+1))p_{x+1}(j) \\ &= (j+x+1)2\pi\hbar\nu(p_x(j+1) + p_{x+1}(j)) \end{aligned}$$

and since

$$p_x(j+1) + p_{x+1}(j) = p'_x(j+1) + p'_{x+1}(j) ,$$

this is conserved; this also shows that the mean total number of particles is conserved.

We conclude that $\{\tau^{n*}p\}$ is an orbit in $\Sigma(\Omega)$, such that the mean total energy and particle number is fixed, moving through local grand canonical states for which entropy is a Lyapunov function. By Lyapunov's theorem the proof of convergence to a fixed point will be complete when we show that there is a unique fixed point. This we now do. To be stationary, a state must be a fixed point of each $T_x(j)$, and so must satisfy

$$p_x(j+1)q_{x+1} = q_x p_{x+1}(j) \text{ for all } x, j \geq 0 \tag{4.19}$$

as well as the LTE conditions, Eqn. (4.12):

$$p_x(j+1)/q_x = \exp\left(-\beta_x\left((j+1)2\pi\hbar\nu + mdgx - \mu_x\right)\right) \tag{4.20}$$

$$p_{x+1}(j)/q_{x+1} = \exp\left(-\beta_{x+1}(j2\pi\hbar\nu + mgd(x+1) - \mu_{x+1})\right) . \tag{4.21}$$

We note that these conditions do not depend on λ, or on whether λ depends on x or j, as long as $\lambda > 0$. This is usual in statistical dynamics, in that the equilibrium state does not depend on $T^{\dagger}$. Combine (4.19) with (4.20) and (4.21) for all j to give $\beta_x = \beta_{x+1}$ and $\mu_x = \mu_{x+1}$, since $mgd = 2\pi\hbar\nu$. Thus our model of convection is fully mixing, in that it ensures a uniform beta and chemical potential at equilibrium. The value of μ is fixed by the total mean number of particles, provided that it is not greater than $N+1$. As usual, beta is fixed by the mean energy, which must be non-negative. We conclude that there is only one fixed point in the set of local grand canonical states with a given mean energy and particle number. This is enough, by Lyapunov's direct method, to establish that the system converges at large time, since the set of local grand canonical states with given total mean energy and particle number is compact with the usual topology on the parameters μ_x, β_x. At equilibrium the density $\rho_x = p_x/d$ varies with height x; we see that

$$p_x = \sum_j p_x(j) = Z^{-1}\exp -\beta\{mdgx - \mu\}\left(1 - \exp\{-2\pi\beta\hbar\nu\}\right)^{-1} .$$

Putting in the grand partition function Eqn. (4.13) and rearranging, we get (Exercise (4.76))

$$p_x = \frac{\exp -\beta\{mdgx - \mu\}}{1 + \exp -\beta\{mdgx - \mu\} - \exp\{-\beta mdg\}} . \tag{4.22}$$

Putting $x = 0$ gives us a relation between the chemical potential μ and the density at $x = 0$, which can be solved to give

$$\exp\{\beta\mu\} = p_0\,(1 - \exp\{-\beta mgd\})\,(1 - p_0)^{-1}.$$

Substituting, we get (Exercise (4.76))

$$p_x = \frac{p_0 \exp\{-\beta mdgx\}}{1 - p_0\,(1 - \exp\{-\beta mdgx\})}\,. \tag{4.23}$$

For small values of p_0 and large values of x this shows an exponential thinning of the density with increasing x. If p_0 is not small there is a correction due to the hard core.

This example allows us to understand how to compute the pressure in statistical dynamics. Often, in equilibrium statistical mechanics, the pressure is defined in terms of the grand partition function. But here the pressure is not an intensive variable, the same throughout the system, but must depend on x. What is holding up the molecules at height x, against the gravitational field? Let $z = xd$ be the height in dimensional units. The difference in pressure at z and $z + \delta z$ provides the force to hold up the weight $mg\rho\delta z$. So we get the equation

$$-\frac{\partial P}{\partial z}\delta z = mg\rho\delta z \tag{4.24}$$

where we write ρ as a function of z. In addition, the pressure must vanish at $z = \infty$. You might not like that remark if Λ is finite; in that case, we note that the pressure must vanish when the density is zero. Either condition fixes the constant of integration. To solve the differential equation, note that

$$Z(z) = 1 + \frac{\exp\{\beta\mu\}}{1 - \exp\{-\beta mdg\}}\exp\{-\beta mgz\} \tag{4.25}$$

from which we can compute

$$\frac{\partial \log Z}{\partial z} = \frac{\beta mg \exp\beta\{\mu - mgz\}}{\exp\beta\{\mu - mgz\} + 1 - \exp\{-\beta mdg\}} = -\beta mgd\rho\,. \tag{4.26}$$

Since $\log Z$ satisfies the boundary conditions, we conclude that

$$P(z) = (\beta d)^{-1}\log Z(z). \tag{4.27}$$

This relation (with d=1) is usually derived for homogeneous systems within the equilibrium theory.

The third criticism of this theory of chemistry, that the forward and backward rates of reaction are equal, contrary to experiment, will be addressed in the next section.

4.2 The Heat-Particle

The sample space chosen in the model of the gas in a gravitational field (4.8) attributes the kinetic energy to the particular particle at x; however, the interaction between close particles is very strong, and it is reasonable, as well as simpler, to share the kinetic energy between neighbours. There is another thing to bear in mind, that in thermal systems the electromagnetic field is also thermalised, and some energy of motion of the charged particles should be attributed to the photon field. We shall attribute all the kinetic energy of the particles to some extra degrees of freedom called heat-particles. Like the electromagnetic field, these will be located on the chemical bonds between molecules, and can be regarded as carrying the energy of bond vibration. This should not be counted twice, once as the kinetic energy of the particles, and a second time as a field vibration. The introduction of the heat-particle instead of the space of kinetic energy of specified particles expresses the idea that heat does not belong to any particular particle. In the early days, the idea that the kinetic energy is quickly thermalised was incorporated in the Smoluchowski approximation. We express the same physical idea in a different way, through the presence of the heat-particle. When we view a body at positive temperature, actual photons are observed as black body radiation, so they are really there.

We now add to the general structure of the space Λ a set $\mathcal{B}$ of pairs $\{x, y\}$ specifying the bonds, so that $\{\Lambda, \mathcal{B}\}$ is a finite graph. The vertices of the graph are the points of Λ, and the edges of the graph are the bonds. Denote a typical bond by $b \in \mathcal{B}$, so that $b = (x, y)$. We choose a sample space Ω_{cx} at each site, which can be classical Fock space of one or more chemical types, or the hard core space of several chemicals, or a mixed product of these. The suffix c indicates that this is the space of chemicals. To each bond b we associate a sample space $\Omega_{\gamma b}$, which is a product of classical Fock spaces, one for each heat-particle; the suffix γ indicates that the space concerns the photon field. For simplicity we usually only introduce heat-particles with energy equal to that of one of the energy transfers in the process, just as the kinetic energy $2\pi\hbar\nu$ in Example (4.8) was chosen equal to mgd to allow transitions to conserve energy. The total sample space thus has the form

$$\Omega = \prod_x \Omega_{cx} \times \prod_b \Omega_{\gamma b}. \tag{4.28}$$

For example, if there are two chemicals A and B and they have hard cores, and can reach every site, then each Ω_{cx} could be chosen to be $\{\emptyset,\ A,\ B\}$ and if there is one heat-particle we choose $\Omega_{\gamma b} = \{0,\ 1,\ 2, \ldots\}$. Then a sample point is the pair $\omega = (\omega_c, \omega_\gamma)$; here, ω_c is an ordered set of symbols $\emptyset, A$ or B, one for each x, and ω_γ is an ordered set of non-negative integers, one for each bond.

The energy of the system is a random variable which is a sum of a potential energy and a kinetic energy. The potential energy $\mathcal{E}_c$ is a function only of the variables ω_c, and represents the chemical energy bound up in the molecule, as well as the physical potential energy such as that of gravity. Electostatic energy is also an important example in chemistry and biology. The kinetic or heat energy, $\mathcal{E}_\gamma$, is a function only of ω_γ; thus

$$\mathcal{E}(\omega_c, \omega_\gamma) = \mathcal{E}_c(\omega_c) + \mathcal{E}_\gamma(\omega_\gamma). \tag{4.29}$$

We do not postulate any contribution to the energy that depends on both ω_c and ω_γ; this is reasonable if we interpret the two parts as potential and kinetic energy (and we omit magnetic forces) but with the new idea that $\mathcal{E}_\gamma$ is the energy of the electromagnetic field, it might seem that molecules should exhibit some interaction. Nevertheless, we stick by Eqn. (4.29), and interpret it as the energy of free, asymptotic molecules and free photons, after they have come apart from their scattering. The scattering is achieved by the stochastic map T, which is bistochastic and so is a random mixture of reversible maps, by Birkhoff's theorem. We interpret these, as did Boltzmann, as scattering by a random short-ranged potential. It is known from scattering theory that the scattering operator conserves the total energy of the asymptotic states. (Of course, it does mix up the various contributions.) This is why we still get reactions occurring even though there is no "interaction" term in Eqn. (4.29). Thus the dynamics is determined by a stochastic map T mapping the energy-shells to themselves; mixing between energy-shells is achieved by the stoss map, which replaces $T^\dagger p$ by $\mathcal{M}_c T^\dagger p \otimes \mathcal{M}_\gamma T^\dagger p$. Here, $\mathcal{M}_c$ and $\mathcal{M}_\gamma$ are the marginal maps onto the chemical space and the heat space respectively. The stoss map is quite within the spirit of statistical methods; more is true here; since heat is, by definition, the thermalised part of the energy, it rapidly causes any disturbance of its thermal state to revert back to a new thermal state. So it is reasonable to follow the stoss map by the map that replaces $\mathcal{M}_\gamma T^\dagger p$ by the canonical state with the same mean energy. This is the LTE-map for heat, say Q_γ. Thus we are coming round to defining the *isolated dynamics with heat-particles* as the map

$$p \mapsto T^\dagger p \mapsto \mathcal{M}_c T^\dagger p \otimes Q_\gamma \mathcal{M}_\gamma T^\dagger p \ .$$

This map conserves mean energy and obeys the second law. Then the aim of the theory is to show that, for a suitable choice of T, the system converges to equilibrium for large time; the equilibrium state will be a canonical state for the heat, and a grand canonical state for the chemicals. The beta and chemical potentials are determined by the initial mean values of the energy and the atomic numbers, not by an external heat-bath. We express this as saying that an isolated system finds its own equilibrium.

It is often hard to handle a model of the above sort, because the state at time t continues to carry details of the distribution of all the chemicals taken together; that is, it carries the information about the correlations between different points of Λ. Great simplifications take place if we add more randomness, namely, we replace the state $\mathcal{M}_c T^\dagger p$ by the grand canonical state with the same energy and atomic numbers. Let us call this map Q_c. One time-step is then

$$\tau^* p = Q_c \mathcal{M}_c T^\dagger p \otimes Q_\gamma \mathcal{M}_\gamma T^\dagger p \,. \tag{4.30}$$

This expresses that a great stirring has rapidly occurred within the time-step, within the region of space represented by x. The resulting map τ^* is called "isolated dynamics with LTE". The map τ^* is highly non-linear, since $Q_c p_c$ is a product of degree $|\Lambda|$ in the grand canonical states at each x. We shall see that the resulting equations are among those postulated by chemists on the basis of the law of mass action.

Example 4.31 Let us reconsider the problem of a particle in a gravitational field, but let us assign the kinetic energy to a heat-particle rather than to the chemical. Thus $\Lambda = \{0,\, 1,\, \ldots,\, N\}$, and

$$\mathcal{B} = \{(0,1),\, (1,2),\, \ldots, (N-1, N)\}$$

is the set of bonds. Also, we have one particle with a hard core, and one heat-particle on each bond, so we choose $\Omega_{cx} = \{\emptyset,\, 1\}$ and $\Omega_{\gamma b} = \{0,\, 1,\, \ldots\}$. We interpret $x \in \Lambda$ as the height above ground. The sample space for the system is

$$\Omega = \prod_{x\in\Lambda} \Omega_{cx} \times \prod_{b\in\mathcal{B}} \Omega_{\gamma b}, \qquad\qquad \omega = (\omega_c,\, \omega_\gamma). \tag{4.32}$$

The energy $\mathcal{E}$ differs from that in Example (4.8) in that a site can possess energy even if it contains no particles; this is just what we want: it is the radiation in the cavity. Thus

$$\mathcal{E}(\omega) = \mathcal{E}(\omega_c) + \mathcal{E}_\gamma(\omega_\gamma) = \sum_{x=0}^{N} mgdx\omega_{cx} + \sum_{b\in\mathcal{B}} 2\pi\hbar\nu\omega_{\gamma b} \,.$$

Here, as defined above, ω_{cx} is either 0 or 1, and $\omega_{\gamma b}$ is one of $0,\, 1,\, 2, \ldots$. As before, we need compatibility between the kinetic energy and the energy jumps in the potential; the simplest choice is the one we made before: $mgd = 2\pi\hbar\nu$. Then convection can occur if a particle at height x absorbs j quanta from the bond $b = (x-1, x)$, jumps to the empty site at height $x+1$, and emits $j-1$ quanta to the bond $(x, x+1)$. One quantum is used up to overcome the gravitational potential. It could equally well absorb j quanta from the bond $(x, x+1)$, jump from x to $x+1$ and release $j-1$ quanta to the bond $(x+1, x+2)$. To get a symmetric process, and hence a bistochastic map, we choose the reverse of these processes to be equally likely. The simplest case is $j = 1$, and we shall give the details for this.

Choose $x \in \Lambda$ and concentrate on the factors $\Omega_{cx} \times \Omega_{c,x+1} \times \Omega_{\gamma b}$, where $b = (x, x+1)$. Introduce the bistochastic map $T^{\dagger}$ which takes the pure state $\{\omega_{cx} = 1,\, \omega_{c,x+1} = \emptyset,\, \omega_{\gamma b} = j\}$ to the state $\{\omega_{cx} = \emptyset,\, \omega_{c,x+1} = 1,\, \omega_{\gamma b} = j-1\}$, with probability λ; the inverse is taken to have the same probability. All other components of ω are unchanged. Following $T^{\dagger}$, we apply the stoss map with LTE. Now, any state (q_x, p_x) on a two-point set is already a grand canonical state with some local value of μ, so the LTE map is

$$p \mapsto \bigotimes_x \mathcal{M}_x T^{\dagger} p \otimes \bigotimes_b Q_{\gamma b} \mathcal{M}_\gamma T^{\dagger} p.$$

We stay with the convention introduced in [24] to write s for the state of the heat particle, and s_β for the canonical state with beta β, while retaining p_x for the marginal state of p on the set Ω_{cx}, and $q_x = 1 - p_x$. Then p_x is the probability of occupation of the site x, so it is the density. Since $T^{\dagger}$ acts only on x, $x+1$ and b, we can ignore the other factors and take $p \in \Sigma(\Omega_{cx} \times \Omega_{c,x+1} \times \Omega_{\gamma b})$, of the form

$$p = p_x \otimes p_{x+1} \otimes s.$$

Put $p' = T^{\dagger} p$. Then one time-step gives

$$\begin{aligned}
p'(0,0,j) &= p(0,0,j) \text{ for all } j \geq 0 \\
p'(0,1,j) &= (1-\lambda)p(0,1,j) + \lambda p(1,0,j+1) \text{ for all } j \geq 0 \\
p'(1,0,j) &= (1-\lambda)p(1,0,j) + \lambda p(0,1,j-1) \text{ for all } j \geq 1 \\
p'(1,0,0) &= p(1,0,0) \\
p'(1,0,0) &= p(1,0,0)\ .
\end{aligned} \tag{4.33}$$

In addition, it is reasonable to add a diffusion of the heat, with a probability κ that an occupied bond sends a heat-particle to a neighbouring bond. The full dynamics is a convex mixture of all these. Let p'_x be the density at x after the time-step. Then

$$\begin{aligned}
p'_x &= \sum_{j\geq 0} p'(1,0,j) + \sum_{j\geq 0} p'(1,1,j) \\
&= p(1,0,0) + \sum_{j\geq 1}\{(1-\lambda)p(1,0,j) + \lambda p(0,\lambda,j-1)\} + p_x p_{x+1} \sum_j s(j) \\
&= p_x q_{x+1} s(0) + (1-\lambda)p_x q_{x+1}(1-s(0)) + \lambda q_x p_{x+1} + p_x p_{x+1} \\
&= p_x - \lambda p_x q_{x+1}(1-s(0)) + \lambda q_x p_{x+1}\ .
\end{aligned} \tag{4.34}$$

This is no surprise: the transition probability from $x+1$ to x is λ times $p_{x+1}q_x$, which is the probability that there is a particle at $x+1$ and a hole at x. The inverse probability is similar, multiplied by the factor $(1 - s(0))$. Now, we have applied the LTE map Q_γ to the state, so s is a canonical state, with beta β say, the instantaneous local beta of the bond $(x, x+1)$. Then $1 - s(0) = 1 - Z_\beta^{-1} = e^{-2\pi\beta\hbar\nu}$. Thus the forward and backward rates are related by the factor $e^{-\beta E}$, where E is the energy difference between the states. This is known as the Boltzmann factor to physicists, and less accurately as the principle of detailed balance; it follows from a simple form of microscopic reversibility, which in our case is that T is symmetric. It will be shown to arise in all reactions that absorb or emit heat. In chemistry it is known as the Arrhenius factor; it gives the typical temperature dependence of the ratio of the forward and backward rates, well confirmed by measurements.

There is a unique fixed point for any initial values of the total energy and total number of molecules. For, the heat-particles cannot be at equilibrium, because of the direct diffusion, unless the temperature is uniform. Then the condition for equilibrium of the molecules

$$p_{x+1}/q_{x+1} = e^{-\beta E} p_x/q_x \tag{4.35}$$

determines the density in sequence, starting at $x = 0$. It can be summarised as saying that the chemical potential is independent of x at equilibrium. Thus we get the same variation of density with height as in Example (4.8). This is Exercise (4.79)

In this example suppose that we omit the LTE map on the space Ω_c, while keeping the map Q_γ which thermalises the heat-particles. By the principle of detailed balance, any fixed point must be a fixed point of the diffusion of the heat particles as well as the convection of the molecules. So there must be a uniform temperature, and the state s_β of the heat-particle must be unchanged by the dynamics. The factors s in the dynamics of the chemicals just appear as constants, the Arrhenius factors. In this case the dynamics of the chemicals reduces to a linear stochastic matrix, which is irreducible on each number shell, since each pair of states is connected by a finite chain of non-zero transition probabilities. The fixed point must therefore be the uniform distribution on each number-shell, by the uniqueness part of the Perron-Frobenius theorem. By theorem (3.26), the system converges to a mixture of uniform distributions on each number-shell, the fraction of probability on each shell being the same as the initial state; the actual number of molecules is conserved, and not only the mean number. Thus the system fails to converge to the grand canonical state; there is not enough mixing for that. We conclude that the chemical LTE-map is needed to get the grand canonical state.

The chemical LTE map produces from a state with a definite number of molecules a state which is a mixture of states with various numbers, but with

the same mean number. This means that if the initial state has a definite number, then after one time-step, there is a non-zero probability of finding a different number. This seems to say that there is a violation of the law of conservation of atomic number. Some people are more worried by this than the same problem that arises in the thermalisation of energy. I was for a time, and I asked 'tHooft about it. His reply was worth keeping:
"The theorems and results in mathematical physics are not theorems and results about Nature, but about our description of it."
In other words, a model (in this case the LTE-map) is only a partial description, and the interpretation of the model might not agree with experiment in all detail. We must treat the number of particles statistically, only the mean values, variance and other higher moments, are predicted by the model. The exact number of atoms is not one of the predictions of the model. If the volume is large, the local fluctuations in equilibrium are almost the same, whichever ensemble is arrived at. This is the result of the important theorem on the equivalence of ensembles. The text-books just say ([8]) we use the grand canonical state for computational convenience, and rely on the fact that the limit as $|\Lambda| \to \infty$ is the same for all ensembles. We state a version of this in Theorem (5.34).

Our theory aims to be more ambitious, and to describe dynamics, not only the equilibrium state. We also want to apply it to *mesoscopic systems*, in which Λ is not infinite. In this case the dynamics with or without chemical LTE is different. Undoubtedly, the dynamics with Q_c is simpler, and converges to a different equilibrium. To avoid the problem of non-conservation of particles, we could just keep the Boltzmann map and leave out LTE; then we get convergence to the canonical state, and do not mix the number-shells. But then we are left with the same philosophical problem with the mixing of energy-shells: a state with a sharp energy can move to a state with mixed energy. I offer the following idea: to say that we have an isolated system is an idealisation, as no system is completely isolated. The walls of the jar are not part of the system, and are designed to conserve the mean number of molecules; we cannot prevent the rapid exchange of adhering molecules with those of the system; this does not alter the mean number, but any given measurement will reveal that the number has changed from that in the initial state (but with small chance of being very different). We choose to model this by the LTE map, on the assumption that the exchange of molecules is one of the fast variables, and so all correlations are destroyed in one time-step.

In the next section we show how to model any chemical reaction, and how the heat-particle represents the energy of reaction.

4.3 Activity-led Reactions

The law of mass action for chemical reactions states that the rate of the reaction

$$A + B + C + \ldots \rightarrow \text{products}$$

is proportional to the product of the concentrations of A, B, $C, \ldots$. The same is claimed for the back-reactions. Let us denote these concentrations by N_A, N_B and so on. Thus the reaction

$$A + B \rightleftharpoons C + D$$

should have a forward rate $K_+ N_A N_B$ and a backward rate $K_- N_C N_D$:

$$\frac{dN_A}{dt} = K_- N_C N_D - K_+ N_A N_B \ . \tag{4.36}$$

Equilibrium holds when

$$N_A N_B / N_C N_D = K_- / K_+, \tag{4.37}$$

a formula taught to generations of students of physical chemistry. Arrhenius's law

$$K_- / K_+ = \exp\{-\beta q\} \tag{4.38}$$

(where q is the heat emitted in the forward reaction) is found to be very accurate for many reactions over a wide range of temperatures. Eqn. (4.36) is approximately true for many dilute solutions; it needs corrections in some experiments. It is found that at high concentration the reaction rate is less than that given by Eqn. (4.36), suggesting some sort of saturation, due perhaps to the shielding of the active molecules from interacting with other than their nearest neighbours. Moreover, there are theoretical reasons why Eqn. (4.36) cannot be exact. Namely, if the sample space is say classical Fock space, with the energy of a simple oscillator, and N is the mean particle number, then the equilibrium state given by Eqn. (4.37) is not a grand canonical state. So it is difficult to justify the dynamical law (4.36) by statistical dynamics. This particular difficulty is avoided if the rate equations were based on the activity of the chemical rather than their concentrations. Thus it is suggested [27] that the rate equation for $A + B \rightleftharpoons C + D$ should be

$$\frac{dN_A}{dt} = K_- a_C a_D - K_+ a_A a_B \tag{4.39}$$

rather than Eqn. (4.36). Here, N is the mean number, $N = \langle \mathcal{N} \rangle$, and a is the activity. For a free particle with energy $\mathcal{E} = 2\pi\hbar\nu\mathcal{N} = E\mathcal{N}$ the grand canonical state is the probability on classical Fock space given by

$$\text{Prob}\{\mathcal{N} = j\} = Z(\beta, \mu)^{-1} \exp\{j\beta(\mu - E)\} \tag{4.40}$$

and the activity is:

$$a = N/(1+N) = \exp\{\beta(\mu - E)\}. \tag{4.41}$$

Then the equilibrium condition becomes, instead of Eqn. (4.37)

$$a_A a_B / a_C a_D = K_- / K_+ \,. \tag{4.42}$$

Let us first consider the case when $K_+ = K_-$, so the heat of reaction is zero. Then Eqn. (4.42) becomes

$$\exp\{-\beta(E_A - \mu_A)\}\exp\{-\beta(E_B - \mu_B)\} = \exp\{-\beta(E_C - \mu_C)\}\exp\{-\beta(E_D - \mu_D)\}$$

which is satisfied if $\mu_A + \mu_B = \mu_C + \mu_D$ since by energy conservation we already have

$$E_A + E_B = E_C + E_D \,.$$

Similarly, activity-led equations for a general reaction $A_1 + A_2 + \ldots + A_m \rightleftharpoons B_1 + B_2 + \ldots + B_n$ can be written down (Exercise 4.80). You are asked to check that if $q = 0$ then a grand canonical state is a fixed point if and only if the sum of the chemical potentials on each side of the reaction is the same.

We conclude that a grand canonical state with balanced chemical potentials is a fixed point of the activity-led dynamics, provided that the heat of reaction is zero. It is easy to modify the dynamical law to cover the cases when some of the chemicals are the same kind of molecule as others. The number of molecules entering the reaction is called the "stoichiometry" of the molecule. If a type of molecule appears on both sides of the reaction, say $A_1 = B_1$, then this chemical is called a catalyst. The general conclusion, that equilibrium occurs when the chemical potentials on both sides are equal, is still true if there are catalysts present, whatever the stoichiometry. Do try Exercise (4.82).

So far, we have taken the heat of reaction to be zero. To include the case where heat is emitted, say in the forward reaction

$$A_1 + A_2 + \ldots + A_m \rightleftharpoons B_1 + B_2 + \ldots + B_n + q$$

we simply introduce $B_{n+1} = \gamma$, a heat-particle with quantum of energy equal to q. The activity of γ is $a_\gamma = e^{-\beta q}$, and the chemical energies on each side of the reaction no longer sum up to the same total, but differ by q:

$$E_{A_1} + \ldots + E_{Am} = E_{B_1} + \ldots + E_{Bn} + q \,.$$

As a result, the grand canonical state with balanced chemical potentials is a fixed point. The backward equation contains the factor $a_\gamma = e^{-\beta q}$, so the Arrhenius relation is predicted to hold. This is even true when heat-particles appear on both sides of the equation, in which case heat is a catalyst that allows the reaction to start. This is what happens in the theory of the transition state. In that theory [28], it is postulated that the reaction does not proceed directly, but first moves to a "transition state", requiring an energy E_a, the activation energy. This is supplied by the kinetic energy, or in statistical dynamics, by a heat-particle. The energy of the heat-particle on the other side of the reaction must then be $q + E_a$; or we could have two or more heat-particles whose total energy is $q + E_a$. In any case, the forward-to-backward rates are in the ratio of $e^{-\beta q}$, as predicted by Arrhenius. So there is good reason to study the activity-led equations. The question arises, can they be derived by statistical dynamics? If so, then they obey the laws of thermdynamics. We now show that we can indeed find a $T^\dagger$ conserving energy so that these equations arise by the application of $T^\dagger$ followed by the LTE-map Q_c. We follow [29] closely. We start with a detailed study of the simplest case, that of a unimolecular reaction.

A unimolecular reaction is a reaction

$$A \rightleftharpoons B\ . \tag{4.43}$$

We assume no heat flows, so that the states A, B must have the same energy, number of atoms, etc. The possible configurations of A are the numbers $\{0, 1, 2, ...\} = \Omega_A$ of particles in the cell, and the configuration space for B is $\Omega_B = \{0, 1, 2, ...\}$, and for the system as a whole we choose $\Omega = \Omega_A \times \Omega_B$. A typical point of Ω will be denoted $\omega = (j, k)$. The numbers of A- and B-particles, $\mathcal{N}_A$, $\mathcal{N}_B$, are the random variables

$$\mathcal{N}_A(\omega) = j \quad , \quad \mathcal{N}_B(\omega) = k \tag{4.44}$$

and the reaction (4.43) conserves $\mathcal{N} = \mathcal{N}_A + \mathcal{N}_B$. A probability p on Ω is described by numbers $p_{jk} \geq 0$, $\sum_{j,k} p_{jk} = 1$.

We start at some time with $p_A \in \Sigma(\Omega_A)$ and $p_B \in \Sigma(\Omega_B)$, and form the independent product $p_A \otimes p_B \in \Sigma(\Omega)$. The reaction in one time-step is then described by a bistochastic map $T^\dagger$:

$$T^\dagger : \Sigma(\Omega) \longrightarrow \Sigma(\Omega) \tag{4.45}$$

conserving $\mathcal{N}$. The new distributions p'_A, p'_B after one time-step are then obtained as the marginal distributions of $T^\dagger(p_A \otimes p_B)$. Explicitly, $\mathcal{M}_A$, $\mathcal{M}_B$, the marginal operators, are

$$(\mathcal{M}_A p)_j = \sum_k p_{jk} \quad , \quad (\mathcal{M}_B p)_k = \sum_j p_{jk} \tag{4.46}$$

for any $p \in \Sigma(\Omega)$.

The operators $\mathcal{M}_A$, $\mathcal{M}_B$ do not alter the mean values of $\mathcal{N}_A$, $\mathcal{N}_B$ respectively.

So far, the dynamics in one time-step is the map

$$
\begin{aligned}
(p_A, p_B) \mapsto p_A \otimes p_B \mapsto T^\dagger(p_A \otimes p_B) \mapsto \\
\big(\mathcal{M}_A(T^\dagger(p_A \otimes p_B)), \mathcal{M}_B(T(p_A \otimes p_B))\big) = (p'_A, p'_B) \,.
\end{aligned} \tag{4.47}
$$

Even if p_A and p_B are local equilibrium states of the form (4.40), in general p'_A and p'_B given by (4.47) will not be.

The average value of $\mathcal{N}_A$ in (4.40) is

$$
N_A = \sum_{j\geq 0} j p_{Aj} = Z_A^{-1} \sum_{j\geq 0} j a_A^j = (1 - a_A)\frac{a_A}{(1-a_A)^2} = \frac{a_A}{1-a_A} \,;
$$

thus

$$
N_A = \frac{a_A}{1-a_A} \;, \quad a_A = \frac{N_A}{1+N_A} \;, \quad Z_A = 1 + N_A = (1-a_A)^{-1}. \tag{4.48}
$$

The local equilibrium state (4.40) can be expressed in terms of N, thus:

$$
\text{Prob}\{\mathcal{N}_A = j\} = \frac{1}{1+N_A}\left(\frac{N_A}{1+N_A}\right)^j \,. \tag{4.49}
$$

Given any state $p \in \Sigma(\Omega)$, we have defined $Q_c p$ to be the local equilibrium state, (4.49), with the values of N_A and N_B determined by

$$
N_A = \sum_1^\infty j p_A(j) \tag{4.50}
$$

and a similar definition for N_B. By definition, Q_c does not alter the mean value of $\mathcal{N}_A$ or $\mathcal{N}_B$. The map Q_c is the mathematical realisation of the hypothesis of local equilibrium.

The dynamics of the model is defined by (4.47) followed by Q_c

$$
\tau^* p = ((\tau^* p)_A, (\tau^* p)_B) = (Qp'_A, Qp'_B) \,. \tag{4.51}
$$

By construction, τ^* takes a pair of local equilibrium states on Ω_A, Ω_B to another pair; we can parametrize this either by the map $(N_A, N_B) \mapsto (N'_A, N'_B)$ or $(a_A, a_B) \mapsto (a'_A, a'_B)$.

It remains to specify the bistochastic map T. We shall give it as a symmetric Markov matrix on $\ell^2(\Omega)$, the Hilbert space of square-summable random variables with scalar product

$$
\langle f, g\rangle = \sum_{j,k=0}^\infty f(j,k) g(j,k) \,.
$$

An orthonormal basis for $\ell^2(\Omega)$ is given by the projection operators onto (j,k), for each $(j,k) \in \Omega$:

$$\left.\begin{aligned} P(j,k) &= 1 \quad \text{if} \quad \omega = (j,k) \\ &= 0 \quad \text{otherwise.} \end{aligned}\right\} \tag{4.52}$$

Then

$$\langle P(j,k), P(j',k')\rangle = \delta_{jj'}\delta_{kk'} \ . \tag{4.53}$$

Any random variable on Ω is the sum of multiples of the $P(j,k)$; thus

$$\mathcal{N}_A = \sum_{j,k=0}^{\infty} jP(j,k) \quad , \quad \mathcal{N}_B = \sum_{j,k=0}^{\infty} kP(j,k) \ . \tag{4.54}$$

A probability $p \in \Sigma(\Omega)$ can also be written in this way; thus the local equilibrium state (a_A, a_B) can be written

$$(a_A, a_B) = \frac{1}{Z_A Z_B} \sum_{j,k=0}^{\infty} a_A^j a_B^k P(j,k). \tag{4.55}$$

Then the expectation of an observable f in the state p is given by

$$\begin{aligned} \sum_{j,k=0}^{\infty} f(j,k)p(j,k) &= \sum_{j,k,j',k'} \langle f(j,k)P(j,k), p(j',k')P(j',k')\rangle \\ &= \langle p, f\rangle \ . \end{aligned} \tag{4.56}$$

Our operator T is defined once it has been specified on each $P(j,k)$. For some $K \in (0, 1/2)$ let us put

$$\begin{aligned} TP(0,0) &= P(0,0) \\ TP(0,k) &= (1-K)P(0,k) + KP(1,k-1) \ ; \quad k \geq 1 \\ TP(j,0) &= (1-K)P(j,0) + KP(j-1,1) \ ; \quad j \geq 1 \\ TP(j,k) &= (1-2K)P(j,k) + KP(j-1,k+1) \\ &\quad +KP(j+1,k-1) \qquad ; \ j,k \geq 1 \ . \end{aligned} \tag{4.57}$$

The first says that if there are no particles, then this remains true after one time step. The second says that if A is absent and there are k B's, then one B is converted to an A with probability K, independent of the number $k \geq 1$. The third is the same with the roles of A and B reversed. The fourth says that if both A and B are present then each makes a transition to the other with rate K depleting the original state with probability $1 - 2K$. (This is why we need $0 \leq K \leq 1/2$; we choose $0 < K < 1/2$ to ensure the ergodicity of T.)

It is by choosing these transition rates to be independent of $j, k \geq 1$ that we get activity-driven reactions.

One easily checks from (4.57) that T is a symmetric operator and $T1 = 1$: this means that T is a symmetric Markov matrix. We now compute the change in the local equilibrium state induced by the map τ^*. Since the state is determined by N_A, N_B it is enough to compute the change in these averages. We also note that the marginal maps $\mathcal{M}_A$ and $\mathcal{M}_B$ and the equilibrating map Q_c do not change the averages N_A, N_B, whose changes are thus exactly those caused by $T^\dagger$:

$$N'_A = \langle \tau^* p, \mathcal{N}_A \rangle = \langle Q_c \mathcal{M}_A T^\dagger p, \mathcal{N}_A \rangle = \langle T^\dagger p, \mathcal{N}_A \rangle. \tag{4.58}$$

Now insert (4.54), (4.55) in (4.58); we get

$$N'_A = \frac{1}{Z_A Z_B} \sum_{j,k=0}^{\infty} a_A a_B \langle T^\dagger P(j,k), \sum_{j',k'=0}^{\infty} j' P(j',k') \rangle \ . \tag{4.59}$$

Looking at (4.57), $T^\dagger$ consists of a term proportional to K, and a constant term. Thus

$$T^\dagger P(j,k) = P(j,k) + O(K) \qquad (K \to 0) \ . \tag{4.60}$$

The sum of terms in (4.59) independent of K must sum to N_A, since $N'_A \to N_A$ as $K \to 0$.

To evaluate the term $O(K)$ in (4.59) it is best to separate the terms in (4.57) that contribute to the forward reaction, $A \to B$, from those that contribute to the reverse, $A \leftarrow B$. This is done for mathematical convenience; only the net flow can be actually measured. The forward terms must have $j \geq 1$ since otherwise $A \to B$ is not possible. Thus the first two lines of (4.57) do not contribute to the forward process. The third line gives the contribution of $j \geq 1$, $k = 0$, and is

$$\frac{1}{Z_A Z_B} \sum_{j \geq 0} a_A^j \langle (1-K) P(j,0) + K P(j-1,1), \sum_{j' \geq 0, k' \geq 0} j' P(j',k') \rangle \ ,$$

of which the contribution to $O(K)$ is

$$K \frac{1}{Z_A Z_B} \sum_{j \geq 1} a_A^j (-j + (j-1)) = -K \frac{1}{Z_A Z_B} \sum_{j \geq 1} a_A^j \tag{4.61}$$

using (4.53). The last line of (4.57) gives the contribution to (4.59) equal to

$$\begin{aligned} &\frac{1}{Z_A Z_B} \sum_{j,k \geq 1} a_A^j a_B^k \langle (1-2K) P(j,k) + K P(j-1,k+1) \\ &\quad + K P(j+1,k-1), \sum_{j',k' \geq 0} j' P(j',k') \rangle \\ &= \frac{1}{Z_A Z_B} \sum_{j,k \geq 1} a_A^j a_B^k [(1-2K)j + K(j-1) + K(j+1)] \ , \end{aligned}$$

using (4.53). The contribution linear in K vanishes, but one can recognise

$$\frac{1}{Z_A Z_B}\sum_{j,k\geq 1} a_{_A}^j a_{_B}^k[-Kj + K(j-1)] = \frac{-K}{Z_A Z_B}\sum_{j,k\geq 1} a_{_A}^j a_{_B}^k \tag{4.62}$$

as being the forward contribution, the reverse contribution being equal and opposite. The total forward contribution is thus the sum of (4.61) and (4.62):

$$\frac{-K}{Z_A Z_B}\left(\sum_{j\geq 1} a_{_A}^j + \sum_{j\geq 1,k\geq 1} a_{_A}^j a_{_B}^k\right) = \frac{-K}{Z_A Z_B}\sum_{j\geq 1,k\geq 0} a_{_A}^j a_{_B}^k = -Ka_{_A} \ .$$

This is exactly the sum over all states of B, and only those states of A that allow the forward reaction.

Similarly, the sum of term in (4.59) linear in K contributing to the reverse reaction $A \leftarrow B$ is

$$\frac{K}{Z_A Z_B}\left(\sum_{k\geq 0} a_{_B}^k + \sum_{j,k\geq 1} a_{_A}^j a_{_B}^k\right) = \frac{K}{Z_A Z_B}\sum_{j\geq 0,k\geq 1} a_{_A}^j a_{_B}^k = Ka_{_B} \ ,$$

and it comes from all states of A and only those states of B that allow the reverse reaction. This proves the dynamics desired:

$$N_{_A}' = N_{_A} + K(-a_{_A} + a_{_B}) \ . \tag{4.63}$$

Obviously, at a fixed point, $a_{_A} = a_{_B}$ and $N_{_A} = N_{_B}$. Since $E_{_A} = E_{_B}$ (energy conservation in $A \rightleftharpoons B$) we must also have $\mu_{_A} = \mu_{_B}$ at equilibrium, which is a very special case of Exercise (4.80).

We now construct a stochastic model for the reaction

$$\begin{aligned} &m_1A_1 + \cdots + m_rA_r + M_1C_1 + \cdots + M_sC_s \\ &\rightleftharpoons l_1B_1 + \cdots + l_tB_t + L_1C_1 + \cdots + L_sC_s \end{aligned} \tag{4.64}$$

such that the dynamics is activity-led, i.e. is given by (4.84). We first do the case of an isolated system, i.e. $K_+ = K_-$.

We take the sample space to be

$$\Omega = \Omega_{A_1} \times \cdots \times \Omega_{A_r} \times \Omega_{B_1} \times \cdots \times \Omega_{B_t} \times \Omega_{C_1} \times \cdots \times \Omega_{C_s} \ . \tag{4.65}$$

We shall denote the numbers of $A_1, A_2, \ldots$ in the sample point by $(j_1, j_2, \ldots) = j$ and the numbers of $B_1, B_2, \ldots$ by $(k_1, k_2, \ldots) = k$, and the numbers of $C_1, C_2, \ldots$ by $(i_1, i_2, \ldots) = i$. Thus the typical sample point in Ω is

$$\omega = (j_1, j_2, \ldots; k_1, k_2, \ldots; i_1, i_2, \ldots) = (j, k, i) \ .$$

Let $P(j,k,i)$ denote the projection onto ω; it is the function that takes the value 1 at ω and zero elsewhere. Then the local equilibrium state is

$$p = Z_{A_1}^{-1} \cdots Z_{C_s}^{-1} \sum_{j,k,i} a_{A_1}^{j_1} \cdots a_{B_1}^{k_1} \cdots a_{C_1}^{l_1} \cdots P(j,k,i) \tag{4.66}$$

where a's are the activities defined in (4.41). The expectation of the random variable

$$\mathcal{N}_{A_\alpha} = \sum_{j,k,i} j_\alpha P(j_1, \ldots, j_\alpha, \ldots, j_r; k_1, \ldots; i_1, \ldots) \tag{4.67}$$

is equal to $\langle p, \mathcal{N}_{A_\alpha} \rangle$. To define the dynamics we must specify the bistochastic map $T^\dagger$ corresponding to (4.64). It is enough to specify $T^\dagger P(j,k,i)$ for all j,k,i.

If any j_α is less than m_α or any i_α less than M_α , then the forward reaction cannot go. Also if any k_β is less than l_β or any i_β is less than L_β, then the backward reaction cannot go. Let

$$\Omega_+ = \{\omega : j_\alpha \geq m_\alpha, \alpha = 1, ..., r \text{ and } i_\alpha \geq M_\alpha, \alpha = 1, ..., s\} \tag{4.68}$$
$$\Omega_- = \{\omega : k_\alpha \geq l_\alpha, \alpha = 1, ..., t \text{ and } i_\alpha \geq L_\alpha, \alpha = 1, ..., s\} . \tag{4.69}$$

These are the configurations for which the forward and backward reactions can occur. We define $T^\dagger$=identity on $P(\omega)$, $\omega \notin \Omega_+ \cup \Omega_-$. In a way similar to the unimolecular case, it will be convenient to define the forward and backward parts of $T^\dagger$, $T_+^\dagger$ and $T_-^\dagger$, separately (although these cancel out on $\Omega_+ \cap \Omega_-$):

$$\begin{aligned} T_+^\dagger P(j,k,i) &= (1/2 - K)P(j,k,i) + KP(j_1 - m_1, ..., j_r - m_r; \\ &\quad k_1 + l_1, ..., k_t + l_t; i_1 + L_1 - M_1, ..., i_s + L_s - M_s) \end{aligned} \tag{4.70}$$

for $\omega \in \Omega_+$ and

$$\begin{aligned} T_-^\dagger P(j,k,i) &= (1/2 - K)P(j,k,i) + KP(j_1 + m_1, ..., j_r + m_r; \\ &\quad k_1 - l_1, ..., k_t - l_t; i_1 + M_1 - L_1, ..., i_s + M_s - L_s) \end{aligned} \tag{4.71}$$

for $\omega \in \Omega_-$. Then $T^\dagger = T_+^\dagger + T_-^\dagger$ is symmetric and stochastic.

The change in N_{A_1} due to the map $\tau^* = Q_c \mathcal{M}_{A_1} T^\dagger$ in the forward direction is the K-dependent part of $\langle T_+ p, \mathcal{N}_{A_1} \rangle$ namely

$$\begin{aligned} & Z_{A_1}^{-1} \cdots Z_{C_s}^{-1} \sum_{\omega \in \Omega_+} a_{A_1}^{j_1} \cdots a_{B_1}^{k_1} \cdots a_{C_1}^{l_1} \cdots a_{C_s}^{l_s} \sum_{\omega' \in \Omega} j_1' \langle \{-KP(j_1, ..., k_1, ..., i_1, ...) \\ & \quad + KP(j_1 - m_1, ..., k_1 + l_1, ..., i_1 - M_1 + L_1, ...)\}, P(j_1', ...) \rangle \\ & = KZ_{A_1}^{-1} \cdots Z_{C_s}^{-1} \sum_{\Omega_+} A_{A_1}^{j_1} \cdots A_{B_1}^{k_1} \cdots a_{C_1}^{i_1} \cdots (-j_1 + (j_1 - m_1)) \\ & = -m_1 K a_{A_1}^{m_1} \cdots a_{A_r}^{m_r} a_{C_1}^{M_1} \cdots a_{C_s}^{M_s} . \end{aligned} \tag{4.72}$$

In this, we have used that $Z_{A_\alpha}^{-1}\sum_{j_\alpha\geq m_\alpha} a_{A_\alpha}^{j_\alpha}$ equals $a_{A_\alpha}^{m_\alpha}$ etc. Similarly the backward rate is

$$m_1 K a_{B_1}^{l_1}\cdots a_{B_t}^{l_t} a_{C_1}^{L_1}\cdots a_{C_s}^{L_s},$$

and we have proved (4.84) for discrete time. The equation for N'_{B_β} is entirely similar. For the autocatalytic ions C_1, we get for $\langle T_+^\dagger p, \mathcal{N}_{C_1}\rangle$ instead of (4.72) the expression

$$KZ_{A_1}^{-1}\cdots Z_{C_s}^{-1}\sum_{\omega\in\Omega_+} a_{A_1}^{j_1}\cdots a_{B_1}^{k_1}\cdots a_{C_1}^{i_1}\cdots(-i_1+(i_1-M_1+L_1))$$

which, together with the backward reaction, leads to the activity-led equation

$$\frac{1}{L_1-M_1}\frac{dn_{C_1}}{dt} = K(a_{A_1}^{m_1}\cdots a_{A_r}^{m_r}a_{C_1}^{M_1}\cdots a_{C_s}^{M_s} - a_{B_1}^{l_1}\cdots a_{B_t}^{l_t}a_{C_1}^{L_1}\cdots a_{C_s}^{L_s})\ .$$

This completes the proof of (4.84).

It follows from this analysis that we can apply the general results of statistical dynamics to these equations. The orbit of an initial point lies in the compact space of states with given mean numbers N_j of various conserved atomic numbers, and entropy is a Lyapunov function. It can be shown that there is a unique fixed point for each possible set of N_j. The system therefore converges to a fixed point, which is the grand canonical state with the same mean values of all the conserved quantities as the initial state. The details of the proof are omitted, as they are very similar to the isothermal case, given in the next chapter.

In order to cope with reactions which produce heat q, we interpret one of the chemicals as a heat-particle with q as its quantum. To cover the cases where there is a non-zero activation energy, we take one of the catalysts to be a heat-particle. In this way the Arrhenius factor is obtained, where the beta of the factor is the instantaneous local beta of the thermalised heat-particle.

4.4 Exercises

Exercise 4.73 Show that in Example (4.2)

$$\begin{aligned} p_1'(B) &= p_1(B) - \kappa\,(p_1(B) - p_2(B)) \\ &- \lambda\,(2p_1(B)p_2(B) - p_1(A)p_2(C) - p_1(C)p_2(A))\,. \end{aligned}$$

Exercise 4.74 Consider Λ with two points, and the space $\{\emptyset,\ A,\ B,\ C\}$ at each site. Set up a model describing the process $A + C \rightleftharpoons B$.

Exercise 4.75 When we quantise a free Schrödinger particle in a box of side ℓ with periodic boundary conditions, the wave-number must be quantised, and the number of states with kinetic energy t is approximately const. t. Find the constant. Modify the model in Example (4.8), with the assumption that the transition probability to each state at a neighbouring point is non-zero for each of these states. Find the dependence of density with height at equilibrium.

Exercise 4.76 Show that in Example (4.8) the density at height x is given by

$$\rho_x = d^{-1}\frac{\exp\{-\beta(mdgx-\mu)\}}{1+\exp\{-\beta(mgdx-\mu)\}-\exp\{-\beta mdg\}} \tag{4.77}$$

and that in terms of the density at $x = 0$ we get

$$\rho_x = \frac{\rho_0\exp\{-\beta mdgx\}}{1-\rho_0 d\,(1-\exp\{-\beta mdgx\})}. \tag{4.78}$$

Note that a factor d was omitted in [26].

Exercise 4.79 We are given any $\beta \neq 0$, and an energy function $\mathcal{E}$ on a two-point set Ω, which is zero on $\omega = \emptyset$, and E on $\omega = 1$; show that any probability $(q = 1 - p,\ p)$ on Ω can be written as a grand canonical state $q = Z^{-1}$, $p = Z^{-1}e^{-\beta(E-\mu)}$ for a uniquely determined μ. Hence show that the fixed point of the model (4.31) corresponds to a constant chemical potential and beta. Show that the density as a function of height at equilibrium is the same as in Example (4.8).

Exercise 4.80 Show that the activity-led equations

$$\frac{dN_j}{dt} = -K(a_1a_2\ldots a_m - b_1b_2\ldots b_n) \tag{4.81}$$

where a_j is the activity of A_j and b_k is the activity of B_k, $1 \leq j \leq m$, $1 \leq k \leq n$, has many fixed points, for example any grand canonical state with the sum of chemical potentials on each side the same.

Exercise 4.82 Consider the reaction

$$m_1A_1 + \ldots + m_rA_r + M_1C_1 + \ldots + M_sC_s \rightleftharpoons l_1B_1 + \ldots l_tB_t + L_1C_1 + \ldots + L_sC_s. \tag{4.83}$$

Here, $m_1 \ldots m_r$ are the stoichiometries of the r $A-$particles, $l_1 \ldots l_t$ are the stoichiometries of the t $B-$particles. The $C-$particles are catalysts, which appear on the left and right of the reaction, with stoichiometries $M_1 \ldots$ and $L_1 \ldots$ respectively. The activity-led dynamics for this reaction is

$$\begin{aligned} \frac{1}{m_\alpha}\frac{dN_{A_\alpha}}{dt} &= -K a_{A_1}^{m_1} \ldots a_{A_r}^{m_r} a_{C_1}^{M_1} \ldots a_{C_s}^{M_s} + K a_{B_1}^{l_1} \ldots a_{B_t}^{l_t} a_{C_1}^{L_1} \ldots a_{C_s}^{L_s} \\ &= K\Delta \\ \frac{1}{L_\alpha - M_\alpha}\frac{dn_{C_\alpha}}{dt} &= -K\Delta. \end{aligned} \tag{4.84}$$

Show that a product of grand canonical states all with the same beta is a fixed point if and only if the sum of chemical potentials, including stoichiometry, is the same on both sides of the reaction.

Chapter 5

Isothermal Dynamics

Ostwald devised and made a thermostat, which enabled him to control the temperature of the apparatus in which a chemical reaction was taking place. When the temperature fell, the gas in a jar contracted and this worked a tap which turned up the gas flame; when the temperature rose, the gas expanded and this turned down the flame. So to within the accuracy available at that time, Ostwald was able to study reactions under *isothermal* conditions. The change from isolated conditions, in which (mean) energy is constant, to isothermal conditions, in which beta is constant, is called a *Legendre transform*, and β is called the Legendre variable dual to $\langle \mathcal{E} \rangle$. We note that energy is an extensive variable, proportional to the volume, whereas beta is an intensive variable, constant throughout the system. In thermostatics the equilibrium state with a given mean energy is obtained as the state maximising the entropy subject to the given mean energy, and the equilibrium state with a given temperature is obtained by maximising, not the entropy S, but the Legendre transform of S, denoted by Ψ, related to the Helmholtz free energy, subject to the beta being kept constant. In the next section we introduce the related transform of the whole dynamics, and not just of the equilibrium state. Thus the isolated dynamics is converted into the isothermal dynamics; we shall see that, unlike the isolated dynamics, the isothermal dynamics is linear, and so defines a stationary Markov chain. The second law, that entropy increases along the orbit, is converted to the free-energy theorem, that the free energy decreases along the orbit. This idea can be generalised to any of the other conserved extensive quantities, such as the atomic number. If we set up physical conditions that ensure that there are exactly the right flows of particles so that the chemical potential of a particular atom is constant, we are doing a Legendre transform to a variable dual to that atomic number; there is a corresponding thermodynamic potential, the Gibbs free energy, that decreases along the orbit. The resulting dynamics will be called *isopotential.* In the second section, we prove the fundamental theorem of chemical kinetics, that any activity-led isothermal, isopotential dynamics converges to equilibrium. There can be no chaos or permanent oscillations. In the third section, we prove that any Markov chain with a strictly positive fixed point can be regarded as the isothermal version of a suitable statistical dynamics. We show that the relative entropy, related to the free-energy difference between the state and the fixed point, decreases to zero, which leads to convergence of the state in a useful norm.

5.1 Legendre Transforms

Suppose we are given a function, S say, of some variables, $p_1, p_2, \ldots, p_n$ say, and we want to find the value of the n-vector p that maximises $S(p)$, subject to a constraint, say $E(p) = \overline{E}$. The case we shall use is when E is a linear function of p, namely the mean energy: $E(p) = \langle p, \mathcal{E} \rangle$. According to the method of Lagrange multipliers, the values of p giving the turning points can be obtained by seeking the turning points of $\Psi = S(p) - \beta E(p)$ without constraints. We can then find the stationary point $\overline{p}$ in terms of β, the Lagrange multiplier; β is found by fitting the solution p, to the constraint, so that β and $\overline{E}$ become functionally related, and the maximum value of S becomes a function of $\overline{E}$. Ψ itself can be written as a function of β, by using the relation between $\overline{E}$ and β to eliminate $\overline{E}$. When written as a function of β, Ψ is called the Legendre transform of S.

It is elementary to justify the method of Lagrange multipliers when S and $E(p) = \langle p, \mathcal{E} \rangle$ are smooth functions of the components of p, and E is a non-constant function, in a neighbourhood of the point of maximum, or, more generally, the turning point. It is easy to generalise from one to many variables, so we illustrate the argument when $n = 2$. So suppose that $S = S(p, q)$ is a function of two variables, and we wish to find its turning points subject to the constraint $E(p, q) = \overline{E}$. Solve the constraint for q as a function of p, E, so that $S = S(p, q(p, E))$. Then by the chain rule, a small change in p gives the change dS:

$$dS = \left(\frac{\partial S}{\partial p}\right)_q dp + \left(\frac{\partial S}{\partial q}\right)_p \left(\left(\frac{\partial q}{\partial p}\right)_E dp + \left(\frac{\partial q}{\partial E}\right)_p dE\right) = 0$$

at a turning point. Since we move on the surface $E(p, q) = \overline{E}$, we put $dE = 0$. This gives the turning points as the solutions to

$$\left(\frac{\partial S}{\partial p}\right)_q + \left(\frac{\partial S}{\partial q}\right)_p \left(\frac{\partial q}{\partial p}\right)_E = 0. \tag{5.1}$$

Now $E = E(p,q)$, so

$$dE = \left(\frac{\partial E}{\partial p}\right)_q dp + \left(\frac{\partial E}{\partial q}\right)_p dq,$$

so putting $dE = 0$ gives the variations at constant E:

$$\left(\frac{\partial q}{\partial p}\right)_E = -\left(\frac{\partial E}{\partial p}\right)_q / \left(\frac{\partial E}{\partial q}\right)_p .$$

Substitute in Eqn. (5.1) to arrive at

$$\left(\frac{\partial E}{\partial q}\right)_p \left(\frac{\partial S}{\partial p}\right)_q - \left(\frac{\partial S}{\partial q}\right)_p \left(\frac{\partial E}{\partial p}\right)_q = 0. \tag{5.2}$$

This is exactly the condition that (S, E) fail to be a valid set of coordinates for the (p,q)-plane; so let $(p_0,\ q_0)$ satisfy Eqn. (5.2). Then at $(p_0,\ q_0)$ the level sets of S and E are parallel: the surface $E(p,q) = E(p_0,q_0)$ touches the surface $S(p,q) = S(p_0,q_0)$. Such points (p_0,q_0) form a curve in the (p,q)-plane parametrised by $\overline{E}$, and S is a function of $\overline{E}$ along the line of turning points. This result can be got by the method of Lagrange multipliers without so many technical difficulties; in Lagrange's method, we seek the turning points of

$$\Psi = S(p,q) - \beta E(p,q) \tag{5.3}$$

without constraints, where β is kept constant. The turning points are given by

$$\frac{\partial S}{\partial p} = \beta\frac{\partial E}{\partial p} \qquad\qquad \frac{\partial S}{\partial q} = \beta\frac{\partial E}{\partial q}. \tag{5.4}$$

When we eliminate β we get the same condition as before, that the turning points of S along the manifold with fixed E are the points where the level surfaces of S and E touch. Moreover, the Lagrange multiplier is

$$\beta = \frac{\partial S}{\partial p}/\frac{\partial E}{\partial p} = \frac{\partial S}{\partial q}/\frac{\partial E}{\partial q} = \left(\frac{\partial S}{\partial E}\right)_q = \left(\frac{\partial S}{\partial E}\right)_p . \tag{5.5}$$

At a turning point, we get the same answer whether we chose to eliminate q, as here, or p. From this we see that

$$d\Psi = \left(\frac{\partial S}{\partial p}\right)_E dp + \left(\frac{\partial S}{\partial E}\right)_p dE - \beta dE - E d\beta = \left(\frac{\partial S}{\partial p}\right)_E dp - E d\beta,$$

from which we read off

$$\left(\frac{\partial \Psi}{\partial p}\right)_\beta = \left(\frac{\partial S}{\partial p}\right)_E$$

and

$$\left(\frac{\partial \Psi}{\partial \beta}\right)_p = -E.$$

By singling out the variable p rather than q, we can define Ψ as a function of p and β when we are away from the turning points; then its Legendre transform with respect to the variable β is

$$\Psi(p, \beta(p, E)) - \beta(-E) = S(p, E).$$

Thus Ψ and S are each other's Legendre transforms. In Exercise (5.48) you are asked to verify all this for the entropy and its Legendre transform Ψ when Ω is the countable space $\{0,\ 1, \ldots\}$ with some energy function $\mathcal{E}$ such that the partition function is finite.

A more versatile theory is obtained by considering all the energy-levels $\mathcal{E}_j$ to be parameters; then β arises as an overall factor which can be put equal to 1, and the thermodynamic potential Ψ becomes a function of all the $\mathcal{E}_j$. This, the analytic theory of Legendre transforms, depends on the existence of derivatives and the implicit function theorem. In statistical mechanics, after the thermodynamic limit has been taken, S is not necessarily differentiable as a function of p; but it remains a concave function. To cover this case, the Fenchel transform can be used. The turning-point condition is replaced by a condition that the plane $\langle p, \mathcal{E}\rangle = \overline{E}$ should touch the convex set given by the level surfaces of S. Thus

$$\Psi(\mathcal{E}_1,\ \mathcal{E}_2, \ldots) = \sup_{p \in \Sigma}\{S(p) - \langle p, \mathcal{E}\rangle\}. \tag{5.6}$$

One shows that Ψ is a concave function of $\mathcal{E}$ and that its Fenchel transform is S as a function of p. See [30, 20, 5].

In the next section we describe how the partial Legendre transform (over the variables of the heat-particle) can transform the whole orbit of the isolated dynamics into the orbit of the system under isothermal conditions.

5.2 The Free-energy Theorem

We have seen that the sample space for a system is of the form $\Omega = \Omega_c \times \Omega_\gamma$ with energy $\mathcal{E}_c + \mathcal{E}_\gamma$, and that the dynamics of an isolated system is determined by an energy-conserving bistochastic map T. The initial state $p \in \Sigma(\Omega)$ is changed in one time-step to

$$p' = \mathcal{M}_c T^{\dagger} p \otimes \mathcal{M}_\gamma T^{\dagger} p = \tau^* p.$$

Suppose now that the state at time $t = 0$ is at a uniform beta β; this is expressed by saying that $p = p_c \otimes s_\beta$, where s_β is the canonical state of the heat-particle.

Unless the state p is invariant under the dynamics τ^*, it is most unlikely that the part of $\tau^* p$ that concerns the heat-particle, namely $s' = \mathcal{M}_\gamma T^\dagger p$, is itself in a canonical state. We can, however, mathematically implement the thermostat by changing this state s' back to s_β, so we define the *isothermal* dynamics to be

$$p \mapsto p' = \tau_\beta^* p = \mathcal{M}_c T^\dagger p \otimes s_\beta. \tag{5.7}$$

That is, the thermostat ensures that each mode of $s' = \mathcal{M}_\gamma T^\dagger (p_c \otimes s_\beta)$ is supplied with just the right amount of heat to restore it to the canonical state s_β.

We see that the map τ_β^* can be regarded as acting on $\Sigma(\Omega_c)$, where it is a linear map, unlike the Boltzmann map of the isolated system. We have done a partial Legendre transform, replacing the dynamics at constant mean energy by one with constant beta (or temperature, $1/\beta$) which is the same as one with constant mean heat $\langle s_\beta, \mathcal{E}_\gamma \rangle$. Although s_β is a well-defined function of $\mathcal{E}_\gamma$, as happens in the Fenchel transform, we cannot really regard $\tau_\beta^* p$ as a function of independent variables $\mathcal{E}_\gamma$, since the whole formalism requires that $T^\dagger$ conserve the energy. We shall content ourselves to regard $\tau_\beta^* p_c$ as a state in $\Sigma(\Omega_c)$ parametrised by β. Then the Legendre transform of the entropy, called Ψ, is a Lyapunov function for the dynamics:

Theorem 5.8 (The Free-energy Theorem) *Let* $\Psi(p_c) = S(p_c) - \beta \langle p_c, \mathcal{E}_c \rangle$; *then*

$$\Psi\left(\tau_\beta^* p_c\right) \geq \Psi\left(p_c\right) + \|s' - s_\beta\|^2/2.$$

Proof

We know that the bistochastic map $T^\dagger$ increases the entropy (or leaves it the same). Also the Boltzmann map, which replaces $T^\dagger p$ by $p' = \tau^* p = \mathcal{M}_c T^\dagger p \otimes \mathcal{M}_\gamma T^\dagger p$, increases the entropy. Let $p = p_c \otimes s$ and $p' = p'_c \otimes s'$. Then we have

$$S(p') = S(p'_c) + S(s') \geq S(p) = S(p_c) + S(s).$$

In isothermal dynamics, we always start with the state of the heat-particle in equilibrium, so $s = s_\beta$. Then

$$S(s) = -\sum_{\omega_\gamma} s_\beta(\omega_\gamma)(-\beta \mathcal{E}(\omega_\gamma) - Z_\gamma(\beta)).$$

Let us write

$$-\sum_{\omega_\gamma} s'(\omega_\gamma) \log s'(\omega_\gamma) = -\sum_{\omega_\gamma} s'(\omega_\gamma) \log s_\beta(\omega_\gamma) + \sum_{\omega_\gamma} s'(\omega_\gamma) \log \left(s_\beta(\omega_\gamma)/s'(\omega_\gamma)\right).$$

By Kullback's inequality,

$$\sum_{\omega_\gamma} s'(\omega_\gamma) \log \left(s_\beta(\omega_\gamma)/s'(\omega_\gamma)\right) \leq -\|s_\beta - s'\|^2/2. \tag{5.9}$$

As we saw in Section (4.1), the mean energy is conserved by the isolated dynamics; thus

$$\sum_{\omega_c} p_c(\omega_c)\mathcal{E}_c(\omega_c)+\sum_{\omega_\gamma} s_\beta(\omega_\gamma)\mathcal{E}_\gamma(\omega_\gamma) = \sum_{\omega_c} p_c'(\omega_c)\mathcal{E}_c(\omega_c)+\sum_{\omega_\gamma} s'(\omega_\gamma)\mathcal{E}_\gamma(\omega_\gamma). \quad (5.10)$$

We put these estimates together in the following chain of inequalities:

$$\begin{aligned}
S(p_c') - \sum_{\omega_\gamma} s'(\omega_\gamma)\log s_\beta(\omega_\gamma) &= S(p_c') - \sum_{\omega_\gamma} s'(\omega_\gamma)\log s'(\omega_\gamma) \\
&\quad -\sum_{\omega_\gamma} s'(\omega_\gamma)\log\left(s_\beta(\omega_\gamma)/s'(\omega_\beta)\right) \\
&\geq S(p_c') - \sum_{\omega_\gamma} s'(\omega_\gamma)\log s'(\omega_\gamma) + \|s' - s_\beta\|^2/2 \\
&= S(p') + \|s' - s_\beta\|^2/2 \\
&\geq S(p) + \|s' - s_\beta\|^2/2 \\
&= S(p_c) + S(s_\beta) + \|s' - s_\beta\|^2/2 \\
&= S(p_c) - \sum_{\omega_\gamma} s_\beta(\omega_\gamma)\log s_\beta(\omega_\gamma) + \|s' - s\|^2/2.
\end{aligned}$$

Hence

$$\begin{aligned}
S(p_c') \quad &- \quad \sum_{\omega_\beta} s'(\omega_\gamma)\left(-\beta\mathcal{E}_\gamma(\omega_\gamma) - \log Z_\gamma(\beta)\right) \geq S(p_c) \\
&- \quad \sum_{\omega_\gamma} s_\beta(\omega_\gamma)\left(-\beta\mathcal{E}_\gamma(\omega_\gamma) - \log Z_\gamma(\beta)\right) + \|s' - s_\beta\|^2/2.
\end{aligned}$$

The log Z terms cancel, and using Eqn. (5.10) we have

$$\begin{aligned}
-dQ &= \sum_{\omega_\gamma}\left(s'(\omega_\gamma) - s_\beta(\omega_\gamma)\right)\mathcal{E}_\gamma(\omega_\gamma) \\
&= \sum_{\omega_c}\left(p_c(\omega_c) - p_c'(\omega_c)\right)\mathcal{E}_c(\omega_c) = U - U' \quad (5.11)
\end{aligned}$$

where U is the chemical energy of the chemicals, and dQ is the heat that has to be added to keep the temperature constant. Thus we get

$$S_c' - \beta U' \geq S_c - \beta U + \|s' - s_\beta\|^2/2. \quad (5.12)$$

This is the free energy theorem [31, 32]. □

It is usually expressed in terms of the internal energy, including the kinetic energy, of the chemicals, and the total entropy, including that of the heat. However, the contributions of the heat-particle to the internal energy and to the entropy are unchanged in the isothermal case, and so our result is not in conflict with the usual claim that the *free energy*, $U - TS$, decreases with time.

We see that in one time-step the system gains entropy at least $dQ/T + \|s' - s_\beta\|^2/2$. We may therefore define the mysterious concept, a *quasi-static* process, as one for which $s' = s_\beta$, that is, one for which the dynamics $T^\dagger$ does not alter the distribution of the heat-particle in any way. This is the same as saying that the change in entropy is dQ/T, the smallest it can be.

There is nothing special about the energy, and the degrees of freedom of the heat-particle, compared with other conserved quantities; for each conserved quantity, say atomic number $\mathcal{N}$, we can define the corresponding *isopotential* dynamics by exchanging particles with the surroundings after each time-step in such a way that the chemical potential of this atom is returned to its initial value. Physically, this means that it is in contact with a bath at a specified chemical potential. In practice it would be difficult to prevent heat escaping with the particles, so isopotential dynamics is usually not considered except at a fixed beta. So the convention has arisen of including a factor β along with μ in the grand canonical state; this gives μ the dimensions of energy. To implement this map mathematically, we need the sample space to be the product of the classical Fock space of the atom, and the rest, and for the dynamical law $T^\dagger$ to map the number-shells into themselves (as well as the energy-shells into themselves). Then, a similar calculation to that in the free-energy theorem shows that the system gains entropy of at least μdN when dN atoms are absorbed in one time-step to keep the chemical potential constant at μ. This applies to any number of atoms that are conserved in the dynamics and whose chemical potential is kept fixed. The dynamical laws are obtained from the isolated dynamics by simply keeping the beta and the chemical potentials fixed for the controlled atoms, and taking the same formulae. In this set-up, the flow of atoms in the isopotential dynamics is taken to be free atoms, not atoms bound in other molecules.

We can prove Arrhenius' law for isothermal dynamics under the assumption that T is a symmetric operator. This means that $T = T^\dagger$ when $\Sigma(\Omega)$ is identified with the subset of ℓ^1 random variables, using the priviledged basis, namely, the points of Ω. For, suppose we consider the general reaction Eqn. (4.81) with some of the symbols on each side representing heat-particles of various energies, ε_j. Then in the activity-led model the forward rate will contain factors $a_{\gamma,j} = e^{-\beta\varepsilon_j}$ for the heat-particles absorbed, and the backward rate will contain factors of the same type for each heat-particle emitted in the reaction. The ratio of these will be $e^{\beta q}$, where $q = \sum \varepsilon_{\text{in}} - \sum \varepsilon_{\text{out}}$ is the heat emitted, the heat of reaction. In fact, in the activity-led equations, the forward rate itself is proportional to $e^{-\beta Q_{\text{in}}}$, where Q_{in} is the activation energy, the heat needed to get the reaction started. Many chemical reactions agree with this prediction to high accuracy over large temperature ranges. When it is found not to be true, chemists tend to look for reasons, such as the occurrence of more than one route for the reaction to occur; we could also look for other models, not leading to the activity-led law; but the Arrhenius factor will always be present, coming from the canonical state of the heat-particle.

5.3 Chemical Kinetics

The question as to whether complex chemical rate-equations always converge to equilibrium was answered in a series of papers by chemists [33, 34, 35, 36]. These authors worked with density-led equations, for which an underlying stochastic model might not exist. Rondoni [37] has proved a similar result for a wide class of equations, which he calls "X-led". Here we give a proof for the density-led and the activity-led cases, which are derived from particular statistical models in the last Chapter. We shall take the space Λ to be a finite set, and each Ω_x, with $x \in \Lambda$, will be the hard-core sample space of the chemicals (including the empty site) for the density-led case, and the set-product of the classical Fock spaces of chemical species for the activity-led case. We shall consider the dynamics to be a convex combination of a finite number, R, of independent reactions in which chemicals $A_1, \ldots, A_r$ react to form chemicals $B_1, \ldots, B_t$, with chemicals $C_1, \ldots, C_s$ acting as catalysts. We take it that the stoichiometries are $m_1 \ldots m_r$ for the A's, $l_1 \ldots l_t$ for the B's, and $M_1 \ldots M_s$ for the initial catalysts, and $L_1 \ldots L_s$ for the final catalysts. In symbols we have the "simple" reaction

$$\begin{aligned} & m_1 A_1 + \cdots + m_r A_r + M_1 C_1 + \cdots + M_s C_s \\ \rightleftharpoons\ & l_1 B_1 + \cdots + l_t B_t + L_1 C_1 + \cdots + L_s C_s. \end{aligned} \tag{5.13}$$

Consider first the density-led case. We include enough points in Λ to allow all the reactants, that are needed for the forward and backward reactions, to be present. The choice of sample $\Omega = \prod \Omega_x$ space ensures that the configuration with one A-particle at $x \in \Lambda$ and no other particle present, is distinct from the configuration where the single A-particle is at $y \neq x$, but the configuration with two A-particles at x and y does not distinguish the particles. Thus we are adopting the field point of view. We can then accommodate any stoichiometry by having a non-zero transition probability from the state with several chemicals of the same type at nearby points in Λ, to the state representing the products that they combine to form. The stoichiometry then enters in the definition of the transition matrix, but not in the choice of sample space, provided that Λ is large enough. Let $\nu = |\Lambda| . |\Omega_x|$. This is the total number of components of all the marginal probabilities p_x with $x \in \Lambda$. Let $\mathcal{K} = [0,1]^\nu$. A point p in $\Sigma = \Sigma(\Omega)$ which is independent over Λ, as after the stoss map, defines a point in $\mathcal{K}$ once we fix on an ordering convention. In the dynamics from a given initial point of Σ, there will be some conserved quantities, and these will say that certain mean atomic numbers are constant. This implies that the motion inside $\mathcal{K}$ must lie in certain affine planes, whose position is fixed by the initial values of these atomic numbers. The conditions that the marginal probabilities add up to 1 also define affine planes. It is easy to see that a boundary point of $\mathcal{K}$ which comes from a point in Σ must come from a boundary point of Σ.

The linear part of the dynamics will be given by a convex sum of primitive symmetric stochastic operators each of which links a certain initial configuration of particles with a final configuration of products. For example, we may wish to include a non-zero probability that the reaction $A + B \rightleftharpoons C$ occurs when A is at x, B is at $x + 1$, and the product C is created at $x + 1$, where B was; then the inverse of this must be included. We may choose to omit the reaction in which the product C is created at x. Then the model expresses that the particle A accretes to B to make C, rather than that B accretes to A to make C. In the hard-core model, with this choice of dynamics, the site x is left empty after the reaction, and must be empty before the reaction for the inverse reaction to be possible. We assume that the inverse rate is the same as the forward rate, (the principle of microscopic reversibility). In artificial ecology, this model represents a hungry predator A at x leaping on a prey B at $x + 1$ to end up with a fat predator C at $x + 1$. In this case the probability of the inverse reaction is zero, and so the following analysis applies only to the chemical case. The initial state, taken to be the product over Λ of the marginal states, is the product of the occupation probabilities of A at x and B at $x + 1$. The probability of the final state will be the product of the occupation probability of C at $x + 1$ and the empty probability at x. After the action of the linear part, we apply the Boltzmann map and replace the final state by the product over Λ of the marginals. We obtain a version of *density-led* dynamics, but modified from the classical law of mass action by the presence of these empty-state probabilities, just as if the empty state were a particle. You have already met this in Exercise (4.74). This modification of the classical density-led law of mass action was suggested in [31]. The full dynamics τ^* of complex reactions will be a convex linear combination of such maps τ_k^*, and by the principle of detailed balance, equilibrium for the complex reaction will necessarily be a fixed point of each of these τ_k^*, by the strict concavity of the entropy.

The fundamental theorem of chemical kinetics says that this version of the law of mass action leads to convergence to equilibrium as time becomes large [31]. We already have most of the ingredients of the proof: the motion takes place within the compact set $\mathcal{K}$, and there is a strict Lyapunov function, the entropy. If there is a unique fixed point, then Lyapunov's theorem (3.23) will ensure convergence. However, typically there is not a unique fixed point in the whole of Σ. If an essential ingredient, say a catalyst or more generally one of the chemicals on each side of the reaction, has zero probability, then the reaction will not proceed. Thus there may be fixed points on the boundary of the state space, in addition to the equilibrium states in the interior. It was shown in [38] that the boundary fixed points are *repellers* at least for the usual law of mass action, which needs balanced reactions. The same proof works here, with the probability of an emply site replacing a chemical concentration from time to time. Thus, if $p \in \Sigma$ is an interior point, all its components are positive; then $p' = \tau_k^* p$ cannot be a boundary point; for if one component, say p'_1, the marginal onto the first factor Ω_1 of Ω, has a zero component, then we get an equation of the form

$$0 = \sum T_{1,\ldots} p_1 \ldots p_N$$

where all the p's are positive. This implies that all the components of $T^\dagger$ involved in the sum are zero, which is impossible, as T is a stochastic matrix. Hence if p is an interior point, so is p', and so no simple reaction τ_k^* can take a point from the interior to the boundary in one fell swoop. The complex reaction is a convex sum of these, and so cannot take an interior point to the boundary in one time-step either.

We now exclude the possibility that a sequence of points on an orbit under time-evolution can lead from an interior point to the boundary. Irrespective of the particular dynamics, each *simple* reaction starting at an interior point always takes place on a line in $\mathcal{K}$. It is a neat remark of Rondoni [38] that any one of the τ_k^* has exactly $\nu - 1$ linear conserved quantities, including the $|\Lambda|$ quantities $\sum p_j$, and so essentially only one of the ν variables changes. The orbit therefore lies on the intersection of $\nu - 1$ hyperplanes, which is a line through the initial state called the *line of reaction*. Such a line of reaction cuts a boundary face of $\mathcal{K}$ at a non-zero angle, since it does not lie entirely in the face. The complex reaction, a convex sum of these, takes place inside the convex hull of the lines of reaction of τ_k^* through the current state p. If the orbit approaches the boundary of $\mathcal{K}$, it must therefore do so at a non-zero angle. The entropy is a function of the marginals, and can be extended to a concave function on $\mathcal{K}$, since it can be defined for any semiprobability. $S(p)$ is increasing as we leave the boundary. So any orbit that approaches the boundary at a non-zero angle must run through a sequence of points of decreasing entropy for large times, impossible as entropy is non-decreasing. This contradiction shows that there is a compact set, at a positive minimum distance from the boundary of $\mathcal{K}$, that is mapped to itself by τ^*. We consider the dynamical system got by iterating τ^* acting on this compact set. We denote a general element of $\operatorname{Span}\mathcal{K} = \mathbf{R}^\nu$ by p.

Now let $p(0)$ lie in the interior of $\mathcal{K}$. Our problem is to show that $\tau^{*n} p(0)$ converges to a limit as $n \to \infty$. If $p(0)$ is a fixed point of τ^* we are done. So we may assume that $\tau^* p(0) \neq p(0)$. Our method, not dissimilar to the original methods [33, 34, 35, 36], is to show that the motion takes place on a certain hyperplane through $p(0)$, and that this hyperplane contains a unique fixed point. We treat the isolated system first.

The simple reaction τ_k^* gives an orbit starting at $p(0)$ and lying on the line of reaction L_k passing through $p(0)$, defined by $\nu - 1$ planes

$$L_k = \{p \in \mathcal{K} : N_j^{(k)}(p) = N_j^{(k)}(p(0)), \quad j = 1, \ldots, \nu - 1\}.$$

Here $N_j^{(k)}$, $j = 1, \ldots, \nu - 1$ are the conserved quantities for τ_k^*, including the conservation of probability of the marginals,

$$\sum_{\omega_x} p_x(\omega_x) = 1. \tag{5.14}$$

Now consider the complex reaction; suppose that R of the lines L_k are linearly independent, but no more. Then $\{L_k\}$ span an R-dimensional convex cone with vertex $p(0)$, and the lines can be made to pass through the origin by subtracting $p(0)$ from each point in the line. Thus, $\{L_k - p(0) : k = 1, \ldots, R\}$ form part of a linear basis in Span $\mathcal{K}$. The conserved quantities $N_j^{(k)}$ are elements of the dual space to Span $\mathcal{K}$; this is obvious for the atomic numbers, as they are random variables. But it is also true that the marginal probabilities in Eqn. (5.14) are given by sums of certain components of p and so are conserved linear functionals. The conserved functionals $N_j^{(k)}$ vanish on the corresponding axis $L_k - p(0)$; for, they are conserved, so $N_j^{(k)}(p) = N_j^{(k)}(p(0))$ for $p = \tau^* p(0) \neq p(0)$ on the line L_k. Then $N_j^{(k)}(p - p(0)) = 0$, so $N_j^{(k)}$ annihilates $p - p(0)$ as well as 0, and being linear, kills all points on the line. Conversely, any linear functional that annihilates all the points on the R lines of reaction is conserved. Thus the conserved quantities, which are not all linearly independent, span a space of dimension exactly $\nu - R$ in the dual to Span $\mathcal{K}$, since they annihilate a space of dimension R. Let us take as linear coordinates in Span $\mathcal{K}$ a certain $\nu - R$ linearly independent conserved functionals N_j, $j = 1, \ldots, \nu - R$, together with R parameters $t_1, \ldots, t_R$, one on each axis $L_k - p(0)$.

The ν labels for the chemicals in the hard-core sample spaces Ω_x for $x \in \Lambda$ can be identified with ν of the corners of $\mathcal{K}$, in which A_x is the point $p_x(A) = 1$ and all other components are zero. This is not a possible point in $\Sigma(\Omega)$, since it violates all except one of the constraints in Eqn. (5.14). But nevertheless these form a basis for Span $\mathcal{K}$. Let $p \in \mathcal{K}$ be a fixed point of the map τ^* with the same values of the conserved quantities as $p(0)$, and let $f = \log p$, meaning that f is a function on the set of labels: $f(A_x(j)) = \log p_x(j)$. Define f to be zero at $0 \in \mathcal{K}$. This f has a unique linear extension to Span $\mathcal{K}$, which we also denote by f. By the principle of detailed balance, p is a fixed point of each τ_k^*, which is the dynamics of the simple reaction

$$A_{k_1} + \cdots + A_{k_K} \rightleftharpoons A'_{k_1} + \cdots + A'_{k_K}$$

whose fixed point obeys

$$p_{k_1} p_{k_2} \cdots p_{k_K} = p'_{k_1} p'_{k_2} \cdots p'_{k_K}$$

which in terms of f can be written

$$f(k_1)+\cdots+f(k_K)=f(k_1')+\cdots+f(k_K'). \tag{5.15}$$

For each given k, each side of Eqn. (5.15) is the value of f evaluated at a point of Span $\mathcal{K}$, the left-hand side being the value of f at $P=A_{k_1}+\cdots+A_{k_K}$ and the right-hand side being the value of f at a different point, $P'=A'_{k_1}+\cdots+A'_{k_K}$. Both P and P' have the same value for each of the $\nu-1$ conserved quantities $N_j^{(k)}$, $j=1,2,\ldots,\nu-1$, so the difference $P-P'$ lies on the axis $L_k-p(0)$, which is the simultaneous null space of $N_j^{(k)}$; hence f, which is linear, vanishes at a non-zero point on this axis, and so must vanish along $L_k-p(0)$. Similarly, f vanishes on all R axes $L_k-p(0)$, $k=1,\ldots,R$, and therefore is a linear function of the chosen $\nu-R$ linearly independent conserved quantities $N_j(p)$, $j=1,\ldots,\nu-R$. So we get

$$\log p_i=\sum_{j=1}^{\nu-R}\lambda_j N_j(\mathbf{p}_i) \tag{5.16}$$

where $\mathbf{p}_i=(0,\ldots,1,\ldots)$ with 1 in the ith place, and λ_j are some real numbers. But this system of equations is identical to what we get by using the method of Lagrange multipliers to maximise $S(p)$ subject to the $\nu-R$ independent linear constraints $N_j(p)=$ constant. Because of the strict concavity of S, there is only one turning point in the interior of Σ, so the uniqueness of the fixed point in density-led dynamics is proved, and consequently, by Lyapunov's theorem, the convergence of the system is established. □

We get a bonus, in that the convergence to equilibrium is established for the dynamics based on $T^\dagger$ followed by the stoss map, in which we replace the state by the product of its marginals at the points in Λ, and we do not need the further randomisation which would be caused if we applied the LTE map as well. The fixed point is a grand canonical state; for each $x\in\Lambda$ there is a conserved quantity N_x, being the total marginal probability, and we could have chosen these to be among the linearly independent conserved numbers. Then the local partition function Z_x is related to the corresponding Lagrange multiplier λ_x by $\lambda_x=\log Z_x$. The other λ's are the various chemical potentials.

We get the fundamental theorem for isothermal and isopotential dynamics in the same way; the only difference is that the state of the heat-particle, and of the chemicals whose chemical potential is kept fixed, is fixed to be the canonical or grand canonical state with given parameters; it does not affect the uniqueness of the fixed point. In place of the entropy of the whole system, the free energy of the chemical subsystem undergoing the dynamics is a Lyapunov function, so we prove convergence by Lyapunov's theorem.

For the activity-led dynamics, the sample space is infinite, since at each x we have the classical Fock space $\Omega_x=\{0,\ 1,\ 2,\ldots\}$. This leads to problems in showing that the limit state has the same energy as the mean energy during the dynamics [39]. However, the LTE-dynamics takes place through a subset of Σ labelled by finitely many parameters; this enables us to find a compact set mapped to itself by the dynamics. We now sketch the argument.

Let us choose a symmetric operator $T^\dagger$ to define the linear part of the dynamics as in Sect. (4.3) to give activity-led dynamics when we apply the LTE-map. A state at time t is then a grand canonical state of each of the $r+s+t$ chemicals at each of the $|\Lambda|$ points x. This is determined by the mean number fields $n(x)$, one for each particle. The reduced state-space is thus finite-dimensional, and labelled by $|\Lambda|(r+s+t)$ non-negative real numbers. In any realistic chemical reaction at fixed temperature, there is a conserved quantity, the total mean number of atoms. If the initial state has a finite mean number of atoms, then the dynamics, which tells us the number of molecules of each type at each point, takes place in a bounded set inside a finite-dimensional space. The R independent reactions give rise to $\nu - R$ conserved numbers, in the same way as in density-led reactions. The region of Σ determined by the initial values of the conserved quantities can be parametrised by a subset of the space of the activities a of each chemical at each point, and these parameters lie between 0 and 1. The boundary where some a's are zero might be fixed points, as they correspond to the absence of a chemical that might be needed for the reaction to go. So if the grand canonical state we get from applying LTE to $T^\dagger p$ has zero activity, this means that $T^\dagger p$ has zero particles of this type, and so has a zero entry at all points except $0 \in \Omega_x$. So p is a boundary point of Σ. As before, we show that boundary points are repellers, as they correspond to local minima of the entropy. Since all the mean numbers remain bounded along the orbit, it follows that the dynamics takes place in a compact submanifold of Σ whenever the initial point is not on the boundary. The fixed points of a simple reaction at the point x is given by the requirement that the product of the activities of the reactants should be equal to the product of the activities of the products. This leads to R independent lines of reaction, as above. In the same way as above, there are $\nu - R$ independent conserved quantities. The equations for the fixed point say that the product of the initial activities is the product of the final activities, for each process. This leads to an equation similar to (5.15), except that the number of terms on each side might not be equal. Then the same argument shows that there is a unique fixed point for each set of initial values of the conserved quantities. Thus we can apply the Lyapunov direct method to prove convergence of the activity-led dynamics.

The upshot of this section is that every chemical system, whether isolated, isothermal or isopotential, converges to chemical and thermal equilibrium, and that permanent oscillations, or chaos, are not possible. Any really interesting chemistry must come from "driven" systems.

5.4 Convergence in Norm

Let T_c be a stochastic map on the algebra $\mathcal{A}_c$, with dual $T_c^\dagger$ having a strictly positive fixed point p_0. We have seen that the relative entropy $S(p_c|p_0)$ is not increased by the isothermal dynamics. Since the relative entropy is bounded below by zero, it follows that $S(p_c|p_0)$ converges along the orbit. Under ergodic conditions, the relative entropy converges to zero. In this section we show that the relative entropy is greater than a certain metric distance of p_c from p_0, given by a useful norm, first studied by Fisher [40], and now the subject of much study [41]. We thus prove that the states along the orbit converge in this metric when the ergodic conditions hold. We follow [32].

We first note that the so-called condition of detailed balance, advocated by Glauber [6] is not the most general condition occurring in isothermal dynamics. Let us choose the natural basis in Span Ω_c, namely, the points of Ω_c itself. Then T_c and its adjoint become infinite matrices. The condition of detailed balance according to Glauber is

$$T_{cij} = e^{\beta(\mathcal{E}_i - \mathcal{E}_j)} T_{cji}. \tag{5.17}$$

This is just like the Arrhenius relation between the forward and backward chemical rates, and it has the same source; the principle of microscopic reversibility as used in the above theory of chemical kinetics is that the bistochastic map T on $\mathcal{A}(\Omega)$ should be a symmetric matrix. In quantum theory an even more general formulation is possible, as we shall see. In any reaction conserving energy, the difference between the energies of the initial and final states is made up by the energies of one or more heat-particles. In isothermal dynamics the heat-particles are in the canonical state at some beta; we denoted this state by s_β. Then we get $T_c^\dagger$ from $T^\dagger$ by taking the marginal distribution of the chemicals after one time-step:

$$p_c' = T_c^\dagger p_c = \mathcal{M}_c\left(T^\dagger(p \otimes s_\beta)\right). \tag{5.18}$$

Then if T is symmetric (in the natural basis) and energy-conserving, we see that the ratio of the forward to the backward rate is s_i/s_j, which gives Eqn. (5.17). In this derivation, it does not matter whether there is one or several heat-particles, on one or on both sides of the equation. Now, the symmetry of T is not at all a natural assumption in applications of statistical dynamics to neural nets, or to artificial ecologies. But to be able to do thermodynamics, we need T to be bistochastic. How is this condition reflected in the properties of T_c?

The first thing to remark is that the canonical state $p_{c\beta}$ is a fixed point of T_c. In fact, we know already that $p_{c\beta} \otimes s_\beta$ is a fixed point of any energy-conserving map $T^\dagger$, since this state is constant on each energy-shell. It will turn out, in the next section, that no further conditions than this, that $p_{c\beta}$ is a fixed point, can be derived from the bistochasticity of T. We therefore take this, or rather the following condition easily shown to be equivalent to it, as the definition of detailed balance. We first introduce a new metric topology on the algebra. To avoid confusion with the free energy, denoted F, in this section we shall use f, g to denote random variables. For f and g in $\mathcal{A}_c$, define the scalar product for each $\beta > 0$ by

$$\langle f, g\rangle_\beta = Z_\beta^{-1} \sum_{j\in\Omega_c} e^{-\beta\mathcal{E}_c(j)} f_j g_j, \tag{5.19}$$

where

$$Z_\beta = \sum_{j\in\Omega_c} e^{-\beta\mathcal{E}_c(j)}.$$

This idea is borrowed from the theory of quantum statistical mechanics [42]. We denote the norm coming from this scalar product by

$$\|f\|_\beta = \left(Z_\beta^{-1} \sum_{j\in\Omega_c} e^{-\beta\mathcal{E}_c(j)} |f(j)|^2\right)^{1/2}. \tag{5.20}$$

Let $\mathcal{A}_\beta$ be the completion of $\mathcal{A}_c$ in the metric given by $\|\bullet\|_\beta$. This contains quite a few unbounded random variables, because of the strongly convergent factors $e^{-\beta\mathcal{E}_c}$. The dual space $\mathcal{A}_\beta^*$ of this Banach space is the completion of $\operatorname{Span}\Omega_c$, of real measures of finite support, in the metric defined by the dual norm

$$\|p\|_{-\beta} = \left(Z_\beta \sum_{j\in\Omega_c} e^{\beta\mathcal{E}_c(j)} |p_j|^2\right)^{1/2}. \tag{5.21}$$

We shall see that this is the natural norm governing the convergence of states evolving under isothermal conditions, rather than the L^1-norm usually used for measures. In this we follow [43]. It is also the norm in which the "equivalence of ensembles" holds. The rising exponential $e^{\beta\mathcal{E}_c(j)}$ forces the high-energy tail of p to converge much faster than if we use the L^1-norm, and leads to a larger norm and stronger topology. It was used by Fisher a long time ago [40].

Let T_c be a bounded operator on $\mathcal{A}_\beta$, which is a Hilbert space with scalar product $\langle\bullet,\bullet\rangle_\beta$, and denote the adjoint of T_c by $T_c^{(\beta)}$. Note that $T_c^{(\beta)}$ acts on $\mathcal{A}_\beta$.

Definition 5.22 (Detailed Balance) *We say that $T_c : \mathcal{A}_\beta \to \mathcal{A}_\beta$ obeys* detailed balance *(relative to $\beta\mathcal{E}_c$) if T_c and $T_c^{(\beta)}$ are both stochastic maps.*

Apart from being positive maps, if T_c satisfies detailed balance, then $T_c 1 = T_c^{(\beta)} 1 = 1$. We note that the canonical measure p_β lies comfortably in $\mathcal{A}_\beta^*$. We have an easy lemma.

Lemma 5.23 *A stochastic map T_c obeys detailed balance if and only if p_β is a fixed point of $T_c^\dagger$.*

For the proof, note that if $T_c^\dagger p_\beta = p_\beta$, then for all $f \in \mathcal{A}_c$,

$$\langle 1, f\rangle_\beta = \langle p_\beta, f\rangle = \langle T_c^\dagger p_\beta, f\rangle = \langle p_\beta, T_c f\rangle = \langle 1, T_c f\rangle_\beta = \langle T_c^{(\beta)} 1, f\rangle_\beta;$$

since $\mathcal{A}_c$ is dense in $\mathcal{A}_\beta$, we get $T_c^{(\beta)} 1 = 1$ as required. For the converse, if $T_c^{(\beta)} 1 = 1$,

$$\langle T_c^\dagger p_\beta, f\rangle = \langle p_\beta, T_c f\rangle = \langle 1, T_c f\rangle_\beta = \langle T_c^{(\beta)} 1, f\rangle_\beta = \langle 1, f\rangle_\beta = \langle p_\beta, f\rangle.$$

Since this is true for all $f \in \mathcal{A}_c$, we have $T_c^\dagger p_\beta = p_\beta$. □

The Lemma says that our formulation of detailed balance is obeyed by the isothermal dynamics τ_β^* used in the first half of this book. Also, it is the most general formulation possible in any classical theory that converges to equilibrium, as in the classical case it imposes no further conditions other than that T_c be stochastic with equilibrium p_β as fixed point. For, the positivity of T_c implies that its matrix elements are non-negative, and so are those of its transpose $T_c^\dagger$. The positivity of $T_c^{(\beta)}$ then follows from the simple relation

Lemma 5.24

$$p_\beta(j) T_{cjk}^{(\beta)} p_\beta(k)^{-1} = T_{jk}^\dagger.$$

Proof.

$$\begin{aligned} \langle p_\beta T_c^{(\beta)} p, f\rangle &= \langle T_c^{(\beta)} p, f\rangle_\beta = \langle p, T_c f\rangle_\beta = \langle p_\beta p, T_c f\rangle \\ &= \langle T_c^\dagger (p_\beta p), f\rangle. \end{aligned}$$

Since this is true for all p and f, we have

$$p_\beta T_c^{(\beta)} = T_c^\dagger p_\beta$$

where p_β denotes the operator of multiplication by p_β. We get the result by taking matrix elements. □

At $\beta = 0$, p_β is the unit operator, and our version of detailed balance reduces to the statement that T_c is bistochastic. This is enough to ensure that the entropy is monotonic increasing in time. We shall see that at $\beta > 0$, detailed balance is enough to ensure that the thermodynamic function $\Psi = S_c - \beta\langle \mathcal{E}_c\rangle$ is monotonic increasing.

We now show that the equilibrium state can be characterised as the state with the smallest $-\beta$-norm.

Theorem 5.25 *Suppose $\beta\mathcal{E}_c$ is such that $Z_\beta = \sum_{j\in\Omega_c} e^{-\beta\mathcal{E}_c(j)} < \infty$. Then the unique probability measure with the smallest $-\beta$-norm is the equilibrium state p_β.*

Proof.
Minimise $\sum_{j\in\Omega_c} e^{-\beta\mathcal{E}_c(j)}p_c(j)^2$ over $p_c \in \Sigma_c$ by introducing a Lagrange multiplier λ:

$$\frac{\partial}{\partial p_k}\left(\sum_j e^{\beta\mathcal{E}_c(j)}p_j^2 + \lambda\sum_j p_j\right) = 2p_k e^{\beta\mathcal{E}_c(k)} + \lambda = 0$$

at a turning point, giving the unique answer

$$p_k = \text{const. } e^{-\beta\mathcal{E}_c(k)}.$$

The second derivative is

$$\frac{\partial^2}{\partial p_k \partial p_l} = 2e^{\beta\mathcal{E}(k)}\delta_{kl}$$

which is a diagonal matrix with positive entries on the diagonal, and so is positive definite. The turning point is therefore a minimum. □
We now come to the main effective result of this section, which is an adaptation of a result of Koseki [17].

Theorem 5.26 *If T_c obeys detailed balance, then $T_c^\dagger$ is a contraction in $\|\bullet\|_{-\beta}$. If in addition T_c is irreducible, with positive diagonal elements, then p_β is the only fixed point of $T_c^\dagger$, and if $p \neq p_\beta$ then $\|T_c^\dagger p\|_{-\beta} < \|p\|_{-\beta}$.*

Note: T_c irreducible means that for any $i, j \in \Omega_c$, there is a chain of finite length $i_1 = i,\ i_2,\ i_3,\ \ldots, i_n = j$ such that $T^\dagger_{i_k, i_{k+1}} > 0$, for $k = 1, \ldots, n-1$. Whereas in statistical dynamics the bistochastic map T is not irreducible on Ω as it maps each energy-shell to itself, the isothermal map T_c is typically irreducible, unless there are conserved quantities such as atomic number in addition to the energy.
Proof.
Obviously for a sum over any finite subset Ω_0 of Ω_c we have for any $p, q \in \Sigma(\Omega_c)$:

$$\sum_{i,j} Z_\beta T^\dagger_{cji}\left(e^{\beta\mathcal{E}_j}q_j - e^{\beta\mathcal{E}_i}p_i\right)^2 \geq 0 \qquad (5.27)$$

and this inequality persists in the limit as Ω_0 increases to Ω_c if the sums converge. Let $q = T_c^\dagger p$. The first part is the sum of non-negative terms

$$Z_\beta \sum_{i,j} T^\dagger_{cji} e^{-\beta(\mathcal{E}_i - \mathcal{E}_j)} e^{\beta\mathcal{E}_j} q_j^2 = Z_\beta \sum_{i,j} T^{(\beta)}_{ji} e^{\beta\mathcal{E}_j} q_j^2.$$

By detailed balance the sum of $T^{(\beta)}_{ji}$ over i converges to 1 as Ω_0 increases to Ω_c, so the first part reduces to

$$Z_\beta \sum_j e^{\beta\mathcal{E}_j} q_j^2 = \|T^\dagger p\|^2_{-\beta}.$$

The cross terms give

$$-2Z_\beta \sum_{i,j} T^\dagger_{cji} e^{\beta\mathcal{E}_j} p_i q_j$$

and the sum over i is

$$\sum_{i\in\Omega_c} T^\dagger_{cji} p_i = q_j = (T^\dagger_c p)_j.$$

The cross-terms thus give us

$$-2\|T^\dagger p\|^2_{-\beta}.$$

The last term is exactly $\|p\|^2_{-\beta}$, since T_c is stochastic. We conclude that

$$-\|T^\dagger_c p\|^2_{-\beta} + \|p\|^2_{-\beta} \geq 0$$

as desired.

We get a proper contraction unless

$$e^{\beta\mathcal{E}_j} = e^{\beta\mathcal{E}_i} p_i \qquad \text{for all } i,\ j \text{ with } T^\dagger_{cji} > 0.$$

So let $p \in \Sigma(\Omega_c)$ be such that $T^\dagger_c$ is not a proper contraction at p. We shall show, using the assumptions of the theorem, that $p = q = p_\beta$. Since we get a proper contraction unless every term is zero, we get

$$e^{\beta\mathcal{E}_j} q_j = e^{\beta\mathcal{E}_i} p_i \qquad \text{for all } i,\ j \text{ with } T^\dagger_{cji} > 0.$$

Since $T^\dagger_{cii} > 0$ for all $i \in \Omega_c$, we have $p = q$. Starting with a component of p, say $i = 0$, with $p_0 \neq 0$, and changing the origin of the energy so that $\mathcal{E}_0 = 0$, we get

$$p_j = e^{-\beta\mathcal{E}_j} p_0$$

for all j linked to 0 by one step. Then we get

$$p_k = e^{-\beta(\mathcal{E}_k - \mathcal{E}_j)} p_j = e^{-\beta\mathcal{E}_k} p_0$$

for all k linked to 0 by two steps. In this way we get

$$p_k = e^{-\beta\mathcal{E}_k} p_0 \qquad \text{for all } k \in \Omega_c.$$

Since p is normalised we see that $p = p_\beta$. □

We can obtain a result similar to the uniqueness part of the Perron-Frobenius theorem, that 1 is a simple eigenvalue. For, if p' is another fixed point of $T^\dagger_c$, it can be taken to be real, since its real and imaginary parts in the priviledged basis are also eigenvectors. Then the equation $T^\dagger p = p$ leads, as in the theorem, to $p = p_\beta$.

We now seek a theorem on the convergence of the isothermal dynamics which will cover the case when the sample space is infinite, such as in the important case when the classical Fock space is present. It is true that the state space $\Sigma(\Omega_c)$ is compact in the weak* topology even when Ω_c has infinitely many points, and that this can lead to a convergence theorem [39]. However, we cannot be sure that the limit state has the same mean energy as those along the sequence, unless some condition is placed on the initial state. The problem is that energy is an unbounded observable, and that some energy might be lost "along the energy ladder" if the notion of convergence is too weak. The following rather strong condition on the initial state leads to convergence in the norm $\| \bullet \|_{-\beta}$, which is good enough for most people.

Definition 5.28 *Let us say that a state $p \in \Sigma(\Omega_c)$ is close to equilibrium (C, β) if there exists a number C such that*

$$p(j) \leq C Z_\beta^{-1} e^{-\beta \mathcal{E}_j} \qquad \text{for all } j \in \Omega_c.$$

This includes all equilibrium states with beta greater than β, all states with a maximum energy, and all states of the form fp_β with $f \in \mathcal{A}_c$. The concept is stable under the dynamics:

Lemma 5.29 *If p is close to equilibrium (C, β), and T obeys detailed balance (β), then $T^\dagger p$ is close to equilibrium (C, β).*

For the proof, note that as $T^\dagger$ is positively preserving,

$$T^\dagger \left(C Z_\beta^{-1} e^{-\beta \mathcal{E}} - p \right)(j) \geq 0.$$

Since $T^\dagger$ leaves p_β fixed, we get

$$C Z_\beta^{-1} e^{-\beta \mathcal{E}_j} - \left(T^\dagger p \right)_j \geq 0$$

which proves the lemma.

Now comes the good estimate:

Theorem 5.30 *Let Ω be a sample space and p and q be non-negative functions in $\ell^1(\Omega)$, and let $S(q|p) = \sum_{j \in \Omega} q(\omega) \left(\log q(\omega) - \log p(\omega) \right)$. Suppose that there exists $C > 0$ such that*

$$p(\omega) \leq C q(\omega) \qquad \text{for all } \omega \in \Omega.$$

Then

$$S(q|p) \geq \frac{1}{2C'^2} \sum_\omega q(\omega)^{-1} |p(\omega) - q(\omega)|^2 - \|p\|_1 + \|q\|_1$$

where $C' = \max\{C, 1\}$.

Remark. If ω is such that $q(\omega) = 0$ then $q(\omega) = 0$ and we interpret the contribution as zero.

Proof.

The Taylor series with Lagrange remainder gives

$$\begin{aligned} S(q|p) &= \sum q(\omega)\,(\log q(\omega) - \log p(\omega)) \\ &= \sum q(\omega)\left(\log q(\omega) - [\log q(\omega) + \frac{p(\omega)-q(\omega)}{q(\omega)} - \frac{(q(\omega)-p(\omega))^2}{2\xi(\omega)^2}]\right) \\ &= \sum (q(\omega)-p(\omega))^2\, q(\omega)/(2\xi(\omega)^2) - \|p\|_1 + \|q\|_1. \end{aligned}$$

Here $\xi(\omega)$ is the intermediate value, obeying

$$q(\omega) < \xi(\omega) < p(\omega) \text{ or } p(\omega) < \xi(\omega) < q(\omega)$$

except if $p(\omega) = q(\omega)$. If ω is such that $\xi(\omega) < q(\omega)$ we get for that term in the sum a contribution greater than

$$\frac{1}{2}|q(\omega) - p(\omega)|^2 \frac{q(\omega)}{q(\omega)^2}.$$

If ω is such that $\xi(\omega) < p(\omega) < C\,q(\omega)$, we get a term greater than

$$\frac{1}{2}\,|q(\omega) - p(\omega)|^2\,\frac{q(\omega)}{C^2 q(\omega)^2}.$$

In either case, summing gives the result. □

So if Ω is the space of chemicals Ω_c and $q = p_\beta$, and if τ_β defines a dynamics such that the relative entropy $S(p_\beta|\tau_\beta^n p)$ converges to zero, then the dynamics also converges in the norm $\|\bullet\|_{-\beta}$, if the initial state is "close to equilibrium". This is not such a convenient result, since the free energy is related to the other relative entropy $S(p|p_\beta)$. For this we have the following

Theorem 5.31 *Suppose that T_c obeys detailed balance (β) and let*

$$F(p) = \langle p, \mathcal{E}_c \rangle - \beta^{-1} S(p)$$

denote the free energy, defined on a subset of Ω_c. Then

$$F(T^\dagger p) \le F(p).$$

If further p is close to equilibrium (C, β), then

$$F(T_c^\dagger p) - F(p) \le -(2\beta C)^{-1}\left(\|p\|_{-\beta}^2 - \|T_c^\dagger p\|_{-\beta}\right).$$

Our proof uses the

Lemma 5.32 *Let $s(x) = -x\log x$ and $x = \sum \lambda_i x_i$, $\lambda_i \geq 0$, $\sum \lambda_i = 1$. Then if $x \neq x_i$, $i = 1, 2, \ldots, n$ we have*

$$\sum \lambda_i s(x_i) = s(x) - \frac{1}{2}\sum \lambda_i (x_i - x)^2/\xi_i$$

where $x_i < \xi_i < x$ or $x < \xi_i < x_i$.

Proof of lemma. By Taylor's theorem with Lagrange's remainder,

$$s(x_i) = s(x) + (x_i - x)[-1 - \log x] + (x_i - x)^2\,(-1/(2\xi_i)).$$

Multiply by λ_i and sum; we get the lemma, since $\sum \lambda_i(x_i - x) = 0$.

Choose j and try this lemma with

$$\lambda_i = T^{\dagger}_{cji} e^{-\beta(\mathcal{E}_i - \mathcal{E}_j)} = T^{(\beta)}_{ji}$$

and with $x_i = e^{\beta\mathcal{E}_i} p_i$. Then

$$x = \sum_i T^{\dagger}_{cji} e^{-\beta\mathcal{E}_i + \beta\mathcal{E}_j} e^{\beta\mathcal{E}_i} p_i = e^{\beta\mathcal{E}_j} q_j$$

where $q = T^{\dagger}_c p$. So for each j

$$\begin{aligned} -\sum_i T^{\dagger}_{cji} e^{-\beta(\mathcal{E}_i - \mathcal{E}_j)} \left(e^{\beta\mathcal{E}_i} p_i(\log p_i + \beta\mathcal{E}_i)\right) &= -e^{\beta\mathcal{E}_j} q_j(\log q_j + \beta\mathcal{E}_j) \\ &\quad -\sum_i T^{\dagger}_{cji} e^{-\beta(\mathcal{E}_i - \mathcal{E}_j)} \left(e^{\beta\mathcal{E}_i} p_i - e^{\beta\mathcal{E}_j}\right)^2 /(2\xi_{ij}). \end{aligned}$$

Cancel $e^{\beta\mathcal{E}_j}$ to get

$$\begin{aligned} \sum_i T^{\dagger}_{cji}\,(-p_i \log p_i - \beta p_i \mathcal{E}_i) &= -q_j \log q_j - \beta q_j \mathcal{E}_j \\ &\quad - \sum_i T^{\dagger}_{cji} e^{-\beta\mathcal{E}_i} \left(e^{\beta\mathcal{E}_i} p_i - e^{\beta\mathcal{E}_j} q_j\right)^2 /(2\xi_{ij}). \end{aligned}$$

Now for non-zero terms, either

$$e^{\beta\mathcal{E}_i} p_i < \xi_{ij} < e^{\beta\mathcal{E}_j} q_j \leq Z_\beta^{-1} C$$

or

$$e^{\beta\mathcal{E}_j} q_j < \xi_{ij} < e^{\beta\mathcal{E}_i} p_i \leq Z_\beta^{-1} C.$$

In either case, $1/\xi_{ij} > Z_\beta C^{-1}$. Summing over j we get

$$\begin{aligned} S(p) - \beta\langle p, \mathcal{E}\rangle &\leq S(q) - \beta\langle q, \mathcal{E}\rangle \\ &\quad - \frac{Z_\beta}{2C} \sum_{ij} T^{\dagger}_{cji} e^{-\beta\mathcal{E}_i} \left(e^{\beta\mathcal{E}_i} p_i - e^{\beta\mathcal{E}_j} q_j\right)^2, \end{aligned}$$

which gives the result in view of the proof of Theorem (5.26). □

This theorem has a useful corollary:

Theorem 5.33 *If T_c has positive diagonals and is irreducible, and p is close to equilibrium, then $T_c^n p \to p_\beta$ as $n \to \infty$, the topology of the convergence being the metric given by $\| \bullet \|_{-\beta}$.*

This is a good result, since it does not depend on the finiteness of Ω_c or the reduction of the number of variables to a finite number by the LTE map.

We now state without proof our version of the theorem on the equivalence of ensembles; the proof is given in detail in [44].

Let $\Omega_x = \{0, 1, 2, \ldots\}$ and let $|\Lambda|$ be a finite set, and take the sample space to be $\Omega = \Omega(\Lambda) = \prod_x \Omega_x$. Let $\mathcal{E}(x) = \varepsilon \omega_x$ and

$$\mathcal{E}(\Lambda_0) = \sum_{x \in \Lambda_0} \mathcal{E}(x).$$

The energy in Λ, the random variable $\mathcal{E}(\Lambda)$, divides Ω into the union of energy-shells Ω_E. Define the microcanonical state for Λ with energy E to be

$$p_\Lambda(\omega) = \begin{cases} |\Omega_E|^{-1} & \text{if } \omega \in \Omega_E \\ 0 & \text{otherwise} \end{cases}.$$

Define the marginal distribution of a probability $p_\Lambda \in \Sigma(\Lambda)$ onto $\Lambda_0 \subseteq \Lambda$ to be

$$p(\Lambda, \Lambda_0, \omega_0) = \sum_{\omega \in \Omega : \omega | \Lambda_0 = \omega_0} p_\Lambda(\omega). \tag{5.34}$$

Let us take Λ to infinity in such a way that the energy per site, $E/|\Lambda|$ remains constant. Then for each fixed Λ_0, the marginal state in Λ_0 of the microcanonical state for Λ converges to the canonical state for Λ_0. Thus

Theorem 5.35 (Equivalence of Ensembles) *As $\Lambda \to \infty$,*

$$p(\Lambda, \Lambda_0, \bullet) \to p_{\Lambda_0,\beta}(\bullet) \quad \textit{in the norm } \| \bullet \|_{-\beta}. \tag{5.36}$$

Here,

$$p_{\Lambda_0,\beta}(\omega_0) = Z(\Lambda_0)^{-1} \exp\{-\beta \mathcal{E}(\Lambda_0)(\omega_0)\}$$

and β is fixed by

$$\sum_{\omega_0} p_{\Lambda_0,\beta} \mathcal{E}(\Lambda_0)(\omega_0) = E|\Lambda_0|/|\Lambda|.$$

There is a similar theorem for an assembly of free Fermions.

This theorem might be used to justify the step between random linear dynamics, which converges to the microcanonical state, and the convergence to the canonical state; the latter needs non-linear dynamics which worries some people (not me).

5.5 Dilation of Markov Chains

We now show that any Markov chain with a finite sample space Ω_c and a strictly positive fixed point can be regarded as the isothermal dynamics of some stochastic map on a suitable sample space, conserving a certain energy-function. We follow [45]. The construction is not unique, but Ω can be taken to be as small as $\Omega_c \times \Omega_c$.

We first note that if $p^{(0)} \in \Omega_c$ is strictly positive, then there exists a stochastic map $T_c^{\dagger}$ with $p^{(0)}$ as fixed point. Indeed,

$$T_c^{\dagger} = \begin{pmatrix} p_1^{(0)} & p_1^{(0)} & \cdots & p_1^{(0)} \\ p_2^{(0)} & p_2^{(0)} & \cdots & p_2^{(0)} \\ \cdot & \cdot & \cdots & \cdot \\ p_n^{(0)} & p_n^{(0)} & \cdots & p_n^{(0)} \end{pmatrix} \tag{5.37}$$

has the much stronger property that $T_c^{\dagger} p_c = p^{(0)}$ for all p_c. That is, it takes any state to equilibrium in one step. The stochastic matrix $T_c^{\dagger}$ is the limit, as $n \to \infty$, of $T_1^{\dagger n}$, where $T_1^{\dagger}$ is any ergodic stochastic matrix with $p^{(0)}$ as fixed point. We can also find a family $\{T_\lambda^{\dagger}\}$ of stochastic matrices, all having $p^{(0)}$ as fixed point, with $T_\lambda^{\dagger}$ arbitrarily close to the identity, namely $T_\lambda^{\dagger} = (1-\lambda)1 + \lambda T_c^{\dagger}$. These remarks mean that the following structure theorem is not empty.

Theorem 5.38 *Let $\Omega_c = \{1, 2, \ldots, n\}$. Let $p^{(0)} \in \Sigma(\Omega_c)$ have positive components: $p_j^{(0)} > 0 \quad j = 1, 2, \ldots, n$. Then there exists a neighbourhood N of the unit $n \times n$ matrix and a sample space Ω_γ, with $|\Omega_\gamma| = n$, and an energy function $\mathcal{E} = \mathcal{E}_c + \mathcal{E}_\gamma$ on $\Omega = \Omega_c \times \Omega_\gamma$, such that for any stochastic matrix $T_c \in N$ with $p^{(0)}$ as fixed point, there exists a bistochastic map T on $\mathcal{A}(\Omega)$ conserving $\mathcal{E}$ such that*

$$T_c p_c = \mathcal{M}_c \left(T^{\dagger} p_c \otimes s \right)$$

for all $p_c \in \Sigma(\Omega_c)$; here, s is the canonical state of $\mathcal{A}(\Omega_\gamma)$ with energy $\mathcal{E}_\gamma$ and $\beta = 1$.
Moreover, if T_c obeys the usual form of detailed balance, then T is symmetric.

Proof.
This is by construction. Consider the set $\{0\} \cup \Omega_c$, the given sample space with one point adjoined. Define the *fan* of Ω_c, denoted Fan Ω_c, to be the graph whose vertices are the points of $\{0\} \cup \Omega_c$, and whose edges are the pairs $(\omega_c, 0)$. We can regard the edges as bonds between the points of the system and a common heat-bath, $\{0\}$. We define the space Ω_γ of the heat-particle to be the set of edges of Fan Ω_c. Then $|\Omega_\gamma| = n$, as promised. Define the energy function $\mathcal{E}_c$ on Ω_c by

$$\mathcal{E}_c(\omega_c) = -\log p^{(0)}(\omega).$$

This is well defined and positive, since $0 < p^{(0)}(\omega_c) < 1$. We have adjusted the lowest energy so that the partition function is 1. Since $p^{(0)}(\omega) = e^{-\mathcal{E}_c(\omega)}$ we see that $\sum e^{-\mathcal{E}_{cj}} = 1$. We define $\mathcal{E}_\gamma(\omega_c, 0) = -\mathcal{E}_c(\omega_c)$, and the energy function on $\Omega = \Omega_c \times \Omega_\gamma$ to be

$$\mathcal{E}(\omega_c, (\omega_c', 0)) = \mathcal{E}_c(\omega_c) - \mathcal{E}_c(\omega').$$

We now construct a bistochastic matrix T on $\mathcal{A}(\Omega)$ which causes hopping from $i \in \Omega_c$ to $j \in \Omega_c$ by picking up a heat-particle from the bond $b_i = (i, 0)$, hopping to j, and releasing a heat-particle to the bond $b_j = (j, 0)$. That is, we have the process

$$i + b_i \rightleftharpoons j + b_j.$$

The other bonds and vertices do not interact. Since the transitions occur only on the energy-shell $\Omega_0 \subset \Omega$ of zero energy, our reaction conserves energy, as claimed. We shall now define an $n \times n$ bistochastic matrix T_0 on $\mathcal{A}(\Omega_0)$. The action of T will coincide with that of T_0 on the subspace, and will be the unit operator (of size $(n^2 - n) \times (n^2 - n)$) on the remaining points. T is bistochastic and conserves energy, as it has the form

$$T = \begin{pmatrix} T_0 & 0 \\ 0 & 1 \end{pmatrix}. \tag{5.39}$$

Let

$$\lambda = \sum_{j=1}^{n} e^{\mathcal{E}(j)} \tag{5.40}$$

and put

$$\text{diag } p^{(0)} = \begin{pmatrix} p^{(0)}(1) & 0 & \dots & 0 \\ 0 & p^{(0)}(2) & \dots & 0 \\ \dots & \dots & \dots & \dots \\ 0 & 0 & \dots & p^{(0)}(n) \end{pmatrix}. \tag{5.41}$$

We define T_0 in terms of T_c by

$$T_0 = 1 + \lambda \,\text{diag}\, p^{(0)}(T_c - 1). \tag{5.42}$$

Let $p^{(1)}$ denote $(1, 1, \dots, 1)^t$. Then $\text{diag}\, p^{(0)} p^{(1)} = p^{(0)}$. A matrix A is stochastic if and only if its elements are non-negative, and $Ap^{(1)} = p^{(1)}$. We now check the second property for T_0 and $T_0^\dagger$.

$$\begin{aligned} T_0 p^{(1)} &= p^{(1)} + \lambda \,\text{diag}\, p^{(0)} T_c p^{(1)} - \lambda \,\text{diag} p^{(0)} p^{(1)} \\ &= p^{(1)} + \lambda \,\text{diag}\, p^{(0)} p^{(1)} - \lambda \,\text{diag}\, p^{(0)} p^{(1)} = p^{(1)} \end{aligned}$$

since T_c is stochastic. Hence the rows of T_0 add up to 1. Also,

$$T_0^\dagger p^{(1)} = p^{(1)} + \lambda(T_c^\dagger - 1)\mathrm{diag}\, p^{(0)} p^{(1)} = p^{(1)} + \lambda(T_c^\dagger - 1)p^{(0)} = p^{(1)} \quad (5.43)$$

since $p^{(0)}$ is a fixed point of $T_c^\dagger$. Hence the columns of T_0 add up to 1. We now check the positivity. The off-diagonal elements of T_0 are non-negative. The k th diagonal element is

$$1 + \lambda p_k^{(0)}(T_{kk} - 1)$$

which is positive if for each k, T_{ckk} is close enough to 1. This defines the required neighbourhood N of the unit matrix, in which T_0 is bistochastic.

It remains to show that we recover the Markov matrix T_c as the isothermal dynamics given by the bistochastic matrix T. Recall that T consists of the matrix T_0 in the block labelled by rows and columns (j, b_j), T being the unit matrix on the rest of the space, as in Eqn. (5.39). This may be expressed in terms of matrix elements as:

$$T_{i,b_j;k,b_\ell} = \delta_{i,j}\delta_{k,\ell}\left(\delta_{i,k} + p^{(0)}(i)(T_{ci,k} - \delta_{i,k})\right) + (1 - \delta_{i,j}\delta_{k,\ell})\delta_{i,k}\delta_{j,\ell}. \quad (5.44)$$

One time-step of the isothermal dynamics defined by T is

$$p_c \mapsto p_c' = \sum_{\omega_\gamma} T^\dagger(p_c \otimes s)(\omega_c, \omega_\gamma). \quad (5.45)$$

Here, ω_c runs over $k \in \{1, 2, \ldots, n\}$ and ω_γ runs over $\ell = b_\ell = (\omega_\ell, 0)$. The heat particle is in the canonical state

$$s(\ell) = \frac{e^{\mathcal{E}(\ell)}}{\sum_j e^{\mathcal{E}(j)}}. \quad (5.46)$$

Then $\sum_j s(j) = 1$ and

$$p^{(0)}(i)s(i) = e^{-\mathcal{E}(i)}e^{\mathcal{E}(i)}\left(\sum_j e^{\mathcal{E}(j)}\right)^{-1} = \lambda^{-1}. \quad (5.47)$$

Both these are used in the following.

$$\begin{aligned}
p_c'(k) &= \sum_{i,j,\ell} p_c(i)s(j)T_{i,j;k,\ell} \\
&= \sum_{i,j,\ell} p_c(i)s(j)\delta_{i,j}\delta_{k,\ell}T_{0,i,k} + \sum_{i,j,\ell} p_c(i)s(j)[\delta_{i,k}\delta_{j,\ell} - \delta_{i,j}\delta_{k,l}\delta_{i,k}\delta_{j,\ell}] \\
&= \sum_i p_c(i)s(i)T_{0,i,k} + p_c(k) - p_c(k)s(k) \\
&= \sum_i p_c(i)s(i)[\delta_{i,k} + \lambda p^{(0)}(i)(T_{c,i,k} - \delta_{i,k})] + p_c(k) - p_c(k)s(k)
\end{aligned}$$

$$\begin{aligned} &= p_c(k)s(k) + \lambda \sum_i p_c(i)p^{(0)}(i)s(i)T_{c,i,k} \\ &\quad -\lambda p^{(0)}(k)p_c(k)s(k) + p_c(k) - p_c(k)s(k) \\ &= \sum_i p_c(i)T_{c,i,k}. \end{aligned}$$

Thus the isothermal dynamics coincides with the given Markov chain.
As for detailed balance, we see that

$$T_0^\dagger = 1 + \lambda(T_c^\dagger - 1)\mathrm{diag}p^{(0)} = T_0$$

if $\mathrm{diag}\, p^{(0)}T_c = T^\dagger \mathrm{diag}\, p^{(0)}$, which is detailed balance, Eqn. (5.17). □

The relative entropy of the Markov chain coincides with the free energy of the isothermal dynamics. By Birkhoff's theorem we can decompose the bistochastic map T into permutations conserving energy. This gives us a natural decomposition of the convex set of stochastic maps, all with the same fixed point $p^{(0)}$, into extremal points. This result seems to be new.

The doubling up of the space from Ω_c to $\Omega_c \times \Omega_\gamma$ used here will be a key method in the quantum case.

5.6 Exercises

Exercise 5.48 Consider the sample space $\Omega = \{0, 1, \ldots\}$ with some energy-function $\mathcal{E}$ with a finite partition function. Define, for any state $p \in \Sigma(\Omega)$ the Legendre transform Ψ of the entropy by $\Psi = S(p) - \beta\langle p, \mathcal{E}\rangle$, in which the variable $\mathcal{E}$ is singled out. Show that β is the dual variable, and that S is the inverse Legendre transform of Ψ.

Exercise 5.49 Show that for 2×2 matrices, the neighbourhood N for which the construction of Section(5.5) works does not include all stochastic matrices with $p^{(0)}$ as fixed point in general.

Chapter 6

Driven Systems

The fundamental theorem on chemical kinetics shows that systems at a fixed temperature and chemical potential converge to equilibrium with a monotonic behaviour of the Lyapunov function, the Helmholtz or Gibbs free energy. As a result, no really interesting behaviour, such as oscillations or chaos, can occur. To get more interesting behaviour from properly constructed statistical models needs a driven system, which is a system in which several of the intensive parameters (beta, chemical potential) are kept fixed at non-equilibrium values by outside intervention. They might then be different at different points of Λ, even at a stationary point. They are then not, strictly speaking, intensive variables any more, but may be called local intensive variables.

6.1 Sources and Sinks

A source is a point in the space Λ at which particles, or heat, is fed into the system from outside; a sink is a point from which particles or heat are withdrawn from the system. The rates are determined by the experimenter, and so these rates are *controlled variables.* Mathematically we can implement the control within statistical dynamics by forcing the state $p' = T^{\dagger}p$ after one time-step to conform to the constraints. Here we follow [46]. Experimentally, it is often a good description of the constraint to say that certain points $x \in \Lambda_0 \subseteq \Lambda$ are maintained at prescribed betas $\beta(x)$ or chemical potentials $\mu_j(x)$ for one or more chemicals j. To be able to express this, the sample space at x should contain the sample space of the constrained chemical as a factor. Thus we must assume that $\Omega_x = \Omega_j \times \Omega_{\neq j}$. For consistency of notation, we shall speak of beta as a chemical potential (of the heat-particle). Let $\mathcal{M}_j$ be the marginal projection onto the algebra of random variables of the j-th chemical. We can presume that the state $\mathcal{M}_j\mathcal{M}_x p'$ at a controlled point is independent of the rest of the system, since the flow of heat and particles into the point x from the outside destroys the correlations. The state can therefore be assumed to be a product state, thus

$$p(\omega) = p_{\mu_j}(\omega_{x,j}) \otimes p(\omega_{\neq x,j}).$$

Here, μ_j is the chemical potential of the jth chemical at which we maintain the point x. This property of factorisation of the state is not preserved by the dynamics $T^\dagger$, and the external control must be used to restore the constraint after one time-step. As promised, we implement this control mathematically by just doing it; the first step is

$$p \mapsto \mathcal{M}_j\left(T^\dagger p\right) \otimes \mathcal{M}_{\neq j} T^\dagger p. \tag{6.1}$$

This is then replaced in the first factor by flows of heat and chemicals from the outside, to enforce the controlled state p_{μ_j}, so that one time-step of the controlled dynamics is

$$\tau^* p = p_{\mu_j} \otimes \mathcal{M}_{\neq j} T^\dagger p. \tag{6.2}$$

We can then repeat the operation, once for each controlled variable, to arrive at the desired model of the driven system.

There are alternative controls that can be modelled. For example, if a site can be occupied by any one of several chemicals, A, B, ... but not more than one (as in the hard-core models), we may control the site by insisting that it always contain only a certain chemical, say A. Then after a time-step, we sweep out the product of the reaction from the site, and replace it by the chemical A. More generally, we may replace the state at the controlled site by a chosen statistical mixture of all possibilities. The grand canonical state p_{μ_j} is just one of these. These more general controls are perhaps not worth studying unless we have in mind some experimental set-up that can implement them. We work out some important cases to illustrate the method.

6.2 A Poor Conductor

In the following model, heat is transmitted along a chain of atoms by local excitations from one to the next. In a good conductor such as copper the bulk of the heat is carried by electrons in the "conduction band" and very little by atomic excitations. We therefore describe the present model as a poor conductor. We take [47] as usual Λ to be a finite lattice, $\{0, 1, \ldots, N\}$ and $\Omega_x = \{0,\ 1\}$. The configuration space is then $\Omega = \prod \Omega_x$. We interpret $\omega_x = 0$ as saying that the atom at x is in its ground state, and $\omega_x = 1$ as saying that the atom at x is in its excited state, with energy ε say. The energy of a configuration ω is thus taken to be

$$\mathcal{E}(\omega) = \varepsilon \sum_x \omega_x. \tag{6.3}$$

We assume that Λ is provided with a set of edges, making it into a graph. These will be taken to define the nearest neighbours of the model. We take it that there are no multiple edges, and denote the set of edges, or bonds $b = (x,\, y)$ by $\mathcal{B}$. For each bond $b \in \mathcal{B}$ we define the transition matrix T_x, which only alters the coordinates $(x,\, y)$ of b, as

$$T_x = \begin{array}{c} \\ 00 \\ 01 \\ 10 \\ 11 \end{array} \begin{array}{c} \begin{array}{cccc} 00 & 01 & 10 & 11 \end{array} \\ \begin{pmatrix} 1 & & & \\ & 1-\lambda & \lambda & \\ & \lambda & 1-\lambda & \\ & & & 1 \end{pmatrix} \end{array} \tag{6.4}$$

In this, the rows and columns are labelled by ω_x, ω_{x+1}, which can take the values 0 and 1. Any state in $\Sigma(\Omega_x)$ is a pair of non-negative numbers whose sum is 1, and is thus determined by one parameter. It can therefore quite generally be written as a thermal state

$$p_x(\omega_x) = (1 + e^{-\beta\varepsilon})^{-1} e^{-\beta\mathcal{E}(\omega_x)} \tag{6.5}$$

for some parameter β, which is uniquely determined. Given any state $p \in \Sigma(\Omega)$, its marginal state $\mathcal{M}_x p$ can be written as in Eqn. (6.5), which defines the local beta of the state p. In this case, where $|\Omega_x| = 2$, the stoss map is the same as the LTE map. We take as the dynamics of the system, due to T_x, to be the combined map

$$p \mapsto T_x p \mapsto \otimes_x \mathcal{M}_x T_x p. \tag{6.6}$$

Since the state is, after each time-step, forced to be a product state, it is determined by $(p_0, \ldots, p_N)$, which in turn is determined by the probabilities that the upper level is occupied at the various points. Put $n(x) = p_x(t)$, the density at time t. This is the mean number of occupation of the upper level at x, and so the state at time t is described by this "density-field". We now take the change in state in one time step to be a convex mixture of these T_x, as x runs from 0 to $N-1$. We again call the new rate constant λ. Let us denote by $n'(x)$ the density field at time $t+1$. We have already seen that this leads to the discrete heat equation

$$n'(x) - n(x) = \begin{cases} \lambda\left(n(x+1) - 2n(x) + n(x-1)\right), \text{ for } 0 < x < N \\ \lambda\left(n(0) - n(1)\right), \text{ for } x = 0 \\ \lambda\left(n(N-1) - n(N)\right), \text{ for } x = N \end{cases}.$$

This, then, describes the dynamics of an isolated system. In the driven system, we maintain the ends, $x = 0$ and $x = N$, at fixed betas; this just means that $n(0)$ and $n(N)$ are returned to their initial values after each time-step. Let us denote these initial values by n_0 and n_N and suppose that $n_0 > n_N$. A stationary state is one which is unchanged in one time-step. Such a state is said to represent *stasis*. It is not in equilibrium unless $n_0 = n_N$. Indeed, a stationary solution can easily be found:

$$n(x) = n_0 - \frac{x}{N}(n_0 - n_N). \tag{6.7}$$

In this solution, the energy-density falls linearly as x increases. This is in contrast to the commonly studied model, where it is the temperature that obeys the heat equation, rather than the energy-density, as here. Although at stasis heat enters at $x = 0$ and leaves at $x = N$, this might not be true at all times. If the initial state is hotter at $x = 1$ than T_0, the temperature of the external control at $x = 0$, then heat will leave the hot end at the beginning of the motion. The entropy of the system may thus either decrease or increase; not even the free energy is a Lyapunov function. This is typical of driven systems: some have Lyapunov functions, and so converge to stasis; others may converge without an obvious Lyapunov function. Yet others might oscillate or diverge to infinity. It is clear that our present model converges to stasis.

We can find the rate of entropy production by the system at stasis. We may treat the ends, $x = 0$, $x = 1$ and $x = N - 1$, $x = N$ separately. Thus, if (n_0, n_1) is the density of the first two points at stasis, from above $n_1 = n_0 - N^{-1}(n_0 - n_N)$. After one time-step in which $T^\dagger$ has acted, but no compensating heat flow has yet flowed, we have a new density,

$$n_0' = n_0 + \lambda(n_1 - n_0) = n_0 - \lambda N^{-1}(n_0 - n_N);$$

it is clear that we must add an amount of heat equal to

$$Q = \varepsilon\lambda N^{-1}(n_0 - n_N)$$

to restore the mean value of the local energy at $x = 0$, and thus the beta, to its original value. The value of n_1 is unchanged (we are at stasis). The entropy at $x = 0$ is

$$S_0 = -n_0 \log n_0 - (1 - n_0)\log(1 - n_0).$$

Passing to the limit of continuous time, we see that

$$\begin{aligned}
\dot{S} &= -\dot{n}_0 \log n_0 - \dot{n}_0 - (1 - \dot{n}_0)\log(1 - n_0) - (-\dot{n}_0) \\
&= \lambda N^{-1}(n_N - n_0)\log\left(\frac{1 - n_0}{n_0}\right) \\
&= N^{-1}(n_N - n_0)\varepsilon\beta = Q/T
\end{aligned}$$

in terms of Q and $T = \beta^{-1}$, the temperature. Thus, in the limit of continuous time the entropy production (cancelled by the feeding in of hot heat) is quasi-static: the entropy loss of the environment is exactly Q/T as given by elementary thermodynamics.

The stationary state of this model defines a stochastic process known as a Markov population process [48]. In 1978, a justification of the construction was given by Davies [49]. He starts with a Hamiltonian system, in which each site of a chain is linearly coupled to an oscillator with coupling strength λ. Thus each site has its own private heat-bath, except that the ends of the chain are coupled to reservoirs whose state is maintained at fixed temperatures. The time-evolution of the system is then obtained by tracing over the private heat-baths. At the end, the weak-coupling limit is taken in the sense of van Hove. The density-matrix of the system then obeys equations very similar to those of the model given here. In particular, the correlations between different points vanishes in the limit, and the LTE hypothesis holds. It can be argued that this justifies the model. This is, however, very much a matter of style rather than substance. It is certainly nice to have at least one Hamiltonian leading to the process of the model in a certain limit. Unfortunately there are many other Hamiltonian schemes whose weak-coupling limit have unphysical features, thus defeating one of the main points of the Hamiltonian method. In particular, energy can be lost to the private heat-baths which are inside the chain, contradicting the first law. This is not the same as losing heat to the reservoirs at the ends, which represents heat flowing out of the system. It is even possible for a limit of Hamiltonian motion to violate the conservation of probability. In Davies' model, the internal heat-loss does not occur, and probability is conserved, because of the clever construction. Thus, the Hamiltonian method must be supplemented by an intuition that avoids these unwanted features. Statistical dynamics is the mathematics behind this intuition.

We now turn to a driven system which leads to oscillations.

6.3 A Driven Chemical System

The following is a simplified version of an oscillating system inspired by the Brussellator [50, 51, 45]. It involves four chemicals, A, B, X and Y, which undergo the unidirectional reactions

$$B + X \stackrel{k_1}{\to} Y\ , \tag{6.8}$$

$$X \stackrel{k_2}{\to} 0\ , \tag{6.9}$$

$$2X + Y \stackrel{k_3}{\to} 3X\ , \tag{6.10}$$

$$A \stackrel{k_4}{\to} X. \tag{6.11}$$

The meaning of Eqn. (6.9) is that X makes a transition to an inert chemical with rate k_2, or else to reactants that are swept away after each time-step. The reaction (6.10) is "autocatalytic"; two molecules of X must be present before the molecule Y can be persuaded to make the transition to X.

The unidirectional reactions violate the laws of chemistry and in particular, the Arrhenius relation. They can be understood by postulating that a fifth particle is produced on the right in each reaction, and that this particle is swept away by the experimenter as fast as it is produced. This prevents the back reaction if the new particle is essential for the inverse reaction to proceed. A natural choice for the new particle is the heat-particle. This is tantamount to saying that all reactions are exothermic, and that the heat produced is swept away by keeping the system at zero temperature.

The system is in a single cell (a "well-stirred reactor") and the sample space will be taken to be

$$\Omega = \Omega_A \times \Omega_B \times \Omega_X \times \Omega_Y.$$

Here, Ω_A etc. are all the same, the classical Fock space

$$\Omega_A = \{0,\ 1,\ 2, \ldots\} = \Omega_B = \Omega_X = \Omega_Y.$$

Let $p = p(\omega_A, \omega_B, \omega_X, \omega_Y)$ be a probability on Ω, and denote as usual the marginals on the factors by p_A and so on. For example,

$$p_A(\omega_A) = \sum_{\omega_B=0}^{\infty} \sum_{\omega_X=0}^{\infty} \sum_{\omega_Y=0}^{\infty} p(\omega_A, \omega_B, \omega_X, \omega_Y).$$

Denote by N_A etc. the mean values of the random variables $\mathcal{N}_A(\omega) = \omega_A$, etc. Thus,

$$N_A = \sum_{\omega_A=0}^{\infty} \omega_A p_A(\omega_A).$$

This is the mean number of A-particles when the state is p. The activity is defined in terms of the mean number by $a = N/(N+1)$. We use the activity-led equations of Section (4.3). The equations with the heat-particle swept away are thus

$$\frac{dN_A}{dt} = -k_4 a_A\,, \tag{6.12}$$

$$\frac{dN_B}{dt} = -k_1 a_B a_X\,, \tag{6.13}$$

$$\frac{dN_X}{dt} = -k_1 a_B a_X - k_2 a_X + k_3 a_X^2 a_Y + k_4 a_A\,, \tag{6.14}$$

$$\frac{dN_Y}{dt} = k_1 a_B a_X - k_3 a_X^2 a_Y. \tag{6.15}$$

We want to consider the driven system in which the activities a_A and a_B are maintained at constant values by feeding the system as they are used up. It is very easy to do this: we omit the first two equations, and regard a_A and a_B as constants in the last two equations. We then get the autonomous equations for the two variables a_X and a_Y (after eliminating N in favour of a):

$$\frac{d\mathbf{a}}{dt} = \mathbf{F}(\mathbf{a}) \tag{6.16}$$

where $\mathbf{a} = (a_X, a_Y)$ and $\mathbf{F} = (F_1, F_2)$, with

$$\begin{aligned} F_1 &= \left(-k_1 a_B a_X - k_2 a_X + k_3 a_X^2 a_Y + k_4 a_A\right)(1 - a_X)^2 , & (6.17)\\ F_2 &= \left(k_1 a_B a_X - k_3 a_X^2 a_Y\right)(1 - a_Y)^2 . & (6.18) \end{aligned}$$

Such a system can be studied by Lyapunov's first method: the fixed point is given by $\mathbf{F} = 0$, which gives two equations

$$\begin{aligned} -k_1 a_B a_X - k_2 a_X + k_3 a_X^2 a_Y + k_4 a_A &= 0 , & (6.19)\\ k_1 a_B a_X - k_3 a_X^2 a_Y &= 0 , & (6.20) \end{aligned}$$

with the unique solution

$$\mathring{a}_X = \frac{k_4}{k_2} a_A , \qquad \mathring{a}_Y = \frac{k_1 k_2 a_B}{k_3 k_4 a_A} . \tag{6.21}$$

For this fixed point to lie in the allowed region $0 \le a < 1$, there are limits on the values that the parameters can take. The next step is to study the stability of the fixed point; we expand F in a Taylor series about $(\mathring{a}_X, \mathring{a}_Y)$, and we keep only the linear terms. The term of zeroth order vanishes, by definition of the fixed point. So, write $a_X = \mathring{a}_X + X$ and $a_Y = \mathring{a}_Y + Y$, and rewrite the equations for the new variables $\mathbf{Z} = (X, Y)$; we get the linear equation

$$\frac{d\mathbf{Z}}{dt} = A\mathbf{Z} \tag{6.22}$$

where A is the 2×2 matrix

$$A = \begin{pmatrix} (k_1 a_B - k_2)(1 - \mathring{a}_X)^2 & k_3 \left(\frac{k_4 a_A}{k_2}\right)^2 (1 - \mathring{a}_X)^2 \\ -k_1 a_B (1 - \mathring{a}_Y)^2 & -k_3 \left(\frac{k_4 a_A}{k_2}\right)^2 (1 - \mathring{a}_Y)^2 \end{pmatrix} . \tag{6.23}$$

Linear equations have exponential solutions, which are sinusoidal with an exponential growing or damping factor. Indeed, if A has eigenvalues λ_1 and λ_2, which are complex in general, with eigenvectors ϕ_1 and ϕ_2, then the solution is

$$\mathbf{Z} = e^{\lambda_1 t}\phi_1 + e^{\lambda_2 t}\phi_2 . \tag{6.24}$$

The number, $\lambda = \max\{Re\,\lambda_1,\ Re\,\lambda_2\}$ is called the Lyapunov index of the fixed point $\mathbf{Z} = 0$. If $\lambda < 0$, then all solutions starting in the neighbourhood of the fixed point converge exponentially to it, at the rate $e^{\lambda t}$. We then say that it is a local attractor. If $\lambda > 0$ then the solutions move away from the fixed point, (except possibly on one orbit) as long as the linear approximation is a good one; then $\mathbf{Z} = 0$ is called a repeller. If both eigenvalues are purely imaginary, then in the linear approximation the solution is periodic about $\mathbf{Z} = 0$, and we see oscillatory solutions of the non-linear equations as well, provided that we start close enough to the fixed point. The present model exhibits these features, for suitable choices of the parameters. For example we may choose $a_A = a_B = 1/2$, and $k_1 = 5$, $k_2 = 11/8 = k_4$, $k_3 = 8$. Then the fixed points become $\mathring{a}_X = 1/2$, $\mathring{a}_Y = 5/8$, and the matrix A becomes

$$A = \begin{pmatrix} 9/32 & 1/2 \\ -45/128 & -9/32 \end{pmatrix}, \tag{6.25}$$

whose eigenvalues are purely imaginary. Figure (1) shows the plot of a_X and A_Y against time, and also the orbit, in which time is eliminated and a_X is plotted against a_Y. The initial point, the pair of activities of X and Y, has been chosen very close to the fixed point (1/2, 5/8). It is clear that oscillations occur for these values of the parameters. Examination of the pictures shows that the period of oscillation, about 20, agrees with the prediction you will make if you do Exercise (6.34).

When the eigenvalues are purely imaginary, and there are two variables as here, then there is a periodic solution to the fully non-linear system; this follows from the Poincaré-Bendixon theorem [52]. Then we say that there is a limit-cycle; Fig. (2) shows that the limit-cycle attracts the orbit even when the initial state is rather close to the boundary. For a range of values of the parameters near these critical values, there will continue to be oscillatory behaviour. This is in contrast to isolated or isothermal systems. For these, we proved in the fundamental theorem of chemical kinetics, that there is no such possibility. Not all driven systems exhibit oscillations, as you will see in Exercise (6.36).

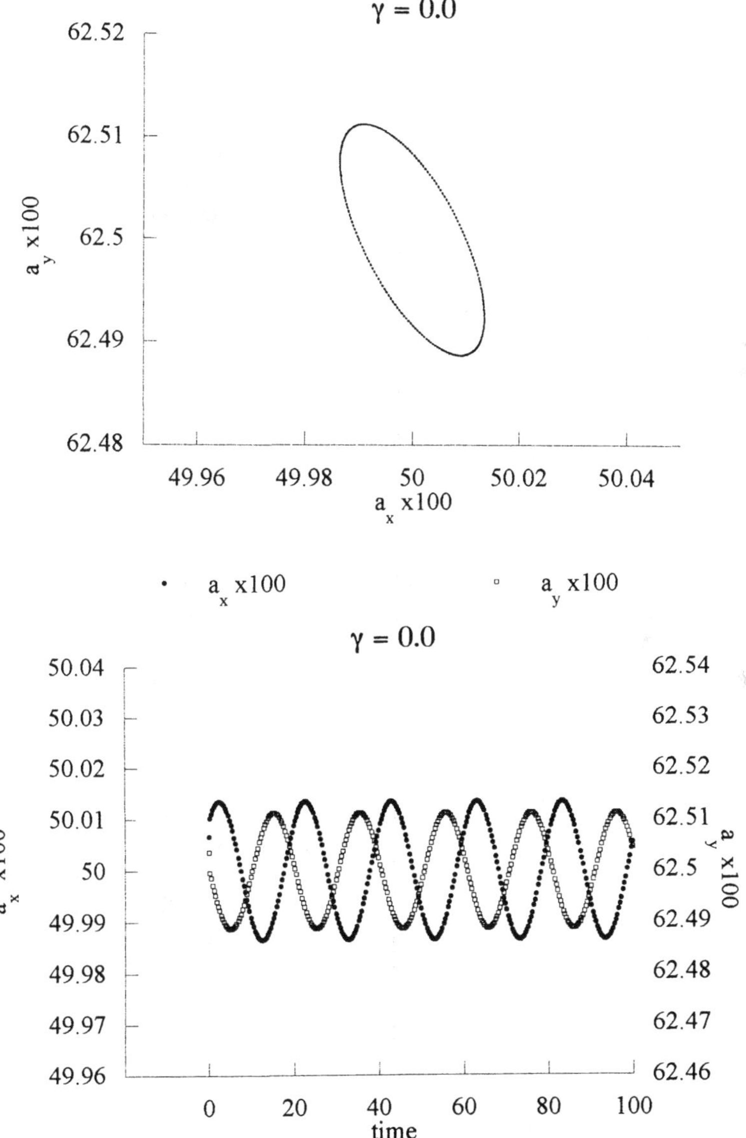

Fig. (1) If the initial point is very close to the critical point, the pair of activities a_X, a_Y executes simple harmonic motion; the upper picture shows the orbit, a_X versus a_Y, and the lower picture shows the time-development of both activities, with the values of a_X on the left ordinate, and the values of a_Y on the right.

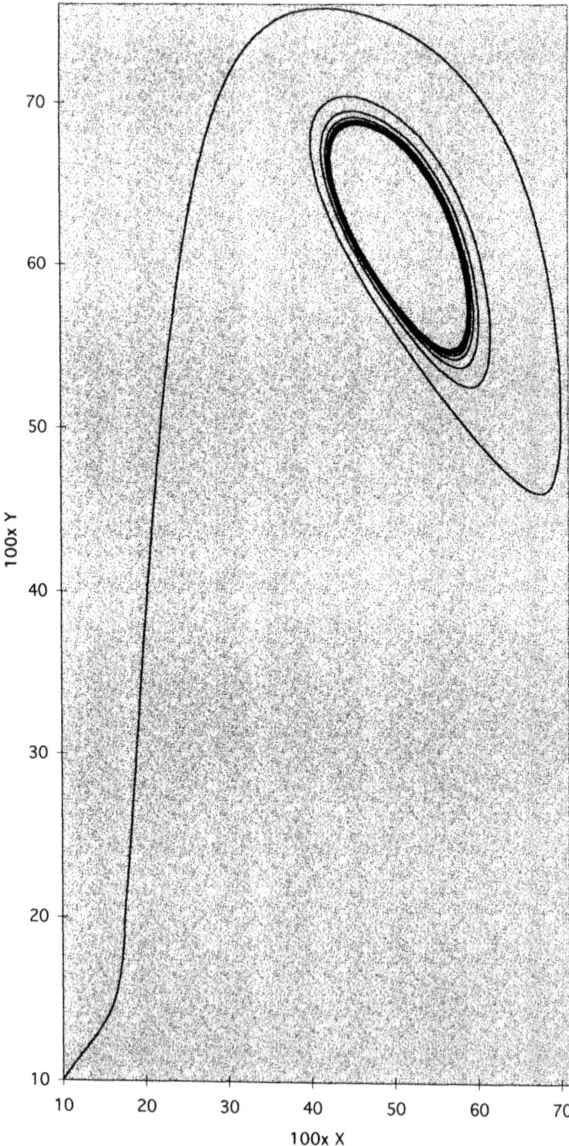

Fig. (2) If the initial point is far from the critical point then the orbit rapidly homes in to the limit cycle.

6.4 How to Add Noise

When a real system is subject to random interference we say that it is subjected to *noise.* A mathematical model of the system without noise must be modified if it is to describe the noisy situation. In a classical theory of particles obeying Newton's laws, it is possible to add a random force to the deterministic force of a particle, whose motion is then governed by a stochastic differential equation, the Langevin equation. The nature of the random interference is modelled by choosing the random force appropriately. Various choices are white noise, coloured noise and thermal noise. One can also try "multiplicative" noise, in which one of the dynamical variables is multiplied by a random term, or the adding of a random term to the Hamiltonian, with its consequent affect on the equations of motion. One of the interesting questions in the theory is then whether the addition of the noise has changed the salient features of the dynamics ([53, 54]). For example, if the system without noise exhibits a bifurcation to periodicity or chaos, is the value (of the control parameter) at which bifurcation occurs altered in the presence of noise? In statistical dynamics we have replaced the particle picture by the field picture, and the Newton's laws by a motion through the state space $\Sigma(\Omega)$, and the traditional description of noisy systems does not appear to be possible. In this section, we present a natural way to introduce noise into statistical dynamics, in a way that is consistent with the first and second laws of thermodynamics. Our construction is not obviously related to a stochastic differential equation; it is, however, similar to the theory of noisy neural nets ([55]), and this might justify our use of the term "noise". Suppose that the system is described by the sample space Ω_c and energy $\mathcal{E}_c$, and a dynamics τ. Our procedure is to extend the system by introducing a heat-particle, and corresponding space Ω_γ, with energy-function $\mathcal{E}_\gamma$, and an energy-conserving bistochastic map T which gives the dynamics on the product space $\Omega = \Omega_c \times \Omega_\gamma$. Let τ_β be the corresponding isothermal dynamics of the combined system. If τ_∞ coincides with the original dynamics, τ, then we say that τ_β is (one model of) the system with thermal noise. The noise allows transitions to occur that cannot go in the original model, because of energy conservation. The probability of such transitions is related to the probability of the inverse by the Boltzmann formula (Arrhenius factor) $e^{-\beta W}$, where W is the energy gap.

We now illustrate the idea with a study of the oscillating chemical reaction of the previous section. The equations for the activity-led isothermal reactions

$$B + X \overset{k_1}{\rightleftharpoons} Y + \gamma\,, \tag{6.26}$$

$$X \overset{k_2}{\rightleftharpoons} \gamma\,, \tag{6.27}$$

$$2X + Y \overset{k_3}{\rightleftharpoons} 3X + \gamma\,, \tag{6.28}$$

$$A \overset{k_4}{\rightleftharpoons} X + \gamma\,, \tag{6.29}$$

are

$$\begin{aligned}
\frac{da_X}{dt} &= \{k_1(a_Y a_\gamma - a_B a_X) - k_2(a_X - a_\gamma) + k_3(a_X^2 a_Y - a_X^3 a_\gamma) \\
&\quad + k_4(a_A - a_X a_\gamma)\}(1 - a_X)^2\,, \qquad (6.30)\\
\frac{da_Y}{dt} &= \{k_1(a_B a_X - a_\gamma a_Y) - k_3(a_X^2 a_Y - a_X^3 a_\gamma\}(1 - a_Y)^2\,, \qquad (6.31)
\end{aligned}$$

which reduce to Eqn. (6.16) when $a_\gamma = 0$. We take the same values of the parameters k_i and a_A, a_B as we had in the previous section, so that with $a_\gamma = 0$ we get a limit cycle. We also denote by $\mathring{a}_X$ and $\mathring{a}_Y$ the same values, 1/2 and 5/8 as before. Note that these are no longer the co-ordinates of a fixed point of the dynamics if $a_\gamma > 0$. The new fixed point will be close to this if a_γ is small, by continuity. The correction to $(\mathring{a}_X, \mathring{a}_Y)$ will be a power series in a_γ beginning with the first power. The matrix A will also be modified, this time exactly by a matrix linear in a_γ, thus:

$$A(a_\gamma) = A + a_\gamma B.$$

This is because the equations of motion are linear in a_γ. We are thus left with problem of finding the new Lyapunov index for the new fixed point. I now show that we can find the Lyapunov index without finding the new fixed point, even up to order a_γ. Indeed, write as before,

$$a_X = \mathring{a}_X + X \qquad\qquad a_Y = \mathring{a}_Y + Y$$

and $\mathbf{Z} = (X, Y)$; then to first order in $\mathbf{Z}$, but exact in a_γ, the equations of motion take the form

$$\dot{\mathbf{Z}} = (A + a_\gamma B)\mathbf{Z} + a_\gamma \mathbf{b}.$$

Clearly, the fixed point has been displaced from $\mathbf{Z} = 0$ to

$$\mathbf{Z} = -a_\gamma A(a_\gamma)^{-1}\mathbf{b} = \mathbf{Z}_0$$

say. Then make a shift of dependent variable from $\mathbf{Z}$ to $\mathbf{Z}' = \mathbf{Z} - \mathbf{Z}_0$ and the resulting equation for $\mathbf{Z}'$ is homogeneous with the same matrix $A(a_\gamma) = A + a_\gamma B$. If a_γ is small, the eigenvalues of $A(a_\gamma)$ will be close to those of A, so they are complex conjugates of each other with the same real part. The Lyapunov index is $Tr\ A(a_\gamma)/2$, as can be seen by finding the solutions of

$$\det\begin{pmatrix} a - \lambda & b \\ c & d - \lambda \end{pmatrix} = 0 \tag{6.32}$$

in the case when there are complex roots. In our case, $TrA = 0$, (the Lyapunov index for the system without noise was zero), so $\mathrm{Tr}\,(A + a_\gamma B) = a_\gamma Tr\, B$. It therefore suffices to find the diagonal elements of B, which means finding, respectively, the coefficient of $a_\gamma X$ on the right-hand side of Eqn. (6.30) and the coefficient of $a_\gamma Y$ on the right-hand side of Eqn. (6.31). Remember that $a_X = \mathring{a}_X + X$, $a_Y = \mathring{a}_Y + Y$. Exercise (6.37) shows that the sought terms are

$$a_\gamma X: \ (1 - \mathring{a}_X)^2(-3k_3 \mathring{a}_X^2 - k_4) - 2(1 - \mathring{a}_X)(k_1 \mathring{a}_Y + k_2 - k_3 \mathring{a}_X^3 - k_4 \mathring{a}_X) ,$$

$$a_\gamma Y: \ (1 - \mathring{a}_Y)[-k_1(1 - \mathring{a}_Y) + 2k_1 \mathring{a}_Y - 2k_3 \mathring{a}_X^3] ,$$

which gives for the matrix B, using the values of the model:

$$B = \begin{pmatrix} -149a_\gamma/32 & \bullet \\ \bullet & 57a_\gamma/64 \end{pmatrix} . \tag{6.33}$$

Thus the Lyapunov index for the noisy system is $\lambda = (-149/32 + 57/64)a_\gamma/2$, which is negative. Thus the new fixed point has been converted into an attractor by the noise. This is illustrated in Fig. (3); we see that the orbit spirals in, and that as functions of time the activities of X and Y decay to their values at the fixed point.

This model shows that we can get a dynamic phase transition: if the parameters were not quite those of the model, but were such that when $a_\gamma = 0$ the Lyapunov index was small and positive, then at zero temperature the fixed point would be a repeller, and the system would spiral out and away for large times. As we increase the temperature, the Lyapunov index would decrease until it reached zero, beyond which it is an attractor. This defines a critical temperature separating two different asymptotic behaviour of the system, at least for motions beginning near the fixed point.

6.5 Exercises

Exercise 6.34 In the driven chemical activity-led reactions given by Eqn. (6.16), show that the condition for a limit-cycle is

$$k_3 k_4^2 a_A^2 k_2^{-2}(1 - \mathring{a}_Y)^2 = (k_1 a_B - k_2)(1 - \mathring{a}_X)^2 \tag{6.35}$$

and that the eigenvalues are then $\pm(1 - \mathring{a}_X)(1 - \mathring{a}_Y)k_4 a_A (k_3/k_2)^{1/2}$.

Exercise 6.36 Consider the model obtained from that in Section (6.3) by replacing $2X + Y \rightleftharpoons 3X + \gamma$ by $X + Y \rightleftharpoons 2X$. Show that the two eigenvalues of the linearised equation are real.

Exercise 6.37 For the noisy model of Section (6.4), show that the matrix B has diagonal elements $-149a_\gamma/32$ and $57a_\gamma/64$.

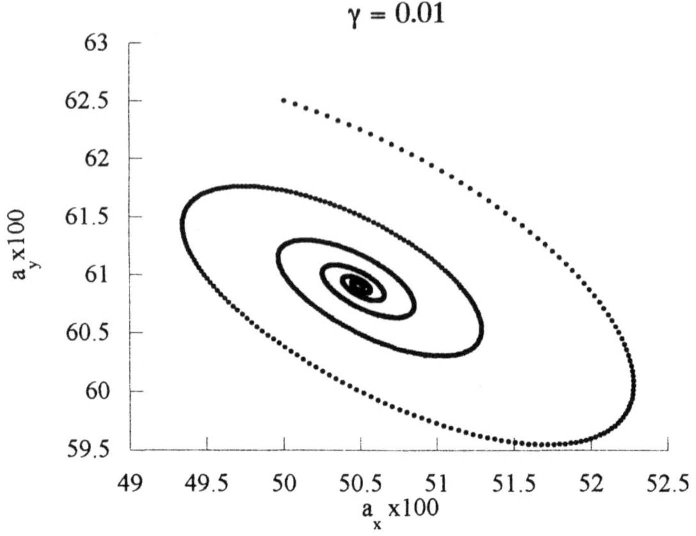

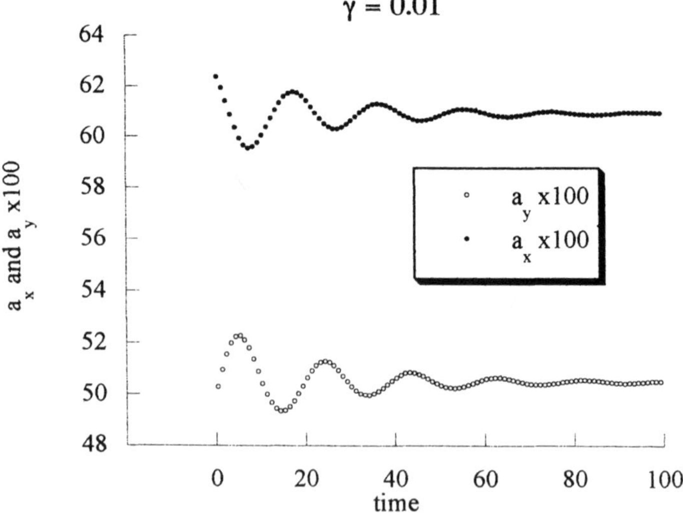

Fig. (3) In the presence of noise, the limit cycle has disappeared and the neutral fixed point has become an attractor. The upper picture shows the orbit, and the lower picture shows the time development. The value of a_γ is 0.01.

Part II

Quantum Statistical Dynamics

Chapter 7

Introduction

The theory described in Part I is based on classical probability theory, and has applications to the theory of neural nets, and to artificial ecologies, as well as to physics and chemistry. However, molecules obey the statistical laws of quantum mechanics, and it is the aim of Part II to develop a parallel theory based on *quantum probability*, which is a true generalisation of probability theory. It cannot be so clearly asserted that quantum *mechanics* is a true generalisation of classical *mechanics*. In Part I we were able to give a good account of thermodynamics by imitating the ideas of quantum mechanics, such as the field point of view, and taking the energy-levels to be discrete. This led to important improvements over the traditional particle theories. The classical theory suffers from two main deficiencies, even when improved in this way. The first is that there is no way to predict the size of atoms, even though there is no problem in putting the observed size into the lattice spacing of Λ. This contrasts with the quantum theory of the atom, whose size is determined by Schrödinger's theory. The second trouble is that the transport equations we get from our theory contain no "flow" term, of the form $\mathbf{v}.\boldsymbol{\nabla}$ which describes convection. This was because we had got rid of the velocity of individual particles as one of the labels, replacing it with kinetic energy. We can get a diffusion term, but not this. Convection requires a close collaboration between material over a large range, and the independent statistical jumps at each $x \in \Lambda$, typical of our theory so far, is ill-adapted to describe this. Any collaboration is destroyed by the stoss map. This problem will find its solution within quantum field theory.

In Chapter 8 we introduce quantum probability, through the concept of a probability algebra, taking the states to belong to the dual space. In the classical case we did it in the other order, and this is a possible way in to the subject [56]. The state-space is a rather complicated object in quantum probability, and it is hard to justify its definition as a primary object. In this we are convinced by the argument of I. E. Segal [57]. The algebra, furnished with a conjugation,

$A \mapsto A^*$ is postulated to be a C^*-algebra with identity, which is a natural generalisation of the function algebras occurring in Part I. Indeed, the algebra of bounded complex functions on a (countable) sample space is an abelian, that is, commutative, C^*-algebra, with * given by complex conjugation, and whose real random variables are self-adjoint elements. To maintain the entirely elementary nature of this book, we shall concentrate on models with discrete space-time and usually choose models based on matrix algebras, though some functional analysis is needed for some infinite models. The idea of combining systems to form larger systems goes through very smoothly by using tensor products. This allows us to define algebras with local structure, and also to define the concepts of independence and marginal states. Traditional quantum mechanics is recovered by choosing $\mathcal{A}$ to be the set $B(\mathcal{H})$ of all bounded operators on a separable Hilbert space, $\mathcal{H}$, say. In any C^*-algebra there is the important subset, the positive cone, which is the convex set of elements of the form A^*A. In the finite-dimensional case the dual space $\mathcal{A}^d$ to the algebra is, as usual, the set of linear functionals on the algebra. Within $\mathcal{A}^d$ can be found the set of states Σ, which are the normalised elements of the cone dual to the positive cone in $\mathcal{A}$. It is a convex set, but is not a simplex unless $\mathcal{A}$ is abelian. We interpret the functional as the expectation of the observable if the system is in the given state. The extremal elements of Σ, the pure states, can be identified with expectations $\langle\psi, A\psi\rangle$ in some normalised vector as in wave mechanics. The non-extremal elements are the mixed states, corresponding to density operators. The desired thermal states will be among these.

In the commutative case the set of normal states Σ can be identified with the set of probability measures on a set, and the functional as the expectation value in the classical sense. The quantum versions of independence and so on then coincide with those already met in Part I. In this way, we arrive at classical probability, but not classical mechanics, as a subtheory of quantum probability. Quantum statistical dynamics is then taken on a similar path to that taken in Part I. We must give an energy, which is a self-adjoint operator in $\mathcal{A}$ in the finite-dimensional case, but could be an unbounded operator H on the Hilbert space $\mathcal{H}$ if $\mathcal{A} = B(\mathcal{H})$ and $\dim \mathcal{H} = \infty$. In this case we make a similar assumption to that needed in the classical case, to ensure thermodynamic stability: $e^{-\beta H}$ has a finite trace. This allows us to define the canonical state. One idea that is essential in quantum theory, but which is not much used in classical probability, is the relation between the states and the cyclic representations of the algebra. This goes under the names of Gelfand, Naimark and Segal, abbreviated to GNS, and is briefly reviewed. The Chapter ends with an account of von Neumann's entropy; we have analogues of several classical estimates for the entropy which will provide the technical results needed to do quantum statistical dynamics. In particular, the stoss map is defined, and is shown to be entropy-increasing.

In Chapter 9 we give a treatment of linear dynamics entirely parallel to that in Part I. It is called linear not because the equations of motion are linear in the dynamical variables, but because it is linear in the state, as in ordinary quantum mechanics. A reversible time-evolution is a homomorphism τ from the group **R**, or **Z** if time is discrete, into the group of automorphisms of the algebra. Clearly, the dynamics given by a continuous one-parameter group commutes with its own generator (the Hamiltonian); so energy is conserved. Naturally, entropy is not changed by reversible dynamics; so we get both laws of thermodynamics, but in a trivial way. The simple act of forming random mixtures of reversible dynamics gives us an irreversible dynamics, in which entropy increases. Thus we get a realistic form of the second law. This is taken as the model for the concept of stochastic map and bistochastic map, and we follow the method of Part I to ensure that the first law holds. The problem is that a random unitary group, generated by $H_V = H_0 + V$, where V is a random potential, will not commute with H_0. We get the first law by a different interpretion; the random unitary operators are scattering operators, corresponding to the evolution of the system, with Hamiltonian H_V, for a very large time (in microscopic units). It is known that for local potentials the scattering operator commutes with the asymptotic energy; it is this operator H that is taken to be the energy that is conserved by the dynamics. The prototype of linear dynamics is thus a mixture of energy-conserving automorphisms. More generally, we consider one time-step to be given by a bistochastic map T leaving each spectral projection of H invariant. We stress the importance of adding one further property, which is automatic in the classical case, and that is complete positivity. We derive Kraus' form for a completely positive map in a simple way if the dimension is finite. This gives a way of constructing satisfactory quantum theories obeying both laws and converging to the microcanonical state at large time.

Chapter 10 starts with the Boltzmann map, which is the composition of the stoss map with a bistochastic map. The resulting dynamics is unusual, in that it is non-linear in the state. This brings in enough randomness to ensure that the system converges to the canonical state rather than the microcanonical state. Two versions are proved; the first uses Lyapunov's theorem, and the second, in which we assume a spectral gap in the dynamics, shows more, in that the rate of convergence is exponential. The next section (10.2) introduces the heat-particle, which is merely an oscillator, or a collection of such. We study the theory of Bosonic oscillators on Fock space, using coherent states. We are able to construct the second quantising map, of which E. Nelson said, "whereas first quantisation is a mystery, second quantisation is a functor". We define the Weyl system, and the generating function of a state. This is the quantum analogue of the characteristic function of a probability measure. The n-point functions are then the analogues of the mixed moments of several random variables. There are also analogues of the cumulants. We define the quasifree states as states with zero cumulants beyond the second; they are therefore the analogues of the Gaussian distributions. We show that the canonical state of a quadratic Hamiltonian is quasifree. The quasifree state is the state of maximum entropy having the given two-point functions. There is also an explicit formula for the entropy of a quasifree state when the modes of the Hamiltonian are known. Section (10.3) does a similar analysis for Fermions, using the real Clifford algebra with complex structure. Again, the quasifree state is the state of largest entropy having the given two-point functions. Thus, in both the Bosonic and Fermionic case the quasifree map, which replaces a state by the quasifree state with the same two-point functions, is entropy-increasing. This provides a natural way to add randomness to a state without changing its particle-density or its energy (if the latter is quadratic). It also leads to a reduced description, since one needs only to parametrise the quasifree states. One can incorporate the conservation of particles by looking at the subset of gauge-invariant states. However, in chemistry the number of molecules is not conserved, and we must introduce fields for the atoms rather than for the molecules. Then the quasifree states form a poor approximation to the canonical state, since even the free Hamiltonian is not quadratic in the atomic fields. This motivates a new algebra, isomorphic to $\mathbf{M}_{n+1}$, which I call the "hard-core algebra", in which only one particle is allowed to occupy any site. This exhibits a more severe exclusion principle than Fermions, since two or more different Fermions are not excluded from being in the same state. Finally, it is shown that a bistochastic map followed by the quasi-free map can lead to the quantum Boltzmann equation for Bosons and Fermions.

Chapter 11 moves from the isolated dynamics to the isothermal dynamics by restoring the state of the heat-particle after each time-step to a canonical state. The resulting dynamics is again linear in the state of the system. The main result is the quantum free-energy theorem. We show that the dynamics is a contraction in a norm, the quantum analogue of the Fisher metric. This enables us to prove a convergence theorem to equilibrium. Driven systems are then easy to formulate (but not to solve). We just restore a certain set of local intensive variables back to their original values after each time-step. We can then add source terms. This is illustrated by the Fröhlich pumped phonon model.

In the final chapters we are more ambitious; we study systems in continuous time and infinite volumes, but still with discrete space. We start with the algebra of the infinite system, defined by the inductive limit of the local algebras. Then we show how the Hamiltonian, defined only locally, can be used to give a reversible time-evolution as a continuous group of automorphisms of the algebra. We describe part of the theory of Haag, Hugenholtz and Winnink, on the thermal states of the infinite volume. I present an argument as to why a state obtained as a small perturbation of the thermal state always converges back to the same thermal state under the reversible dynamics. This is both a triumph and tragedy; a triumph because we have obtained irreversibility from a reversible dynamics (by the idealisation of the infinite-volume); a tragedy because we have lost the first law: added energy is lost to infinity, and does not lead to any change in beta. We then study dissipative systems in continuous time, and give a simple derivation of the form of the generator of a completely positive semi-group, due to Gorini, Kossakowski and Sudarshan, and Lindblad. I give a possible class of dissipative dynamics for the Heisenberg ferromagnet in infinite volume, which should converge to the canonical state. There is a section on the Jaynes-Ingarden theory of information, and Ingarden's concept of temperatures of higher order. This enables us to invent entropy-increasing maps which are much more versatile than the quasi-free map, and which lead to a reduced description. These are used to define a dissipative dynamics for the Ising model, which is proved to return to equilibrium if disturbed by a small amount. The dynamics is non-linear, however, and a finite system in a canonical state will converge to a different beta if it is disturbed. This is an improvement over the theory in an infinite volume, which is physically ambiguous because of the lack of a boundary. We end with a dissipative dynamics for the Heisenberg ferromagnet in which the reversible dynamics is modified by lost information. The rate of convergence is then determined by the Hamiltonian and the level of description alone; there is no other dissipation. This leads to a differential equation determined by the initial state, whose solution might be called a non-linear quantum stochastic process.

Chapter 8

Quantum Probability

8.1 Algebras of Observables

It is best to start with the idea of an observable, and the quantum analogue of the random variables, and to define the states as a derived concept, rather than the other way round. This is because the state space in quantum theory is very complicated. In place of the algebra of random variables on the chosen sample space, quantum theory uses a C^*-algebra. We shall denote the typical C^*-algebra by $\mathcal{A}$, and the typical element by A or B. Thus $\mathcal{A}$ is a vector space over the complex field, obeying the following axioms:

Definition 8.1

1. *$\mathcal{A}$ is furnished with a product, which is associative, and distributive relative to addition, and obeys the usual laws of algebra except possibly commutativity. The elements of $\mathcal{A}$ commute with the number field.*

2. *$\mathcal{A}$ is furnished with a norm, which makes it into a Banach space.*

3. *$\mathcal{A}$ is furnished with a conjugation $*$; this is a map $A \mapsto A^*$ which is antilinear and obeys $(AB)^* = B^*A^*$ for all A and B in $\mathcal{A}$.*

4. *The norm obeys the "C^*-identity" $\|AA^*\| = \|A\|^2$.*

The example that will be used in the next few chapters is $\mathcal{A} = \mathbf{M}_n$. This is the set of $n \times n$ complex matrices; the algebraic rules of addition and multiplication are those of matrix algebra, and the *-operation is that of taking the Hermitian conjugate. The norm is the largest singular value, that is, the largest eigenvalue of $|A| = (A^*A)^{1/2}$. Exercise (8.25). This algebra is also called the full matrix algebra of dimension n. More generally, if $\mathcal{H}$ is a Hilbert space, then the set of all bounded linear operators on $\mathcal{H}$ is a C^*-algebra, denoted $B(\mathcal{H})$. The *

operation is the adjoint, also known as Hermitian conjugate, and the norm is the operator norm

$$\|A\| = \sup\{\|A\psi\| : \|\psi\| = 1\}.$$

If $\dim \mathcal{H} = n < \infty$ then $B(\mathcal{H})$ is isomorphic to $\mathbf{M}_n$; see Exercise (8.26).
We shall always take it that our algebra contains the identity, denoted by 1. An element A of $\mathcal{A}$ is said to be Hermitian if $A = A^*$. The Hermitian elements will represent observables, and they take the place of the (real-valued) random variables in classical probability theory. If $A \in \mathbf{M}_n$, and is Hermitian, then A has real eigenvalues, which form the spectrum of A. This is denoted $\operatorname{Spec} A$. The statistical interpretation of quantum mechanics forces us to the conclusion that any measurement of A will lead to a value in $\operatorname{Spec} A$. In a more general C^*-algebra, the spectrum is defined as follows. The *resolvent set* of any element A is the set of complex numbers λ such that $(A - \lambda 1)$ is invertible. If λ lies in the resolvent set, then $(A - \lambda 1)^{-1}$ exists and is an element of $\mathcal{A}$, called the *resolvent* of A. The complement in $\mathbf{C}$ of the resolvent set is called the spectrum of A. This agrees with the set of eigenvalues when A is a matrix, since the eigenvalues are exactly the values of λ for which $\det(A - \lambda 1) = 0$, so no inverse exists. In the general C^*-algebra, an operator may have a continuous part to the spectrum, as well as some isolated points. The interpretation of Hermitian elements as observables is supported by a theorem which says that the spectrum of a Hermitian element is a subset of $\mathbf{R}$. It is possible, even at this abstract level, to define certain analytic functions of an element. The definition of a function of several elements is however only possible if they mutually commute. Thus, if $A \in \mathcal{A}$ and $f = \sum f_n z^n$ is a function of a complex variable, analytic in a circle $\{z : |z| < r\}$, then we may define $f(A) = \sum f_n A^n$ if $\|a\| < r$. We use the fact that the series for $f(z)$ converges for $|z| < r$ to show that the partial sums in the series for $f(A)$ form a Cauchy sequence in the norm if $\|A\| < r$. Then the series converges to an element of $\mathcal{A}$, by the axiom of completeness. We note that $(z - \lambda)^{-1}$ is an analytic function of z in the circle $\{z : |z| < |\lambda|\}$. In this way we can define the inverse of $A - \lambda 1$ if $|\lambda| > \|A\|$, proving that $\operatorname{Spec} A$ is a subset of $\{\lambda : |\lambda| \leq \|A\|\}$. A C^*-algebra always contains Hermitian elements; indeed, any element A can be written as $A = (A + A^*)/2 + i(A - A^*)/(2i)$, just as any complex number can be written as the sum of its real and imaginary parts. Another important class is the set of unitary elements; an operator $U \in \mathcal{A}$ is said to be unitary if it is invertible, and $U^{-1} = U^*$. The set of unitary elements in a C^*-algebra is a group (Exercise(8.27)). If $A \in \mathcal{A}$ is Hermitian, then e^{iA} is always defined, and is unitary, with inverse e^{-iA}. Thus any C^*-algebra always contains many unitary elements. The same cannot be said about (Hermitian) projections, that is elements $P \in \mathcal{A}$ obeying $P = P^*$ and $P^2 = P$. The spectrum of a projection consists of the two points, $\{0, 1\}$, unless $P = 0$ or $P = 1$. It therefore represents a "question", that is, an observable with a yes-no answer.

The simplest C^*-algebra after $\{0\}$ and $\mathbf{C}$ is $\mathbf{M}_2$, the set of 2×2 complex matrices. This is four-dimensional, with a basis given by the matrix units

$$E_{11} = \begin{pmatrix} 1 & 0 \\ 0 & 0 \end{pmatrix} \quad E_{12} = \begin{pmatrix} 0 & 1 \\ 0 & 0 \end{pmatrix} \quad E_{21} = \begin{pmatrix} 0 & 0 \\ 1 & 0 \end{pmatrix} \quad E_{22} = \begin{pmatrix} 0 & 0 \\ 0 & 1 \end{pmatrix}. \tag{8.2}$$

Any 2×2 matrix is a sum of these with complex coefficients. An important alternative to this basis is that comprising the Pauli matrices and the unit matrix:

$$\sigma_0 = \begin{pmatrix} 1 & 0 \\ 0 & 1 \end{pmatrix} \quad \sigma_x = \begin{pmatrix} 0 & 1 \\ 1 & 0 \end{pmatrix} \quad \sigma_y = \begin{pmatrix} 0 & -i \\ i & 0 \end{pmatrix} \quad \sigma_z = \begin{pmatrix} 1 & 0 \\ 0 & -1 \end{pmatrix}. \tag{8.3}$$

The operators $s_x = \hbar\sigma_x/2$, $s_y = \hbar\sigma_y/2$ and $s_z = \hbar\sigma_z/2$ are used to represent the three components of the vector observable known as the spin, attributed to particles such as electrons, protons and neutrons, and some atoms. These matrices do not commute, but have the commutation relations of the Lie algebra $su(2)$: $[\sigma_x, \sigma_y] = i\hbar\sigma_z$. Let $\boldsymbol{\theta} \in \boldsymbol{R}^3$. Then the exponential $e^{i(\theta_x\sigma_x+\theta_y\sigma_y+\theta_z\sigma_z)}$ is unitary, forming the group $SU(2)$. The rest of the unitary group of $\mathbf{M}_2$ consists of multiples of the identity $e^{i\theta_0}1$. Since the eigenvalues of any of the components is $\hbar/2$, we say that these particles "have spin 1/2" (in units of $\hbar$).

There are many beautiful results within the theory of C^*-algebras; for example, it can be shown that any isomorphism is continuous, and that any derivation is bounded. However in two respects the theory of C^*-algebras is too general as a framework for quantum mechanics: first, a C^*-algebra may contain no projections except 0 (and 1 if it has a unit), whereas Hermitian projections play an important role in quantum probability; and secondly, the tensor product of two C^*-algebras can sometimes be ambiguous, in that there are several norms on the algebraic tensor product all with the C^*-property. Tensor products play an important role in constructing compound systems. Neither of these problems arises in the special class of C^*-algebras known as von Neumann algebras: such an algebra has so many projections that it is generated by them; and there is a natural choice for the norm of a tensor product. However spin systems on an infinite space Λ, as well as Bosons and Fermions, are best described by certain C^*-algebras that are not von Neumann algebras. We do not meet these two problems in these special cases.

An element A of a C^*-algebra of the form $A = B^*B$ is said to be positive. The spectrum of a positive element lies inside the set $\{x \geq 0\}$. The converse also holds (Kaplansky's theorem, [58]). It can be proved that the set of positive elements is convex. It is obviously a cone, since if $A = B^*B$ is positive, then $\lambda A = (\lambda^{1/2}B)^*(\lambda^{1/2}B)$ is also positive, for any $\lambda > 0$. The positive cone in a C^*-algebra plays the role of positive random variables in the theory of probability.

There are two constructions with C^*-algebras which allow us to combine two algebras and get another; these are the direct sum and the tensor product. A C^*-algebra is a vector space (with further structure) and the *direct sum* of two, $\mathcal{A}_1$ and $\mathcal{A}_2$ is the set of pairs (A_1, A_2), with multiplication by a number λ given by $\lambda(A_1, A_2) = (\lambda A_1, \lambda A_2)$. We say that this is "componentwise". Addition and multiplication are also componentwise:

$$\begin{aligned}(A_1, A_2) + (A_1', A_2') &= (A_1 + A_1', A_2 + A_2') \\ (A_1, A_2).(A_1', A_2') &= (A_1 A_1', A_2 A_2').\end{aligned}$$

We denote the element (A_1, A_2) of the direct sum by $A_1 \oplus A_2$. The norm of an element $A_1 \oplus A_2$ is defined to be $\max\{\|A_1\|, \|A_2\|\}$. The direct sum of two vector spaces $\mathcal{A}_1$ and $\mathcal{A}_2$, which in our discussion happen also to be algebras, is denoted $\mathcal{A}_1 \oplus \mathcal{A}_2$. It can be shown that any finite-dimensional C^*-algebra with an identity is the direct sum of full matrix algebras. If the algebra is an algebra of matrices, this means that any element of the algebra can be written in block diagonal form.

The vector space underlying the *algebraic* tensor product of two C^*-algebras is just the tensor product of the underlying vector spaces. *The* tensor product is the completion of the algebraic tensor product in a suitable norm. If one or more of the algebras is of finite dimension, the completion is not needed, as the algebraic tensor product is already complete. So let us now give the definition of the tensor product (over the complex field) of two vector spaces of any dimension. We start with vector spaces V_1 and V_2, and form the set $V = V_1 \times V_2$ of ordered pairs of elements. Next we form the space $\operatorname{Span} V$, of formal finite sums of elements of V, with complex coefficients. We have already met this construction in Sect. (2.1). This space is a vector space of enormous dimension, one for each pair of vectors, taken from V_1 and V_2, and not just from a chosen basis in each. We get the space we want, the tensor product, by forming the quotient vector space of $\operatorname{Span} V$ by a rather large subspace, formed by combinations in $\operatorname{Span} V$ that should be zero by the rules of the tensor calculus. Under these rules, the following elements of $\operatorname{Span} V$ are equivalent to zero, for any vectors v_i and numbers λ:

$$\begin{aligned}(\lambda v_1, v_2) &- \lambda(v_1, v_2)\,, \\ (v_1, \lambda v_2) &- \lambda(v_1, v_2)\,, \\ (v_1 + v_1', v_2) &- (v_1, v_2) - (v_1', v_2)\,, \\ (v_1, v_2 + v_2') &- (v_1, v_2) - (v_1, v_2')\,. \end{aligned} \tag{8.4}$$

Let V_0 be the vector subspace of $\operatorname{Span} V$ spanned by all elements of the form of one of the four lines in Eqn. (8.4). Then we define the *algebraic tensor product* to be the quotient space:

$$V_1 \otimes V_2 = \text{Span}\,(V_1 \times V_2)/V_0 \ . \tag{8.5}$$

(I did not believe it either, when I first saw it written, even though it was in a book.) The equivalence class containing (v_1, v_2), and thus all elements equivalent to it, is denoted $v_1 \otimes v_2$. As usual, we add equivalence classes by adding representative elements, one from each, the addition here being in V, and then we find the equivalence class of the sum. This procedure is independent of the choice of representative. Similarly, we can define the multiplication of an equivalence class by a complex number. That this all works out is part of the theory of the quotient space V/V_0. Because of the definition of V_0 by Eqn. (8.4), the tensor product satisfies corresponding relations

$$\begin{aligned}
(\lambda v_1) \otimes v_2 - \lambda(v_1 \otimes v_2) &= 0 \ , \\
v_1 \otimes (\lambda v_2) - \lambda(v_1 \otimes v_2) &= 0 \ , \\
(v_1 + v_1') \otimes v_2 - v_1 \otimes v_2 - v_1' \otimes v_2 &= 0 \ , \\
v_1 \otimes (v_2 + v_2') - v_1 \otimes v_2 - v_1 \otimes v_2' &= 0 \ .
\end{aligned}$$

If now we have two C^*-algebras, $\mathcal{A}_1$ and $\mathcal{A}_2$, then the structure of $\mathcal{A}_1 \bigotimes \mathcal{A}_2$ as a vector space is just that of the tensor product of the vector spaces $\mathcal{A}_1$ and $\mathcal{A}_2$. In particular, every element of $\mathcal{A}_1 \bigotimes \mathcal{A}_2$ is a finite sum of dyadic tensors, that is, elements of the form $A \otimes B$. To furnish the tensor product with a *-operation and a product, we define them first on dyadic elements:

$$(A \otimes B)^* = A^* \otimes B^*$$

$$(A \otimes B)(A' \otimes B') = AA' \otimes BB'$$

by choosing representatives. We extend these by anti-linearity (for *) and bi-linearity (for the product) to all elements of the tensor product. The easy but rather long calculations needed to verify that the product is the same, whichever representatives are chosen, will be done (by you) in Exercise (8.29).

To put a norm on $\mathcal{A}_1 \bigotimes \mathcal{A}_2$, we encounter an ambiguity unless one of the algebras is of finite dimension. We shall only discuss here the case when $\mathcal{A}_1 = \mathbf{M}_n$ and $\mathcal{A}_2 = \mathbf{M}_m$. Then it can be shown that $\mathbf{M}_n \bigotimes \mathbf{M}_m$ is isomorphic to $\mathbf{M}_{nm}$, and also to the algebra of $n \times n$ matrices whose entries are $m \times m$ matrices, which is denoted $\mathbf{M}_n(\mathbf{M}_m)$. By symmetry, all these are isomorphic to $\mathbf{M}_m(\mathbf{M}_n)$. These algebras have a natural C^*-norm, $\|A\|$ = the maximum eigenvalue of $(A^*A)^{1/2}$. Now any C^*-algebra has a unique norm, so this must be it. The reason why this does not work when both $\mathcal{A}_1$ and $\mathcal{A}_2$ are infinite-dimensional is that the algebraic tensor product is not complete in any C^* norm, and the uniqueness theorem mentioned only holds for C^*-algebras. Remember, Axiom

(2) requires the algebra to be a Banach space, which includes the completeness axiom. Finite-dimensional vector spaces are always complete.

The direct sum and tensor product can be done for more than two algebras, and is associative. This means that $(\mathcal{A}_1 \oplus \mathcal{A}_2) \oplus \mathcal{A}_3$ is isomorphic to $\mathcal{A}_1 \oplus (\mathcal{A}_2 \oplus \mathcal{A}_3)$, and that $(\mathcal{A}_1 \otimes \mathcal{A}_2) \otimes \mathcal{A}_3$ is isomorphic to $\mathcal{A}_1 \otimes (\mathcal{A}_2 \otimes \mathcal{A}_3)$. Distributivity also holds: $\mathcal{A}_1 \otimes (\mathcal{A}_2 \oplus \mathcal{A}_3)$ is isomorphic to $\mathcal{A}_1 \otimes \mathcal{A}_2 \oplus \mathcal{A}_1 \otimes \mathcal{A}_3$.

The physical meaning of the tensor product of algebras is the combination of the two systems they represent into one system. The two systems might be mixed up, or might still be well separated in distance; it is just that we wish to consider them both as part of the same theory. This construction is possible, whether or not the energy of the combined system is the simple sum of the energies of its parts (non-interacting systems) or whether a further energy of interaction must be added. The tensor construction does not imply that the two systems are statistically independent; this depends on whether the state (to be defined in the next section) factorises. But what we do need is that the observer, in contemplating making a measurement on the combined system, must have available all the measurements that are possible on say the first system. This is certainly true of the tensor product; the operators of the form $A \otimes 1$ as A runs over $\mathcal{A}_1$ make up a sub-algebra of $\mathcal{A}_1 \otimes \mathcal{A}_2$ that is isomorphic to $\mathcal{A}_1$. Similarly, the set of elements of the form $1 \otimes A$ with $A \in \mathcal{A}_2$ form a subalgebra of the tensor product isomorphic to $\mathcal{A}_2$. The mapping, $A \mapsto A \otimes 1$ is an ampliation, similar to that of regarding a given function of one variable, say x, as a function of two variables, say x, y, which happens to be independent of the second variable. This construction is common in classical probability theory. It is this tensor construction that does not seem to be possible in certain attempts to construct even more general theories than quantum mechanics; for example, there is no tensor product in the general theory of Jordan algebras, which have been advocated as a possible generalisation of C^*-algebras. Without the ampliation, a theory fails to be *local* as used in the philosophical debate over Bell's inequalities; a more recent and better word is *non-contextual*. If the mathematical object used to describe a certain measurement on a system, S_1, while not considering another system S_2, which happens to be far away, is A_1, but the mathematical object used to describe the *same* measurement on S_1, while contemplating both systems S_1 and S_2, is essentially different, say A_1', then the assignment of A_1 to that particular measurement is called "contextual". In a contextual assignment, the choice between A_1 and A_1' might depend on what observable is being measured on S_2, even though it is miles away. It is intended in these theories that A_1 and A_1' have the same statistical predictions for the first observable, that is, the same probability distribution. Even so, contextual theories are very odd: it is universal in mathematical modelling that the physical acts of measuring something are replaced at an early stage by a well-defined mathematical object. The theory is a theory of these objects, and is expressed in terms of these objects. If the assignment of mathematical objects to physical processes is allowed to slide around, and change with the context, then the theory is capricious; no definite conclusions can be drawn in such theories. Contextual variables were introduced into the debate on quantum mechanics as a result of Bell's inequalities, which say that the predictions of quantum mechanics (in a theory on a Hilbert space of dimension at least four) cannot be obtained from *any* classical probability theory. Instead of admitting that a more general probability theory (quantum mechanics) is needed, Bohm argued that we should use contextual random variables in an otherwise classical theory; this is an even more radical departure from probability theory than quantum mechanics itself. Indeed, contextual assignments of random variables have a drastic effect on the interpretation of the sample space: a certain outcome is represented by different sample points depending on what measurements are being made at other parts of space. This hardly meets the criterion of *reality* laid down by Einstein, Podolsky and Rosen [59], which was one of the original motives for this type of work. This is referred to as the EPR-experiment. In this sense, quantum mechanics is non-contextual, and realistic: the same mathematical object, the matrix A_1, or its ampliations, is always assigned to a given measuring process, and is called the "observable" in question. For example (and this is what is used in Bell's inequalities), if we have two far-separated particles of spin 1/2, the C^*-algebra used is $\mathbf{M}^2 \otimes \mathbf{M}^2$; the z-component of the spin of the first particle is described by the element $\hbar\sigma_z/2 \otimes 1$ irrespective of whether an observer is measuring the spin of the second particle in the z-direction, the x-direction, or not measuring it at all.

Quantum probability is a true generalisation of classical probability, and is not merely a different theory. (It cannot be said so clearly that quantum *mechanics* is a true generalisation of classical *mechanics*.) We shall show this for any countable sample space Ω. Indeed, the set of complex bounded functions on Ω, the complex random variables, form the Hermitian elements of a (commutative) C^*-algebra, and so form a special case of the set-up given here. In the next section we introduce quantum states, and show that any probability measure on Ω is a special type of quantum state, known as "normal", acting on a special type of algebra (an abelian one).

In quantum statistical dynamics, we have a finite set Λ of points in space, and to each $x \in \Lambda$ we are to assign a C^*-algebra $\mathcal{A}_x$, representing the observables that pertain to x. To the region Λ we associate the algebra

$$\mathcal{A}(\Lambda) = \bigotimes_{x\in\Lambda} \mathcal{A}_x \ .$$

If each algebra is the same full matrix algebra, this is the C^*-algebra of the lattice spin system. We shall see later that the existence of the ampliations between partial tensor products allows us to define the C^*-algebra for spins on an infinite lattice.

8.2 States

A given system can be in many different states, just as in classical probability the same sample space admits many probability measures. The state, then, is the description of the statistical properties of the system when prepared repeatedly in the same way. Following I. E. Segal [57] a *state* on a C^*-algebra is an assignment of the expectation value for every element of the algebra. This should obey the natural laws of linearity and positivity. Thus, let $\mathcal{A}$ be a C^*-algebra with identity; then a state on $\mathcal{A}$ is a linear map $\rho: \mathcal{A} \to \mathbf{C}$ obeying

- $\rho(1) = 1$.
- $\rho(\lambda A) = \lambda\rho(A)$ for all $\lambda \in \mathbf{C}$ and all $A \in \mathcal{A}$.
- $\rho(A + B) = \rho(A) + \rho(B)$ for all A and B in $\mathcal{A}$.
- $\rho(A) \geq 0$ for all A in the positive cone of $\mathcal{A}$.

For example, the expectation or mean in a classical theory is a state: let Ω be a countable sample space and let $\mathcal{A}$ be the C^*-algebra of all bounded complex-valued functions F on Ω. Then any probability $p \in \Sigma(\Omega)$ defines a state by $\rho(F) = \langle p, F\rangle$, of which the definition is $\sum_\omega p(\omega)F(\omega)$. The concept of state is more general than that of probability measure, since a state need not be countably additive. As another example, let $\mathcal{H}$ be a Hilbert space with scalar product $\langle \bullet, \bullet\rangle$, and put $\mathcal{A} = B(\mathcal{H})$; let $\psi \in \mathcal{H}$ be a normalised vector. Then

$$\rho_\psi(A) = \langle \psi, A\psi\rangle \tag{8.6}$$

is a state on $\mathcal{A}$. Such a state is called a *vector state*. A third example is that of density matrix; let $\mathcal{A} = \mathbf{M}_n$ and let δ be an $n \times n$ positive semidefinite matrix with unit trace. It is easy to see that the set of density matrices is a convex subset of $\mathbf{M}_n$. We get a state, called ρ_δ, by the map $A \mapsto \rho_\delta(A) = Tr\,(\delta A)$ (Exercise (8.31)). By the cyclic property of the trace, we can also write $\rho_\delta(A) = Tr\, A\delta$. This example can be generalised to infinite dimensions. Let $\mathcal{H}$ be a separable Hilbert space, and $\mathcal{A} = B(\mathcal{H})$. Let us say that a bounded operator δ is of trace class if the operator $(\delta^*\delta)^{1/2}$ has a complete set of normalised eigenfunctions ψ_i with eigenvalues $\{\lambda_i\}$ of finite multiplicity (except that 0, if an eigenvalue, might have infinite multiplicity), and if $\sum_i \lambda_i < \infty$, where we have counted each eigenvalue with its multiplicity. Then we define the trace of an operator δ of trace class to be

$$Tr\,\delta = \sum_i \delta_{ii} \qquad \text{where } \delta_{ii} = \langle \psi_i, \delta\psi_i \rangle.$$

It can be proved that this always converges absolutely, and that the answer is independent of the basis in which the calculation is done. A *density operator* is a positive semi-definite operator of trace class, whose trace is equal to 1. The set of density operators form a convex subset of $B(\mathcal{H})$. One can show that if A is bounded and δ is of trace class, then $A\delta$ and δA are of trace class, and their traces are equal. We can now use a density operator δ to define a state: $\rho_\delta(A) = Tr\,(\delta A)$. Not all states are of this form unless $\dim \mathcal{H} < \infty$, in which case they are. See Exercise (8.32). We are then able to define the inverse of the map $\delta \mapsto \rho_\delta$, and associate a density matrix, denoted by $\Delta\rho$, to the state ρ. In infinite dimensions, a state given by a density operator is said to be *normal*. So if ρ is normal, we denote by $\Delta\rho$ the (unique) density operator such that $\delta\Delta\rho = \rho$.

Let $\mathcal{A}$ be a C^*-algebra with identity and let $\Sigma(\mathcal{A})$ denote the set of states on $\mathcal{A}$. We can regard $\Sigma(\mathcal{A})$ as a subset of the space $\mathcal{A}^d$ of all linear functionals on $\mathcal{A}$ which has a natural linear structure. Then $\Sigma(\mathcal{A})$ is a convex subset of $\mathcal{A}^d$. Thus, if $0 < \lambda < 1$, and ρ_1 and ρ_2 are states, then $\lambda\rho_1 + (1-\lambda)\rho_2$ is a state (Exercise (8.33)). We say that the state space has an *affine* structure. We interpret this combination as the incoherent statistical mixing of the two states, in the proportion λ to $1-\lambda$. It is easy to do this in practice: we choose a sample in the state ρ_1 with probability λ and a sample with state ρ_2 with probability $1-\lambda$. In the same way, we can set up the mixture $\rho = \sum_i^n \lambda_i\rho_i$, where $0 < \lambda_i < 1$ and $\sum_i \lambda_i = 1$. Clearly, the set of normal states has the same affine structure as the set of density operators, when they are related as above.

We say that a state ρ of Σ is a *mixed* if ρ can be written as $\lambda\rho_1 + (1-\lambda)\rho_2$ where $0 < \lambda < 1$ and ρ_1 is different from ρ_2. If a state is not mixed, it is said to be pure. The pure states are thus the extreme points of the convex set Σ. In [58] a different definition of pure is given: ρ is pure if and only if it does not dominate any other state; it is shown to be equivalent to ours (Theorem 2.3.15). According to Minkowski's theorem, in finite dimensions any point in a convex set is a mixture of the extreme points of the set. In general, the pure states generate the whole of state space, in that the smallest closed convex set containing all pure states is $\Sigma(\mathcal{A})$. This is the Krein-Milman theorem. Closed here means closed in the weak* topology.

We meet a significant difference between quantum probability in general and the special case of probability measures: the classical state space (of probability measures on a sample space) is a simplex, while $\Sigma(\mathcal{A})$ is not, provided that $\mathcal{A}$

is not abelian. Recall that a convex subset Σ of $\mathbf{R}^m$ is said to be a simplex if the expression of any element $\rho \in \Sigma$ as a convex sum of extreme points $\rho_i \in \Sigma$ is unique. If $\mathcal{A}$ is abelian, then it can be identified with the set of continuous complex-valued functions on a compact space, and $\Sigma(\mathcal{A})$ is the set of finitely additive measures on it. This is the Gelfand isomorphism theorem. See Theorem 2.1.11A of [58], where the more general case of C^*-algebra without identity is included. The case of a finite-dimensional abelian algebra is not hard; try Exercise (8.34).

Let us concentrate on the case of $\mathbf{M}_n$. In this case an extremal state is given by an extremal density matrix, and this must be a one-dimensional projection; see Exercise (8.35). It follows from this that the pure states are exactly the vector states in this case, the one-dimensional projection giving the same state as any vector in the subspace it defines; see Exercise (8.36). We can get the projection onto the subspace spanned by a single column vector ψ by forming the matrix product $\psi\psi^*$ of the matrix with its transposed complex conjugate. This can also be regarded as a tensor product $\psi\otimes\psi^*$. A projection of dimension $r \leq n$ also defines a density matrix, when normalised by dividing by r, so that its trace is 1. In particular, the matrix $n^{-1}1$, where 1 is the $n\times n$ identity matrix, is a density matrix. This is the quantum analogue of the uniform distribution, also known as the microcanonical state when the energy is a constant. This state can be written as

$$n^{-1}1 = \sum_i^n (n^{-1})E_{ii}$$

in terms of the matrix units E_{ii}. We can write the matrix 1 as a sum of one-dimensional projections in many different ways: we just need to choose any orthogonal basis, and use the projections onto its elements. So the decomposition is not unique, and $\Sigma(\mathbf{M}_n)$ is not a simplex. In the case of $n = 2$ the uniform state describes the spin of a completely unpolarised beam of electrons, say. The density matrix E_{11} describes a beam of particles all completely polarised in the "up" direction, in that they are all eigenstates of s_z with eigenvalue $+\hbar/2$. The density matrix E_{22} similarly describes a beam completely polarised in the "down" direction. We now mix these together, taking a particle from one beam or the other with probability 1/2. The resulting beam is described by the density matrix diag (1/2, 1/2), the uniform state. Exactly the same result would have been obtained if we had similarly mixed two other vector states which are orthogonal, such as the eigenstates of s_x with eigenvalues $\pm\hbar/2$. There is no way that anyone can discover from the mixture what were the constituents by any experiment on the mixed beam: the classical mixing, with probability 1/2, gets confused with the quantum superposition, which allows us to write the eigenstates of s_z as coherent mixtures of those of say s_x. The rules of matrix or tensor products do the rest.

It is sometimes necessary to admit that two different observers should assign two different states to the same exemplar of the system. This leads to the "paradox" of Wigner's friend. In this paradox, an unpolarised beam of electrons is subjected to a Stern-Gerlach measurement by Wigner's friend, but he does not tell Wigner the result. What state should Wigner assign to the exemplar that resulted from the experiment? A betting man in Wigner's position would have to say that it is still unpolarised, even though his friend would rightly assign it an eigenstate. This has been thought to be a paradox, since the state appears to be a subjective entity. However, the same "problem" arises in classical probability, when some information due to conditioning is available to one person and not another. The formal way to handle this is through *information sets*, which keeps track of who knows what in the theory of games. We can generalise this to the *quantum game*; this is a game with several players, each with a strategy, and some chance elements; the difference with the usual game is that the probabilities with which pure strategies are chosen, and the outcomes of the chance elements, are governed by quantum, not classical, probability. Several elaborations of the *EPR* experiment are of this type; paradoxes arise if we do not treat the information sets correctly.

Apart from the uniform or microcanonical state, we shall meet the canonical and grand canonical states, which are defined as follows.

Definition 8.7 (Canonical State) *Let H be a self-adjoint operator on the Hilbert space $\mathcal{H}$ such that $e^{-\beta H}$ is of trace class; then the density matrix on $B(\mathcal{H})$:*

$$\Delta\rho_\beta = Z_\beta^{-1} \exp\{-\beta H\} \text{ with } Z_\beta = Tr(\exp\{-\beta H\}) \tag{8.8}$$

is called the canonical state with beta, or inverse temperature, equal to β and Hamiltonian H. Z_β is called the partition function.

As in the case of classical statistical dynamics, we want to show why many systems tend to converge to the canonical state for large times. Sometimes there is one or more conserved quantities, whose presence in the theory preclude states from converging to the canonical state. In such models we need a generalisation:

Definition 8.9 (Grand Canonical State) *Let H and N be a self-adjoint operators whose spectral projections commute, and are such that $e^{-\beta(H-\mu N)}$ is of trace class, Then the state*

$$\Delta\rho_{\beta,\mu} = Z_{\beta,\mu}^{-1} \exp\{-\beta(H-\mu N)\} \text{ with } Z_{\beta,\mu} = Tr(\exp\{-\beta(H-\mu N)\}) \tag{8.10}$$

is called the grand canonical state with beta β and chemical potential μ, Hamiltonian H and number operator N. $Z_{\beta,\mu}$ is called the grand partition function.

We shall see that as in the classical case, these states maximise the entropy subject to certain constraints.

One large industry in C^*-algebras which is not an important part of classical probability is representation theory. A representation of $\mathcal{A}$ is a *-homomorphism π from $\mathcal{A}$ into $B(\mathcal{H})$ for some Hilbert space $\mathcal{H}$. The representation is said to be faithful if $\pi(A) = 0$ only if $A = 0$. A state ρ is said to be faithful if $\rho(A^*A) > 0$ unless $A = 0$. Gelfand showed that every C^*-algebra has at least one faithful representation, and so can be regarded as an operator algebra without loss of generality. Two representations, π_1 on $\mathcal{H}_1$ and π_2 on $\mathcal{H}_2$ are said to be equivalent if there exists a unitary operator $U : \mathcal{H}_1 \to \mathcal{H}_2$ such that $U\pi_1(A) = \pi_2(A)U$ for every $A \in \mathcal{A}$. We shall need the concepts of irreducible representation, and cyclic representation. Let $\mathcal{X}$ be any set of bounded operators on a Hilbert space $\mathcal{H}$. We say that a subspace $\mathcal{H}_1$ is invariant under $\mathcal{X}$ if $X\psi \in \mathcal{H}_1$ for all $\psi \in \mathcal{H}_1$ and all $X \in \mathcal{X}$. We say that $\mathcal{X}$ is irreducible if the only subspaces invariant under $\mathcal{X}$ are the zero subspace and $\mathcal{H}$ itself. There is then the useful lemma of Schur. Suppose that whenever $\mathcal{X}$ contains X it also contains X^*; then we say that $\mathcal{X}$ is self-adjoint. We define the *commutant* of $\mathcal{X}$, denoted by $\mathcal{X}'$, as the set of operators in $B(\mathcal{H})$ that commute with all elements of $\mathcal{X}$. Schur's lemma says that a self-adjoint set of operators is irreducible if and only if its commutant consists of multiples of the identity. This is Proposition 2.3.8 of [58]. The theorem also says that these are equivalent to the fact that every vector is cyclic (unless $\mathcal{A} = 0$ and $\mathcal{H} = \mathbf{C}$). We say that a vector $\psi \in \mathcal{H}$ is cyclic for $\mathcal{X}$ if the set $\{X\psi : \; X \in \mathcal{X}\}$ is dense in $\mathcal{H}$. We say that a representation π of a C^*-algebra $\mathcal{A}$ is *cyclic* with cyclic vector ψ if ψ is a cyclic vector for $\mathcal{X} = \{\pi(A) : \; A \in \mathcal{A}\}$. Cyclic representations are very common in quantum mechanics: we build up all states by acting on ψ, often the vacuum or equilibrium state, with the operators at our disposal.

We now come to the most important idea, the Gelfand-Naimark-Segal construction.

Theorem 8.11 *Let $\mathcal{A}$ be a C^*-algebra and ρ a state on $\mathcal{A}$. Then there exists a Hilbert space $\mathcal{H}_\rho$, a representation π_ρ and a cyclic vector ψ_ρ for π_ρ, such that*

$$\rho(A) = \langle \psi_\rho, \pi(A)\psi_\rho \rangle \quad \text{for all } A \in \mathcal{A}.$$

The representation is unique up to cyclic unitary equivalence. Moreover, π is irreducible if and only if ρ is pure.

For the proof, see [58], Theorems 2.1.16 and 2.1.19. The great thing about the Gelfand-Naimark-Segal theorem is that it is constructive. We shall show this in the case when ρ is a faithful state; this will be so for the canonical states of interest to us. Thus let $\mathcal{A}$ be a C^*-algebra with identity and let ρ be a faithful state. We now define a scalar product on the vector space $\mathcal{A}$ by

$$\langle A, B\rangle_\rho = \rho(A^*B).$$

This is linear in the second variable and antilinear in the first; it is also definite, in that $\|A\|^2 = \langle A, A\rangle_\rho > 0$ unless $A = 0$. This holds since it expresses that ρ is faithful. The Hilbert space $\mathcal{H}_\rho$ mentioned in the theorem is then the completion of $\mathcal{A}$ in the metric defined by the norm. The representation π_ρ mentioned in the theorem is the action on $\mathcal{A}$ by itself by left multiplication. This can be shown to give rise to bounded operators, which have unique extensions to $\mathcal{H}_\rho$ by continuity. The cyclic vector ψ_ρ is the identity element of $\mathcal{A}$. Naturally, if $\dim \mathcal{A} < \infty$ then the completion and extensions are not necessary.

If $\mathcal{A}$ is a subalgebra of $B(\mathcal{H})$ for some Hilbert space $\mathcal{H}$, and ρ is a normal faithful state (given by density matrix $\Delta(\rho)$), then we may define a norm on certain elements in the vector span of states by

$$\|\sigma\|_\rho = [Tr\,(\Delta(\rho)^{-1}\sigma^*\sigma)]^{1/2}. \tag{8.12}$$

This is the dual to the norm on $\mathcal{A}$ given by the state ρ. It will turn out to be a useful norm which may be used to define a metric on the space of states.

There is a natural definition of independence for quantum probability. The analogue of Boolean subring is subalgebra $\mathcal{A}_1 \subseteq \mathcal{A}$. Let $\mathcal{A}$ be finite dimensional and let $\mathcal{A}_1$ and $\mathcal{A}_2$ be two C^*-algebras. Suppose that $\mathcal{A} = \mathcal{A}_1 \otimes \mathcal{A}_2$ and suppose we are given $\rho_1 \in \Sigma(\mathcal{A}_1)$, $\rho_2 \in \Sigma(\mathcal{A}_2)$. We write $\rho = \rho_1 \otimes \rho_2$ for the unique bilinear extension to $\mathcal{A}$ of the definition on dyads:

$$\rho(A_1 \otimes A_2) = \rho_1(A_1)\rho_2(A_2). \tag{8.13}$$

Then we say that the subalgebras, the ampliations of $\mathcal{A}_1$ and $\mathcal{A}_2$ are independent in the state ρ. The ∞-dimensional case needs a more careful treatment of the tensor product. We shall deal with that problem if it arises.

Of great importance in classical statistical dynamics was the concept of marginal distribution, and we need an analogue of this. So let $\mathcal{A}$ be a tensor product $\mathcal{A}_1 \otimes \mathcal{A}_2$, and let $\rho \in \Sigma(\mathcal{A})$. The marginal state of ρ, $\mathcal{M}_1\rho$ on the first factor, often described as the "trace of ρ over the second factor" is defined, for each element A_1 of the first algebra, as

$$\mathcal{M}_1\rho(A_1) = \rho(A_1 \otimes 1). \tag{8.14}$$

The marginal map obviously obeys

$$\mathcal{M}_1(\rho_1 \otimes \rho_2) = \rho_1.$$

Similarly the marginal state on the second factor is defined. So the marginal map $\mathcal{M}_1 : \Sigma(\mathcal{A}_1 \otimes \mathcal{A}_2) \rightarrow \Sigma(\mathcal{A}_1)$ is the dual of the ampliation, which maps the algebras in the reverse direction.

Quantum probability is a more general model than probability for modelling systems subject to chance; it cannot be usefully described by hidden variables in a classical theory; it would be like trying to model non-Euclidean geometry by using figures of a strange shape in Euclidean space. It can't be done.

8.3 Quantum Entropy

Let ρ be a normal state on $B(\mathcal{H})$ and $\Delta\rho$ the corresponding density operator; in this section, we shall use the symbol ρ to denote $\Delta\rho$. In his book [1], von Neumann defined the entropy of ρ to be

$$S(\rho) = -Tr(\rho \log \rho). \tag{8.15}$$

In this definition, $\log \rho$ is defined by the functional calculus for self-adjoint operators. If $\dim \mathcal{H} = \infty$ then there are some operators ρ, even of trace class, for which this does not converge. Some authors write $S(\rho) = \infty$ in this case. This entropy, like its classical version, the Shannon entropy, is extensive. Thus, suppose that the algebra of observables $\mathcal{A}$ has a local structure over a space Λ, and is a tensor product: $\mathcal{A} = \bigotimes_x \mathcal{A}_x$. Suppose too that the state ρ is independent over Λ, by which we mean that

$$\rho = \otimes_{x\in\Lambda}\rho_x.$$

Then

$$S(\rho) = \sum_{x\in\Lambda} S(\rho_x).$$

The proof is very similar to that for classical probability, and you may do it yourself in Exercise (8.37). Thirring [20] p.58, gives conditions under which S is unique. We see that the entropy of a pure state is zero. This is because we use the convention in the definition of S that $0 \log 0 = 0$ and $\log 1 = 0$; this means that $P \log P = 0$ as an operator if P is a one-dimensional projection. This is in spite of the fact that in quantum mechanics any pure state gives rise to some uncertainty (Heisenberg's uncertainty principle). Thus, $S(\rho)$ measures the lack of information that is in addition to the intrinsic quantum uncertainty. Note that the classical entropy of the probability distribution $|\psi(x)|^2$ of a normalisable wave-function in one dimension is infinite, if it is defined as the limit of the Shannon entropy of the coarse-grained approximations to it. Since it represents a pure state, its quantum entropy is zero. This is one of many ways in which quantum probability improves on classical probability.

In the finite-dimensional case, the microcanonical state is the state that maximises the entropy. To see this, use a Lagrange multiplier to express that the trace of a state is 1. Thus we seek to maximise

$$-Tr\,(\rho \log \rho) + \lambda Tr\,\rho$$

as we roam over the set of positive matrices ρ. The trace is invariant under unitary transformations; let us diagonalise ρ by a unitary matrix. So it suffices to consider the same problem where ρ, and therefore also $\log \rho$, is diagonal. But then the product $\rho \log \rho$ involves only the product of these diagonal elements, and the problem reduces to the classical theorem. The result is that all components along the diagonal are equal, so that ρ is a multiple of the identity matrix. All the unitary transforms are then equal to each other, so the answer is unique. □

The crux is the following

Theorem 8.16 *Let $\mathcal{H}$ be a Hilbert space and $H \geq 0$ an unbounded self-adjoint operator with eigenvalue 0 such that $e^{-\beta H}$ is of trace class for all $\beta > 0$. Let ρ_β be the canonical state defined by the density operator $Z_\beta^{-1} e^{-\beta H}$. Let $E > 0$ be given. Then ρ_β is the unique state that maximises $S(\rho)$ among all normal states ρ on $B(\mathcal{H})$ satisfying the conditions that ρH is of trace class and $Tr(\rho H) = E$. This equation then determines β uniquely.*

This theorem will easily follow from a quantum version [60] of Kullback's lemma:

Lemma 8.17 *Let ρ and σ be density operators. Then*

$$S(\rho|\sigma) = Tr\,(\rho(\log \rho - \log \sigma)) \geq \frac{1}{2} Tr(\rho - \sigma)^2.$$

Note. The left-hand side is called the relative entropy of ρ, given σ.

Proof. Let (ϕ_i, a_i) and (ψ_i, b_i) be the orthonormal eigenvectors and eigenvalues of ρ and σ respectively. Then $0 \leq a_i,\, b_j \leq 1$. Let $f(x) = x \log x$ and $c_{ij} = \langle \phi_i, \psi_j \rangle$. Then $\sum_j |c_{ij}|^2 = \langle \phi_i, \phi_i \rangle = 1$ and

$$\begin{aligned}
& \langle \phi_i, \{f(A) - f(B) - (A-B)f'(B) - (1/2)(A-B)^2\}\phi_i\rangle \\
= \; & \langle \phi_i, f(a_i)\phi_i\rangle - \sum_j \langle \phi_i, \psi_j\rangle\langle \psi_j, f(b_j)\phi_i\rangle \\
- \; & \sum_j \langle \phi_i, (a_i - b_j) f'(b_j)\psi_j\rangle\langle \psi_j, \phi_i\rangle \\
- \; & \frac{1}{2}\left\{ \langle \phi_i, a_i^2 \phi_i\rangle - 2\sum_j a_i b_j \langle \phi_i, \psi_j\rangle\langle\psi_j, \phi_i\rangle + \sum_j \langle \phi_i, b_j^2 \psi_j\rangle\langle \psi_j, \phi_i\rangle \right\} \\
= \; & \sum_j |c_{ij}|^2 \left\{ f(a_i) - f(b_j) - (a_i - b_j) f'(b_j) - \frac{1}{2}(a_i - b_j)^2 \right\}
\end{aligned}$$

and this is non-negative by Taylor's theorem with remainder. Summing over i gives

$$Tr\left\{A\log A - B\log B - (A-B)(\log B + 1) - \frac{1}{2}(A-B)^2\right\} \geq 0 \qquad (8.18)$$

that is

$$Tr\,(A(\log A - \log B)) \geq \frac{1}{2}Tr\,(A-B)^2. \quad \square$$

Note. This is a sharpened form of a result in [61], proved by the same method. There is a more general form with the same proof, in which A and B are positive operators with eigenvalues in [0, 1]. Then the result is adjusted by $Tr\,A - Tr\,B$, as can be seen from Eqn. (8.18).

Proof of theorem. Let $\sigma = \rho_\beta$ in the lemma. Then

$$Tr\,(\rho(\log\rho - \log\rho_\beta)) = Tr\,(\rho\log\rho) + Tr\,(\rho\beta H) + \log Z_\beta \geq \|\rho - \rho_\beta\|^2/2$$

by the lemma. Rearranging gives

$$\begin{aligned} -Tr\,(\rho\log\rho) + \|\rho - \rho_\beta\|^2/2 &\leq \beta Tr\,(\rho H) - \log Z_\beta \\ &= \beta Tr\,(\rho_\beta H) - \log Z_\beta \\ &= S(\rho_\beta). \end{aligned}$$

To show that β is uniquely determined, note that the mean energy is infinite if $\beta = 0$, so we can choose β so that the mean energy is larger than the given E. But then differentiation shows that

$$\frac{\partial}{\partial\beta}Tr\,(\rho_\beta H) = -Tr(\rho_\beta(H - \overline{H})^2) < 0$$

where $\overline{H}$ is the mean energy in the state ρ_β. So it is strictly decreasing. Since the mean energy at infinite β is zero, the energy of the ground state, we can find a unique β with $\overline{H} = E$. $\square$

We see from the proof that under the conditions of the theorem, $S(\rho)$ is finite. A small elaboration on this proof shows that the grand canonical state is the unique state maximising the entropy subject to a given mean energy and a given mean particle number. In fact we can include any finite number of constraints, in terms of the given means of Hermitian operators $X_1, \ldots, X_m$, not necessarily commuting. We have

Theorem 8.19 *Suppose that $X_1, \ldots, X_m$ are positive operators with a common domain, such that for all positive β_i, $\exp\{-\sum_i \beta_i X_i\}$ is of trace class. Then the great grand canonical state*

$$\Delta\rho = Z^{-1}\exp\{-\sum_i \beta_i X_i\} \qquad (8.20)$$

is the unique state that maximises $S(\rho)$ among all normal states such that ρX_i is of trace class and

$$\langle \rho, X_1 \rangle = E_1, \qquad \ldots \qquad \langle \rho, X_m \rangle = E_m.$$

Proof. First, solve a problem we know how to do: let $\lambda_1, \ldots, \lambda_m$ be any positive numbers, and put $H = \sum \lambda_i X_i$. Now maximise $S(\rho)$ subject to one condition, $\langle \rho, H \rangle = E$. The result is unique: $\rho = \rho_\beta$, where $\Delta\rho_\beta = Z^{-1}e^{-\beta H}$, and β is determined by $\langle \rho_\beta, H \rangle = E$. Now the observables X_i will have certain expectations in this state, say $\langle \rho_\beta, X_i \rangle = E_i$; these must satisfy the condition $\sum \lambda_i E_i = E$. Conversely, put $\beta_i = \beta\lambda_i$; then for a given possible set of values E_i, the β_i are unique (by the fact that entropy is a strictly decreasing function of each variable). Moreover, ρ_β is the state of maximum entropy with these values of E_i; for if not, we would get a state with the same value of E but a bigger entropy, contrary to the definition of ρ_β. □

We also see that quantum entropy is not decreased by the stoss map:

Theorem 8.21 *Let $\mathcal{H}$ be a Hilbert space and $\mathcal{A} = \mathcal{A}_1 \otimes \mathcal{A}_2 \subseteq B(\mathcal{H})$, with marginal maps $\mathcal{M}_1$ and $\mathcal{M}_2$. Then for any normal state $\rho \in \Sigma(\mathcal{A})$ with finite entropy we have*

$$S(\mathcal{M}_1\rho \otimes \mathcal{M}_2\rho) \geq S(\rho) + \frac{1}{2}\|\rho - \mathcal{M}_1\rho \otimes \mathcal{M}_2\rho\|_2^2. \tag{8.22}$$

Proof. Just put $\sigma = \mathcal{M}_1\rho \otimes \mathcal{M}_2\rho$ in Kullback's lemma, (8.17) and use the idempotent property of the partial trace:

$$Tr\,(\rho \log \mathcal{M}_1\rho) = Tr_1 \mathcal{M}_1\rho \log \mathcal{M}_1\rho$$

for example. □

We can use Kullback's lemma (quantum version), (8.17) to estimate the concavity of the entropy; we get the same result as Theorem (2.42):

Theorem 8.23 (Concavity of the Quantum Entropy) *Let ρ and σ be density operators of finite entropy on a Hilbert space. Then*

$$S(\lambda\rho + (1-\lambda)\sigma) - \lambda S(\rho) - (1-\lambda)S(\sigma) \geq \frac{1}{2}\lambda(1-\lambda)\|\rho - \sigma\|_2^2 \tag{8.24}$$

for all $\lambda \in [0, 1]$.

Proof. Kullback's lemma says that

$$-Tr\,\rho \log \sigma \geq -Tr\,\rho \log \rho + \frac{1}{2}\|\rho - \sigma\|_2^2;$$

try this with σ replaced by $\lambda\rho + (1-\lambda)\sigma$. It gives

$$\begin{aligned} -Tr\,\rho\log(\lambda\rho + (1-\lambda)\sigma) &\geq -Tr\,\rho\log\rho + \frac{1}{2}\|\rho - (\lambda\rho + (1-\lambda)\sigma)\|_2^2 \\ &= S(\rho) + \frac{1}{2}(1-\lambda)^2\|\rho - \sigma\|_2^2. \end{aligned}$$

Now use it with ρ replaced by σ and σ replaced by $\lambda\rho + (1-\lambda\sigma)$:

$$\begin{aligned} -Tr\,\sigma\log(\lambda\rho + (1-\lambda)\sigma) &\geq -Tr\,\sigma\log\sigma + \frac{1}{2}\|\sigma - (\lambda\rho - (1-\lambda)\sigma\|_2^2 \\ &= S(\sigma) + \frac{1}{2}\lambda^2\|\rho - \sigma\|_2^2. \end{aligned}$$

Take λ times the first estimate and $(1-\lambda)$ times the second, and add, to get

$$\begin{aligned} S(\lambda\rho + (1-\lambda\sigma)) &- \lambda S(\rho) - (1-\lambda)S(\sigma) \\ &\geq \frac{1}{2}(\lambda(1-\lambda)^2 + (1-\lambda)\lambda^2)\|\rho - \sigma)\|_2^2 \\ &= \frac{1}{2}\lambda(1-\lambda)\|\rho - \sigma\|_2^2. \quad \square \end{aligned}$$

Corollary. Any mixture $\tau = \sum_i \lambda_i \tau_i$ of automorphisms τ_i gives rise to a dual action τ^* which is entropy non-decreasing.

Many further properties of quantum entropy can be found in [20] and [62].

8.4 Exercises

Exercise 8.25 Show that $\mathcal{A} = \mathbf{M}_n$ obeys the axioms of a C^*-algebra.

Exercise 8.26 Let $\mathcal{A}_1$ and $\mathcal{A}_2$ be C^*-algebras. A linear map, T, from $\mathcal{A}_1$ to $\mathcal{A}_2$ is called a *homomorphism* if it preserves the structure of C^*-algebra. This means that

- $T(AB) = T(A)T(B)$ for all A, B in A;
- $T(A^*) = (TA)^*$ for all $A \in \mathcal{A}$.

We say that T is an isomorphism if it is bijective; in this case we say that $\mathcal{A}_1$ and $\mathcal{A}_2$ are *isomorphic*.

Show that if $\dim \mathcal{H} = n < \infty$, then $B(\mathcal{H})$ and $\mathbf{M}_n$ are isomorphic.

Exercise 8.27 Show that the set of unitary elements in a C^*-algebra with identity forms a group.

Exercise 8.28 Show that the star operation defined on dyads and extended by antilinearity gives the same answer whichever representatives are chosen.

Exercise 8.29 Show that the definition of product on $\mathcal{A}_1 \otimes \mathcal{A}_2$ is independent of the representatives chosen.

Exercise 8.30 Let G be a finite group, with n elements. Define a multiplication on Span G by

$$(\sum_i \lambda_i g_i) \times (\sum_j \mu_j g_j) = \sum_{ij} (\lambda_i \mu_j)(g_i g_j).$$

In this, the symbol $g_i g_j$ denotes the product in the given group. Define a conjugation by $g^* = g^{-1}$, extended by anti-linearity to Span G. Show that this makes Span G into a *-algebra, $\mathcal{A}(G)$ say. Now let $\mathcal{A}(G)$ act on Span G by left multiplication, (the *regular* representation). Show that in this way, each element of $\mathcal{A}$ defines a linear operator on Span G. Take the elements of G as a basis in Span G, and for each $A \in \mathcal{A}$ let $M(A)$ be the matrix of this linear map in this basis. Show that $A \mapsto M(A)$ is an injective homomorphism from $\mathcal{A}$ into $\mathbf{M}_n$. Use this fact to furnish $\mathcal{A}(G)$ with a C^*-norm.

Exercise 8.31 Let $\mathcal{A} = \mathbf{M}_n$ and let ρ be a positive semi-definite $n \times n$ matrix with unit trace. Show that the map $A \mapsto Tr\,(\rho A)$ defines a state on $\mathcal{A}$.

Exercise 8.32 Show that any state on $\mathbf{M}_n$ is of the form $Tr\,(\rho A)$ for some density matrix ρ. [Hint: Try defining ρ as the matrix whose i, j matrix-elements are the expectations of the matrix units E_{ji}.]

Exercise 8.33 Show that the state space of any C^*-algebra is a convex set.

Exercise 8.34 Let $\mathcal{A}$ be an abelian C^*-algebra of finite dimension, containing the identity. Show that $\mathcal{A}$ is isomorphic to the set of functions on a finite set. [Hint: let Ω be the space of homomorphisms from $\mathcal{A}$ into $\mathbf{C}$.]

Exercise 8.35 Show that a density matrix is extremal in $\Sigma(\mathbf{M}_n)$ if and only if it is a projection of dimension 1. [Hint: show that the spectral resolution of a density matrix expresses it as a mixture of projections.]

Exercise 8.36 Show that a vector state on $B(\mathcal{H})$ given by a vector ψ is the same as that defined by the density operator P_ψ, the projection onto the subspace spanned by ψ.

Exercise 8.37 Let $\rho = \otimes_{x \in \Lambda} \rho_x$ be a state on $\mathcal{A} = \otimes_{x \in \Lambda} \mathcal{A}_x$. Show that $S(\rho) = \sum_x S(\rho_x)$. [Hint: the logarithm of a density matrix, defined by the spectral theorem, has all the usual properties.]

Chapter 9

Linear Quantum Dynamics

9.1 Reversible Dynamics

The dynamics of a quantum system is either expressed in the Heisenberg picture or in the Schrödinger picture. The first concerns the C^*-algebra and the second the state space. The first seems to be the more basic. One time-step in the algebra, like any symmetry transformation, is expressed as a *-algebraic automorphism of the algebra of observables; this is simply an isomorphism τ say, from $\mathcal{A}$ onto $\mathcal{A}$. We write the action on the right: $A \mapsto A\tau$. Because an isomorphism preserves the * operation, τ must take Hermitian elements to Hermitian elements. Any automorphism has an inverse, which is also an automorphism. Obviously, the identity map is an automorphism. Also, the composition of two automorphisms is an automorphism, We denoted the composition by $\circ$: $(A\tau_1)\tau_2 = A(\tau_1 \circ \tau_2)$ for any $A \in \mathcal{A}$. Thus the set of automorphisms of a C^*-algebra form a group denoted $AUT\mathcal{A}$. We may throw the action of the automorphism group onto the states, by duality. Thus, given τ and a state ρ, we define the transformed state $\tau^*\rho$ by

$$\tau^*\rho(A) = \rho(A\tau) \text{ for all } A \in \mathcal{A}.$$

Given the transformation τ for one time step, we define the transformation for n time steps as $\tau^n = \tau \circ \tau \ldots \tau$ (n factors). A time-evolution (with discrete time) is then a homomorphism (here $n \mapsto \tau^n$) from the group $\mathbf{Z}$ into $AUT\mathcal{A}$. In the same way, we can argue that a time-evolution with continuous time is a homomorphism $t \mapsto \tau_t$ from $\mathbf{R}$ into $AUT\mathcal{A}$. To be a homomorphism means that the group law $\tau_s \circ \tau_t = \tau_{s+t}$ holds for all real s and t. In this case it is natural to postulate some sort of continuity for the homomorphism.

Because we are led to a dynamical group, and not just a semi-group, automorphisms lead to reversible dynamics: there is no dissipation.

In a C^*-algebra $\mathcal{A}$, a unitary element $U \in \mathcal{A}$ defines an automorphism by the map $A\tau_U = UAU^{-1}$ for all $A \in \mathcal{A}$. Such an automorphism is called *inner*; it is often denoted by $\mathrm{Ad}\,U$. One can prove that $\tau_U \circ \tau_V = \tau_{UV}$, $\tau_1 = Id$ and $(\tau_U)^{-1} = \tau_{U^{-1}}$. See Exercise (9.23). Indeed the set of inner automorphisms form a group, denoted $INN\mathcal{A}$. One can prove that every automorphism of $\mathbf{M}_n$ is inner. If θ is a real number, and $V = e^{i\theta}U$, then $\tau_U = \tau_V$. More generally, let $\mathcal{Z}(\mathcal{A})$ denote the *centre* of $\mathcal{A}$, that is, the set of elements of $\mathcal{A}$ that commutes with all elements of $\mathcal{A}$, and let $Z \in \mathcal{Z}$ be unitary. Then $\tau_{ZU} = \tau_{UZ} = \tau_U$. In fact this is the extent of the ambiguity in U; see Exercise (9.24). Obviously for a commutative algebra only the identity automorphism is inner. Thus there are C^*-algebras for which not all automorphisms are inner; in infinite dimensions, this is usually the case, even for non-abelian algebras. In that case the following question becomes interesting: given a representation π of $\mathcal{A}$ on a Hilbert space $\mathcal{H}$, and an automorphism τ of $\mathcal{A}$, can we find a unitary operator $U \in B(\mathcal{H})$ such that

$$\pi(A\tau) = U\pi(A)U^{-1} \text{ for all } A \in \mathcal{A}?$$

If so, we say that τ is *spatial in* π. If τ is inner, say $\tau = \tau_V$, then τ is spatial in all representations, since we may take $U = \pi(V)$. There is an important case when we can show that an automorphism τ is spatial in a representation; this is when the representation is of the form π_ρ, where $\rho \in \Sigma(\mathcal{A})$ is invariant under τ^*. For then the operator on $\mathcal{H}_\rho$ given by $\pi_\rho(A)\psi_\rho \mapsto \pi_\rho(A\tau)\psi_\rho$ is unitary; see and do Exercise (9.25). The same remarks apply to any automorphism of $\mathcal{A}$, whether or not it represents time-evolution. In particular, in a non-relativistic theory, a *symmetry* of the system is an automorphism α commuting with the time-evolution; this might or might not be spatial in a representation π, even if τ_t is spatial in π. If α is a symmetry which is not spatial in π, we say that it is spontaneously broken in π. Note that the concept of spontaneously broken symmetry depends on the representation chosen, and so is not a purely algebraic property. Our definition of spontaneously broken symmetry [63] differs from that used by some authors, who use the term to describe the situation where the state ρ used in the GNS construction is invariant under the time-evolution but not under the symmetry automorphism α. This leaves open the question as to whether α is spatial in π_ρ or not. The Heisenberg ferromagnet exhibits a spontaneous breakdown of symmetry in both senses. The ground state is not invariant under rotations; and also the rotation-automorphisms commute with time-evolution, but are not spatial in the vacuum representations. This model is studied in Section (12.2). In [64] I give a strange example; time is continuous and space is discrete, and the space-time automorphisms commute. I construct a state ρ which is invariant under time-evolution but not space translations, but where nevertheless space-time translations are spatial in π_ρ, being given by $U(t)$ and $V(n)$ say. The remarkable fact is that in this example there is no choice of implementing operators $U(t)$ and $V(n)$ that commute with each other. This situation is called an *anomaly*.

In a relativistic theory, with relativity group G, we are given a homomorphism τ from G into $AUT\mathcal{A}$, so that τ_g is an automorphism for each $g \in G$, and $\tau_g\tau_h = \tau_{gh}$; if ρ is invariant under the dual action of the group on the states, then each τ_g is spatial in π_ρ, and the implementing operators $U_g : g \in G$ say, also obey the group law: $U_gU_h = U_{gh}$. In this case there are no multipliers or anomalies. In particular, the time-evolution forms a one-parameter group, U_t, which may under reasonable assumptions be taken to be continuous. Thus, U_t satisfies

Definition 9.1 (One-parameter Group) .

1. $U(0) = 0$
2. $U_t\psi \to \psi$ *in* $\mathcal{H}$ *as* $t \to 0$.
3. $U_sU_t = U_{s+t}$ *for all* s, $t \in \mathbf{R}$.

Stone's theorem asserts that $U_t\psi$ is differentiable for a dense set $\mathcal{D}$ of ψ, with self-adjoint generator, called the energy, H. Thus, in the sense of norm convergence in Hilbert space,

$$iH\psi = \lim_{t\to 0} t^{-1}(U_t - 1)\psi \text{ if } \psi \in \mathcal{D}. \tag{9.2}$$

It is clear that the energy operator depends on the representation in which we are working (unless τ_t is inner, which is rare if $\dim \mathcal{A} = \infty$). Even the property of being bounded below is not algebraic; usually, the generator of the time-evolution, the energy, is positive in the representation generated by a certain state, called the ground state. It will turn out that for infinite systems, the energy is never bounded below in the thermal states (a general term for one of the canonical states). Even when the algebra is of finite dimension, it will be convenient to choose the zero-point of energy to be in the centre of the range of eigenvalues, when we are in a thermal state. We see the beginnings of this in the dilation theorem for Markov chains, Theorem (5.38).

Just as in the classical case, it is easy to see that the entropy is unchanged by any spatial automorphism: the trace is invariant under any unitary transformation, since it is independent of the base used in its calculation. Thus let ρ be a normal state and $\Delta\rho$ the density operator corresponding to it; the functional calculus for self-adjoint operators implies that $\log(U\Delta\rho U^{-1}) = U\log(\Delta\rho)U^{-1}$. Thus if ρ' is such that $\Delta\rho' = U\Delta\rho U^{-1}$, we have

$$\begin{aligned} S(\rho') &= -Tr\,[U\Delta\rho U^{-1}\log(U\Delta\rho U^{-1})] = -Tr\,[U\Delta\rho U^{-1}U\log(\Delta\rho)U^{-1}] \\ &= -Tr\,[U(\Delta\rho\log\Delta\rho)U^{-1}] = S(\rho). \end{aligned}$$

Thus we have the same problem as in classical statistical dynamics, that the microscopic theory does not lead to dissipation. We therefore follow the same steps as we used there, and next consider random unitary dynamics.

9.2 Random Quantum Dynamics

In this section, we will take a representation of the algebra of observables $\mathcal{A}$ in which time-evolution is spatial, with a continuous unitary representation U_t of the group $\mathbf{R}$, with generator H. We will assume that the spectral projections of H lie in $\mathcal{A}$. The idea of quantum ergodic theory is that a measurement of an observable takes a time very much longer than the time-scale for microscopic dynamics. Instead of measuring an observable represented by an operator $A \in \mathcal{A}$, we measure the mean of very many time-translates of A; thus we measure $A_N = N^{-1}\sum_{n=0}^{N} A\tau^n$. (This is still an operator in $\mathcal{A}$, not its quantum mean in some state.) Here, τ is the time translation over a small time-step, typical of the microscopic system. We then let N tend to infinity. On the other hand, N is to be a small time on the scale of the slow variables, whose dynamics we wish to study. We note that for each term in the sum, $U(n)$ commutes with the energy operator H, and so τ and τ^{-1} leave fixed the spectral projections of H (which lie in the algebra). The map $A \mapsto A_N$ is linear, and can be thrown by duality onto $\Sigma(\mathcal{A})$. It is easy to see that the resulting motion through the states conserves the mean energy. See Exercise (9.26). By the corollary to Theorem (8.23), this dynamics increases entropy; it therefore obeys both the first and second laws of thermodynamics.

We are not obliged to adopt the ergodic theorist's interpretation of measurement. These ideas can be incorporated into a more general scheme; we let H be a self-adjoint operator, taken to be the energy, and consider N unitary operators $U_1, \ldots, U_N$, each of which commutes with every spectral projection of H. Let $A\tau_n = U_n A U_n^{-1}$, and consider the mixture $\tau = \sum_i \lambda_i \tau_i$, where $0 < \lambda_i < 1$ and $\sum_i \lambda_i = 1$. Then the dynamics defined by τ obeys the first and second laws of thermodynamics. We call τ a *random reversible dynamics*, (and also limits of such, as the number of terms goes to infinity). In the classical case, conservation of mean energy was a result of the property that the transformation leaves the energy-shell invariant. Equivalently, its action on the algebra of random variables leaves invariant the indicator functions of the energy-shells. The quantum version of this is that each τ_i from which τ is composed, and the inverses, should leave the spectral projections P_E of H fixed. Equivalently, they should map the subalgebras $P_E\mathcal{A}P_E$ to themselves, for all eigenvalues E of H. It will follow from the assumption that e^{-H} is of trace class that the spectrum of H is discrete and that each eigenvalue has finite multiplicity. There is also a restriction on the number of eigenvalues E_n as n becomes large. Similar restrictions were found to be necessary in the classical case, where for example, we needed to assume that each energy-shell was finite.

The interpretation of τ_i as the dynamics due to a random potential causes problems, since there seems no physical reason to expect all these random Hamiltonians to commute with the energy represented by H. We therefore prefer to regard the τ_i as the scattering automorphisms caused by the random interactions; on the time scale of the fast variables, one time-step of our dynamics is indeed very large, and the observables at the end of this time-span are related to the same at the beginning of the time-span by the scattering operator for the fast variables: $A_+ = SA_-S^{-1}$. It is a fact from scattering theory that the scattering operator conserves the energy of the asymptotic particles, for a wide range of potentials; so we are not out of line in asssuming that H represents the asymptotic energy, and that each τ_i leaves its spectral projections invariant. Indeed, the collision term in the Boltzmann equation described scattering and it conserves the asymptotic energy.

We shall be interested in finding examples of dynamics for which $\tau^{*n}\rho$ converges to a limit as $n \to \infty$ for every state ρ. For this, we need that τ mix up each algebra $\mathcal{A}_{E} = P_{E}\mathcal{A}P_{E}$ thoroughly. Since τ is a mixture of automorphisms τ_i, we can always write it as

$$\tau = \tau_1[\lambda_1 1 + \sum_{i=2} \lambda_i \tau_1^{-1}\tau_i];$$

here, the choice of τ_1 was arbitrary. The prefactor τ_1 does not contribute at all to the mixing, since entropy is invariant under automorphisms; this part of the dynamics is reversible. It is the remaining part that must cause the convergence to equilibrium. This idea leads us to the following formulation:

Definition 9.3 *Suppose that a random dynamics τ is a mixture of automorphisms*

$$\tau = \lambda_0 1 + \sum_i \lambda_i \tau_i, \quad \textit{where } 0 < \lambda_i < 1, \qquad \sum_0^n \lambda_i = 1$$

each τ_i leaving the spectral projections of H invariant. We say that τ is ergodic relative to H *if the only elements of $\mathcal{A}$ invariant under τ are functions of H.*

Being a "function of H" is in the sense of the functional calculus for operators, and is equivalent to lying in the C^*-algebra generated by the spectral projections P_{E}. The ergodic condition then expresses that there is no non-trivial subalgebra of $\mathcal{A}_{E}$ that is mapped to itself by the dynamics, so each "energy-shell" really does get fully mixed up.

Theorem 9.4 *Let $\dim \mathcal{A}_{E} = N < \infty$ and suppose that τ is a mixture as in the definition, and is ergodic relative to H. Then for any initial state $\rho \in \Sigma(\mathcal{A}_{E})$,* *we have*

$$\lim_{r\to\infty} \tau^{*r}\rho \to N^{-1}1.$$

Proof. In finite dimensions the set of states is compact. The entropy is a strict Lyapunov function for the dynamics, since by Theorem (8.23),

$$\begin{aligned} S(\tau^*\rho) &= S\left((\lambda_0 1 + \sum_{i=1}^n \lambda_i\tau_i^*)\rho\right) = S\left(\lambda_0\rho + (1-\lambda_0)\sum_{i=1}^n \frac{\lambda_i}{1-\lambda_0}\tau_i^*\rho\right) \\ &\geq \lambda_0 S(\rho) + (1-\lambda_0)S\left(\sum_{i=1}^n \frac{\lambda_i}{1-\lambda_0}\tau_i^*\rho\right) + \varepsilon \end{aligned} \tag{9.5}$$

where

$$\varepsilon = \frac{\lambda_0(1-\lambda_0)}{2}\|\rho - \sum_{i=1}^n \frac{\lambda_i}{1-\lambda_0}\tau^*\rho\|_2^2. \tag{9.6}$$

We note that $\varepsilon = 0$ only if $(1-\lambda_0)\rho = \sum \lambda_i\tau_i^*\rho$, which reduces to $\tau^*\rho = \rho$. By the ergodic condition, this is possible only if ρ is a multiple of the identity, that is, ρ must be the microcanonical state. We can therefore afford to leave out this sharp form of the inequality in the further entropy estimates. We note that

$$\sum_{i=1}^n \frac{\lambda_i}{1-\lambda_0}\tau_i^*\rho = \frac{\lambda_1}{1-\lambda_0}\tau_1^*\rho + \frac{1-\lambda_0-\lambda_1}{1-\lambda_0}\sum_{i=2}^n \frac{\lambda_i}{1-\lambda_0-\lambda_1}\tau_i^*\rho. \tag{9.7}$$

It follows from this and Theorem (8.23) that

$$S\left(\sum_{i=1}^n \frac{\lambda_i}{1-\lambda_0}\tau_i^*\rho\right) \geq \frac{\lambda_1}{1-\lambda_0}S\left(\tau_1^*\rho\right) + \frac{1-\lambda_0-\lambda_1}{1-\lambda_0}S\left(\sum_{i=2}^n \frac{\lambda_i}{1-\lambda_0-\lambda_1}\tau_i^*\rho\right).$$

Put this in Eqn. (9.5) to get

$$S\left(\tau^*\rho\right) \geq \lambda_0 S(\rho) + \lambda_1 S\left(\tau_1^*\rho\right) + (1-\lambda_0-\lambda_1)S\left(\sum_{i=2}^n \frac{\lambda_i}{1-\lambda_0-\lambda_1}\tau_i^*\rho\right). \tag{9.8}$$

Proceeding in the same way, and using the fact that τ_i are automorphisms, we arrive at

$$S\left(\tau^*\rho\right) \geq (\lambda_0 + \lambda_1 + \ldots + \lambda_n)S(\rho) + \varepsilon = S(\rho) + \varepsilon.$$

This shows that S is a strict Lyapunov function, since $\varepsilon = 0$ only at a fixed point. Since the fixed point is unique, and the space is compact, the theorem follows from Lyapunov's direct method. □

9.3 Quantum Dynamical Maps

A random dynamics τ generated by a mixture of automorphisms τ_i of a C^*-algebra $\mathcal{A}$ is a *stochastic map*, that is, it obeys

1. τ is linear;

2. τ is positive (takes positive elements to positive elements);

3. $1\tau = 1$ (τ preserves the identity).

Note that it is (3), (in which τ acts on the right, remember) that ensures that the dual action on the states preserves the normalising condition $\langle \rho, 1 \rangle = 1$. This dual action can be taken as the starting point of a theory of quantum stochastic processes [56]. We shall adopt the notation T for a general stochastic map, and write it on the left, thus: $A \mapsto T(A)$. We shall reserve the right-action $A \mapsto A\tau$ for the special case of random dynamics. The set of stochastic maps on an algebra forms a convex set. A nice result of Størmer [65] shows that a stochastic map T is a contraction in the norm, that is $\|T(A)\| \leq \|A\|$. This is the analogue of the classical result, Theorem (3.14). To prove it, note that if T is stochastic and A commutes with its adjoint, then $\|T(A)\| \leq \|A\|$; for we can simultaneously diagonalise both A and A^*, and the problem reduces to an abelian algebra, to which we can apply Theorem (3.14). In particular, T is contractive on unitary operators, and therefore on the convex span of the unitaries. But any operator on the unit sphere in $\mathcal{A}$ is the norm limit of a convex sum of unitaries; see [58], Vol. I, Cor. 3.2.6. □

The dynamics generated by a stochastic map is a very general concept, similar to the general Markov chain in classical theory, in which we lose both laws of thermodynamics. A random dynamics has a further property, that of preserving the trace: $Tr\, A\tau = Tr\, A$. This makes sense directly if $\dim \mathcal{A} < \infty$; in general, we can look at this condition only for operators of trace class. A stochastic map that also preserves the trace is called *bistochastic*. There is no problem in formulating the trace condition if we are interested only in maps that leave the spectral projections P_E of the energy invariant, and these are of finite dimension. We can then impose the condition for bistochasticity directly on the action of the map on the algebras $\mathcal{A}_E$. More generally, let $\mathcal{A}$ be any algebra of finite dimension; the trace can be used to provide it with a scalar product

$$\langle A, B \rangle = Tr\,(A^* B). \tag{9.9}$$

The corresponding norm is called the Hilbert-Schmidt norm. Indeed, let $A \in \mathcal{A}$; then the map $A \mapsto (\dim \mathcal{A})^{-1} Tr\, A$ is a state ρ on $\mathcal{A}$, the microcanonical state, and the scalar product (9.9) on $\mathcal{A}$ is that given in the GNS theory. The state

ρ is faithful, since if A is a positive matrix and $Tr\,A = 0$, then $A = 0$. The stochastic map T then defines a map $\pi_\rho(A) \mapsto \pi_\rho(T(A))$ on the GNS Hilbert space $\mathcal{H}_\rho$. The Hermitian conjugate of this map then coincides with the action $T^\dagger$ on the states, identified as density matrices. It is therefore positive as well as trace preserving. Thus we may say that a map T is bistochastic if and only if both T and $T^\dagger$ are stochastic. In finite dimensions this is equivalent to the condition that T is stochastic and $T^\dagger$ has the microcanonical state as a fixed point. If $\dim \mathcal{A} = \infty$ we shall limit discussion to normal T; we denote the subspace of Hilbert-Schmidt operators in $\mathcal{A}$ by $\mathcal{A}_2$; this is the set of operators with finite Hilbert-Schmidt norm. The set of bistochastic maps on an algebra is clearly convex (and obviously closed if the dimension is finite). Any element is therefore a mixture of the extreme points (Minkowski's theorem). The reversible dynamics, given by unitary conjugation, are extreme points of this set. One might conjecture that an analogue of Birkhoff's theorem holds, but we are out of luck: not every bistochastic map on an algebra of dimension bigger than two is a mixture of unitary conjugations [66, 67]. The general form of the extreme points is not known. Nevertheless, the main property of random dynamics, that it increases the entropy, does hold for the more general case of bistochastic maps [60]. We shall show that the increase in entropy is strict, using Theorem (8.23) and the following theorem, to be found in [16], Theorem 2-2:

Theorem 9.10 *Let $A \in \mathcal{A}$ be given, with $\dim \mathcal{A} < \infty$, let T be a bistochastic map acting on $\mathcal{A}$, and let $B = T(A)$. Then there exists a random dynamics $\tau = \sum_i \lambda_i \tau_i$ such that $B = A\tau$.*

Remark: This does not mean that any bistochastic map is a mixture of unitary conjugations. In the theorem, the random dynamics τ only does the job of T for the single matrix A, and not all matrices.

Theorem 9.11 *Let T be a bistochastic map on $\mathcal{A}$ and let ρ be a density matrix such that there is no unitary operator U with $T^\dagger\rho = U\rho U^{-1}$. Then $S(T^\dagger\rho) > S(\rho)$.*

Proof. Write
$$T^\dagger\rho = \sum \lambda_i \tau_i^* \rho = \tau^* \rho$$
with at least two non-zero terms, as $T^\dagger$ is not unitary. Choose one of the τ_i to play the role of τ_0, and write $\tau = \tau_0\tau'$. We can drop the factor τ_0 since it does not change the entropy. The dynamics τ' has the form $\lambda_0 1 + \sum \lambda_i \tau_i'$, and Theorem (8.23) gives the result. □.

The hypothesis that the dynamics is ergodic relative to H means that there is a unique invariant state within each $\Sigma(\mathcal{A}_E)$, and since the dynamics maps each such space, the energy-shell, to itself, we can in the ergodic case apply Lyapunov's direct method to show that the dynamics generated by $T^\dagger$ converges to equilibrium for any initial state. The limit state is a mixture of microcanonical states.

Before we give you yet another estimate [60] for the gain in entropy under a bistochastic map, we note that a bistochastic map T, and its adjoint $T^\dagger$ relative to the Hilbert-Schmidt scalar product, are contractions in the operator norm. This is because they are stochastic maps, so we can apply the result of Størmer [65]. We now show that they are contractions in the Hilbert-Schmidt norm too.

Theorem 9.12 *For any bistochastic map T on an algebra $\mathcal{A}$, with* $\dim \mathcal{A} = n < \infty$, *and $A \in \mathcal{A}$, we have $\|T(A)\|_2 \leq \|A\|_2$.*

Proof. The Hilbert-Schmidt norm is related to that given by the microcanonical state ω_0 by a factor $n^{-1/2}$, thus:

$$\|A^*A\|_2 = (Tr\,(A^*A))^{1/2} = n^{-1/2}\,(\omega_0(A^*A))^{1/2} = n^{-1/2}\|A\|_0.$$

It is therefore enough to show the contractive property for $\|\bullet\|_0$. We find

$$\begin{aligned}
\langle T(A),\, T(A)\rangle_0 &= \langle A,\, T^\dagger(T(A))\rangle_0 \\
&\leq \|A\|_0\, \|T^\dagger(T(A))\|_0 \\
&= \left(\|A\|_0 \langle T^\dagger(T(A)),\, T^\dagger(T(A))\rangle_0\right)^{1/2} \\
&= \|A\|_0 \langle A,\, T^\dagger\left(T(T^\dagger(T(A)))\right)\rangle_0^{1/2} \\
&\leq \|A\|_0\, \|A\|_0^{1/2} \|T^\dagger\left(T(T^\dagger T(A))\right)\|_0^{1/2} \leq \ldots \\
&\leq \|A\|_0^{2-2^{-n}} \|(T^\dagger T)^{2^n} A)\|_0^{2^{-n}}.
\end{aligned}$$

Now, both $T^\dagger$ and T are norm contractions, and so is their product any number of times, and putting $B = (T^\dagger T)^{2^n} A$, we see that $\|B\|_0 \leq \|B\| \leq \|A\|$. Letting $n \to \infty$, we get

$$\|T(A)\|_0^2 \leq \|A\|_0^2$$

since $\lim \|A\|^{2^{-n}} = 1$. □

In infinite dimensions, we are interested in T which map the finite-dimensional algebras $\mathcal{A}_E = P_E \mathcal{A} P_E$ to themselves. Then we can define an energy-conserving bistochastic map as a stochastic map T which maps the density operators $(\dim P_E)^{-1} P_E$ to themselves. On the operators in each $\mathcal{A}_E$ we can apply the above theorem, showing that $T^\dagger$ is continuous in the Hilbert-Schmidt norm, and can be extended to all Hilbert-Schmidt operators.

We now prove our gap estimate.

Theorem 9.13 *[60] Let $B = T^{\dagger}(A)$, where T is bistochastic, and suppose that the operator $TT^{\dagger}$, acting on the Hilbert space of Hilbert-Schmidt operators on a Hilbert space, has a spectral gap $\gamma > 0$ (so that the spectrum of $TT^{\dagger}$ lies in $[0,\, 1-\gamma] \cup \{1\}$), then*

$$S(B) - S(A) \geq \frac{\gamma}{2}\|P^{\perp}A\|_2^2. \tag{9.14}$$

Here, $P^{\perp} = 1 - P_1$, where P_1 is the projection, in the Hilbert space of Hilbert-Schmidt operators, onto the eigenvalue 1 of $TT^{\dagger}$.

Proof. Write $B = \sum \lambda_i A_i$, with $A_i = U_i A U_i^{-1}$. Then for each i, by the quantum Kullback inequality,

$$Tr\, A_i\left(\log A_i - \log B\right) \geq \frac{1}{2} Tr\, \left(A_i - B\right)^2.$$

Multiply by λ_i and sum, noting that $Tr\,(A_i \log A_i) = Tr\, A \log A$, we get

$$\begin{aligned}
S(B) - S(A) &= Tr\,(A\log A - B\log B) \\
&\geq \frac{1}{2}\sum_i \lambda_i Tr\,\left(A_i^2 - A_iB - BA_i + B^2\right) \\
&= \frac{1}{2}\left(\sum_i \lambda_i Tr\, A^2 - \sum_i Tr\,(\lambda_i A_i B) - Tr\, B\sum_i \lambda_i A_i + Tr\, B^2\right) \\
&= \frac{1}{2}\left(\langle A,\, A\rangle - \langle B,\, B\rangle\right) \\
&= \frac{1}{2}\left(\langle A,\, A\rangle - \langle T^{\dagger}A, T^{\dagger}A\rangle\right) \\
&= \frac{1}{2}\langle A,\, \left(1 - TT^{\dagger}\right) A\rangle.
\end{aligned}$$

Now writing P_x for the spectral resolutions of $T^{\dagger}T$, we have

$$\begin{aligned}
1 - TT^{\dagger} &= 1 - P_1 - \int_0^{1-\gamma} x\, dP_x \geq 1 - P_1 - (1-\gamma)\int_0^{1-\gamma} dP_x \\
&= 1 - P_1 - (1-\gamma)P^{\perp} = \gamma P^{\perp}.
\end{aligned}$$

Thus

$$S(T^{\dagger}A) - S(A) \geq \frac{\gamma}{2}\langle A,\, P^{\perp}A\rangle$$

which gives the result. □

This result means that if the dynamics is ergodic on each energy-shell $\mathcal{A}_E$, with a gap $\gamma_E > 0$, for all E, then the entropy increases to that of the microcanonical state and the system converges in Hilbert-Schmidt norm to the microcanonical state. It is sometimes more useful than Theorem (9.4) in that the rate of convergence is determined by γ, which is independent of A.

Another property true of random dynamics but not true for the general bistochastic map is complete positivity. This concept was introduced into analysis by Stinespring [68]; its importance in quantum mechanics was discovered by Kraus [69]. We have mentioned that an essential property of quantum probability is the existence of the ampliation, which is a map which takes an observable $A \in \mathcal{A}$ of a system into the observable $A \otimes I \in \mathcal{A} \otimes \mathcal{B}$ where I is the unit operator of a second system, described by the algebra $\mathcal{B}$, but which plays no role. Now a dynamical map τ must be positivity preserving, or else its dual action on the states will not map states to states. The ampliation expresses that we can regard our system as a part of the compound system, and for this to be true of its dynamics, we must require that $\tau \otimes I$ should also be positive, for any choice of the second algebra. This leads to

Definition 9.15 (Complete Positivity) *We say that a linear map τ on an algebra $\mathcal{A}$ is n-positive if $\tau \otimes I$ is a positive map on $\mathcal{A} \bigotimes \mathbf{M}_n$.*
We say that τ is completely positive if it is n-positive for all n.

In classical probability, every positive map is completely positive, so every stochastic process can be expanded by ampliation. One can see this by writing the time-step as the multiplication (in the natural basis in $\operatorname{Span}\Omega$) of the probability by a matrix with non-negative entries. The matrix of the map $\tau \otimes I$ in the product basis also has non-negative entries, and so is positive. Hence the original matrix was completely positive. We can also see that, in quantum probability, any implemented reversible dynamics is completely positive. For, if $\tau = \operatorname{Ad} U$, then put $V = U \otimes I$, and $\operatorname{Ad} V$ implements $\tau \otimes I$ and is positive. Moreover, any random reversible dynamics is also completely positive. To see this, simply note that the set of completely positive maps is convex; see Exercise (9.28). It is annoying that there are completely positive bistochastic maps that are not expressible as random dynamics, if the dimension is greater than 2. This was shown in dimension greater than 3 by Maassen and Kümmerer, [70], and by L. J. Landau and the author in dimension 3 [67]. This puzzling result may be due to the fact that we define a mixture of two maps using classical probability, whereas it might be more natural to use quantum probability in the definition.

If $X \in \mathcal{A}$, then the map $T_X : A \mapsto X^*AX$ is completely positive; for, $T_X \otimes 1_m$, acting on $\mathbf{M}_n \bigotimes \mathbf{M}_n$, is the map

$$A \mapsto (X \otimes 1)^* A (X \otimes 1), \text{ with } A \in \mathbf{M}_n \bigotimes \mathbf{M}_m$$

which is positive. Thus any convex sum of such maps is completely positive. The harder converse in due to Kraus [69]. We [67] shall now show more, that any n-positive map has this form, thus proving Choi's result [71] on the way (that any n-positive map on $\mathbf{M}_n$ is completely positive). We first introduce some notation.

Let m be any positive integer. To any $A \in \mathbf{M}_m$ we associate the linear functional $f_A : \mathbf{M}_m \to \mathbf{C}$ given by

$$f_A(B) = Tr\,(AB) = Tr\,(BA) = (A, B) \text{ say,} \tag{9.16}$$

and any linear functional is uniquely represented in this manner. We shall write Δ for the inverse of the map $A \mapsto f_A$, so that if f is a linear functional, Δf is the matrix such that $f(B) = Tr\,(B\Delta f)$. We have already met this for states in Exercise (8.32). We shall use this for $m = n^2$. If T is a linear map on $\mathbf{M}_n$ we denote by T^t the transpose, defined as the unique map obeying

$$(T^t(A), B) = Tr\,(T^t(A)B) = Tr\,(AT(B)) = (A, T(B)) \text{ if } A, B \in \mathbf{M}_n. \tag{9.17}$$

We note that $T^{tt} = T$ and that T^t is Hermitian if T is Hermitian, and that T^t is positive if T is positive (Exercise (9.29)). Moreover, T^t coincides with the adjoint $T^\dagger$ relative to the scalar product $\langle A, B\rangle = n^{-1} Tr\, A^*B = \rho_0(A^*B)$ where ρ_0 is the tracial state on $\mathbf{M}_n$. See Exercise (9.30). We also need the correspondence between linear maps T from $\mathbf{M}_n$ to itself and linear functionals L on $\mathbf{M}_n \otimes \mathbf{M}_n$, based on the definition

$$L_T(A \otimes B) = n^{-1} Tr\,\left(B^t T(A)\right) \tag{9.18}$$

where B^t denotes the transpose of the matrix B. Eqn. (9.18) defines L_T by linear extension to the whole of $\mathbf{M}_n \otimes \mathbf{M}_n$.

We have seen in the Gelfand-Naimark-Segal construction that an algebra acts on itself by left multiplication. This idea can be extended to both left and right actions, which always commute. Define an action of $\mathbf{M}_n \otimes \mathbf{M}_n$ on $\mathbf{M}_n$ by $(A \otimes B)C = ACB^t$, extended by linearity to sums of elements of the form $\sum A \otimes B$. Define the linear functional $L_{C,D}$ on $\mathbf{M}_n \otimes \mathbf{M}_n$ by the formula

$$L_{C,D}(M) = \langle C, MD\rangle$$

for $M \in \mathbf{M}_n \otimes \mathbf{M}_n$ regarded as an operator on $D \in \mathbf{M}_n$. Any linear functional on $\mathbf{M}_n \otimes \mathbf{M}_n$ can be expressed as a linear combination of such functionals, and defines the matrix $\Delta L \in \mathbf{M}_n \otimes \mathbf{M}_n$. We note that $L_{C,D} = L_T$ if and only if $T(A) = C^*AD$. It is enough to check this for $M = B \otimes B$. Then $\langle C, MD\rangle = \rho_0(C^*ADB^t) = n^{-1} Tr\,(B^t(C^*AD)) = L_T(A \otimes B)$. Clearly, the functional

$$M \mapsto L_{1,1}(M) = \langle 1, M1\rangle$$

is a state on $\mathbf{M}_n \otimes \mathbf{M}_n$, being the expectation in the unit vector $1 \in \mathbf{M}_n$. It is therefore a positive functional. The corresponding density matrix, $\Delta L_{1,1}$ is, as always for pure states, the projection in $B(\mathbf{M}_N)$ onto the unit vector 1.

Definition 9.19 *Let us say that a linear functional L on $\mathbf{M}_n \bigotimes \mathbf{M}_n$ is*

1. weakly positive *if $L(A \otimes B) \geq 0$ whenever $A \geq 0$ and $B \geq 0$.*
2. positive *if $L(M) \geq 0$ for all positive $M \in \mathbf{M}_n \bigotimes \mathbf{M}_n$.*

We note that if T is a positive map, then L_T is weakly positive, as follows from the fact that the trace of the product of two positive operators is non-negative. The main part of the proof of Kraus' formula is contained in the following

Theorem 9.20 *L_T is positive if T is n-positive.*

Proof. Let $\hat{T} = T \otimes \mathrm{Id}$. Then $\hat{T}$ is a positive map. Also

$$L_T(A \otimes B) = \rho_0(B^t T(A)) = L_{1,1}(T(A) \otimes B) = L_{1,1}\left(\hat{T}(A \otimes B)\right). \tag{9.21}$$

Thus

$$L_T = L_{1,1} \circ \hat{T}. \tag{9.22}$$

Hence L_T is positive, being the composition of two positive maps. □
If T is an n-positive stochastic map then L_T is a state. It can therefore be written as a convex combination of pure states, each of which is of the form $L_{C,C}$. Otherwise, if T does not preserve the identity, we can normalise L_T and then apply this decomposition. Thus any n-positive map T is a convex combination of maps $A \mapsto C^*AC$, with at most n^2 terms in the sum. It follows that T is completely positive (Choi's theorem [71]), and has the form given by Kraus. We can also regard the decomposition, which is not unique, into the decomposition of the density matrix ΔL_T into a spectral resolution of one-dimensional projections. We can therefore arrange to do this into elements C_i that are mutually orthogonal (in the sense of Hilbert-Schmidt).

Given algebras $\mathcal{A}_i$ with completely positive dynamics τ_i, then the combined system $\bigotimes \mathcal{A}_i$ can be given the dynamics $\otimes\tau_i$. This is positive. For it is the product of the positive maps $\tau_i \otimes I$, where I denotes the identity on the rest of the factors, other than $\mathcal{A}_i$. We shall take the theory further in connection with continuous semigroups of completely positive maps in Chapter 12.

9.4 Exercises

Exercise 9.23 Show that the map $U \mapsto \tau_U$ is a homomorphism from *UNI* $\mathcal{A}$ onto *INN* $\mathcal{A}$. Here, *UNI* $\mathcal{A}$ denotes the group of unitary elements of $\mathcal{A}$.

Exercise 9.24 Let τ_U be an inner automorphism of $\mathcal{A}$, and let $V \in \mathcal{A}$ be such that $\tau_U = \tau_V$. Prove that there exists a unitary element Z of the centre $\mathcal{Z}(\mathcal{A})$ such that $U = ZV$.

Exercise 9.25 Let $\mathcal{A}$ be a C^*-algebra and τ an automorphism of $\mathcal{A}$. Let $\rho \in \Sigma(\mathcal{A})$ be invariant under τ^*, and define a sesquilinear form on $\mathcal{A}$ by $\langle A, B\rangle = \rho(A^*B)$. Show that $\langle \bullet, \bullet\rangle$ is invariant under τ.

Exercise 9.26 Let U be a unitary operator commuting with the spectral projections of the self-adjoint operator H. Let $\rho \mapsto U^{-n}\rho U^n$ be the time-evolution in the Schrödinger picture. Show that any mixture of this dynamics for various n conserves the mean value of H.

Exercise 9.27 Suppose that $\dim \mathcal{A} < \infty$. Show that the set of random dynamics is a convex set. Show that its extreme points are the automorphisms. Hint: use the strict concavity of the entropy.

Exercise 9.28 Show that the set of completely positive maps on a C^*-algebra is convex.

Exercise 9.29 Let $T : \mathbf{M}_n \to \mathbf{M}_n$ be a linear map. Show that T^t, defined in Eqn. (9.17) obeys $T^t(A^*) = (T^t(A))^*$, and that T^t is positive if T is positive.

Exercise 9.30 Show that $\langle T^t(A), B\rangle = \langle A, T(B)\rangle$, the scalar product being that given by the tracial state.

Chapter 10

Isolated Quantum Dynamics

10.1 The Quantum Boltzmann Map

Suppose that our system consists of two parts, described by algebras $\mathcal{A}_1$ and $\mathcal{A}_2$, and we wish to consider the combined system. As in classical theory, we must distinguish three levels of separateness: if there is no functional relationship between the elements of the two algebras, it is traditional to say that they are "independent" degrees of freedom, and we can use the algebra $\mathcal{A} = \mathcal{A}_1 \bigotimes \mathcal{A}_2$ as the algebra of the combined system. If this holds, we can discuss the second level of separateness, which is whether the systems are "non-interacting" or interacting. This involves the form of the energy. If they are non-interacting, then we can write the total energy as a sum of operators, one for each subsystem:

$$H = H_1 \otimes 1 + 1 \otimes H_2. \tag{10.1}$$

If the algebra is a tensor product, and whether or not the systems are non-interacting, we can discuss whether they are statistically independent. This concept depends on the state, and not on the energy. The systems are statistically independent in the combined system if the state is a tensor product $\rho = \rho_1 \otimes \rho_2$. In this chapter, we will assume that the algebra $\mathcal{A}$ is a tensor product, and the energy has the form given in Eqn. (10.1); one step in the time-evolution will be given by a bistochastic map on $\mathcal{A}$; it can mix up states with the same total energy, and a state which is a tensor product at time $t = 0$ will not in general be such after one time-step. This happens even when the total energy is a sum, as in Eqn. (10.1), because the dynamical map is only required to conserve the total energy, and not the values of the energy of the two parts separately. The map that replaces a state $\rho \in \Sigma(\mathcal{A})$ by the tensor product of its marginals, $\mathcal{M}_1\rho \otimes \mathcal{M}_2\rho$ will be called the quantum stoss map, or just the stoss map. By Theorem (8.21), the map is entropy-increasing, and the increase is strict. This means that the entropy increases unless ρ is already a product state. For non-interacting systems, the stoss map does not change the mean energy:

Theorem 10.2 *If H has the form (10.1), then*

$$\langle \rho, H \rangle = \langle \mathcal{M}_1 \rho, H_1 \rangle + \langle \mathcal{M}_2 \rho, H_2 \rangle. \tag{10.3}$$

Proof. We have $\langle M_1 \rho, H_1 \rangle = Tr_1 Tr_2 \left(\Delta\rho(H_1 \otimes 1) \right) = \langle \rho, H_1 \otimes 1 \rangle$, and the same for the second term. □

We have seen that a dynamics generated by an ergodic bistochastic map τ acting on $\mathcal{A}$, in a way that conserves energy, will converge to a mixture of microcanonical states. It is not able to mix the different energy-shells; for this, some more stochasticity is needed. Just as in the classical case, in quantum statistical dynamics this is provided by the stoss map. Thus, let τ be an energy-conserving bistochastic map and define for $\rho \in \Sigma(\mathcal{A})$ the map

$$\rho \mapsto (\mathcal{M}_1 \tau^* \rho) \otimes (\mathcal{M}_2 \tau^* \rho). \tag{10.4}$$

This we call the Boltzmann map defined by τ. A similar idea was introduced and used by Alicki and Messer [72], who are also responsible for some of the basic properties. The Boltzmann map is non-linear; this might be why it was not considered by Thirring [20], who discusses (p.148) only the linear map $\rho_1 \mapsto \mathcal{M}_1 \tau^* (\rho_1 \otimes \rho_2)$. We use such maps in the next Chapter; there, ρ_2 is kept constant in time. This is ruled out in [20], on the grounds that in real systems the first system has an influence on the second system; for example, some energy might be transferred to it. We shall see that it is nevertheless a good model for isothermal systems, in which the second system is rapidly restored to its former state after each time-step.

The dynamics given by iterating a Boltzmann map obeys the first and second laws of thermodynamics: each of the two steps conserves the mean energy and does not decrease the entropy. Any canonical state ρ_β is given by a density operator $\Delta\rho_\beta$:

$$\Delta\rho_\beta = Z_\beta^{-1} e^{-\beta H} = Z_{1\beta} e^{-\beta H_1} Z_{2\beta} e^{-\beta H_2}.$$

It is a fixed point of a Boltzmann map, since it is a function of H as well as being a tensor product of states on each of the two algebras. We have already got some of the ingredients allowing us to prove that the iterated dynamics takes any initial state to equilibrium. The remaining properties that we need are that the fixed point be unique, that the increase in entropy be strict, and that the orbit be compact.

We are interested in theories with local structure. Thus, we have a finite set Λ, with $x \in \Lambda$ denoting the typical point. It may be that correlations between different points x are destroyed by random effects, which we can model by the local stoss map:

$$\rho \mapsto \otimes_x (\mathcal{M}_x \rho) \ , \tag{10.5}$$

which increases the entropy unless ρ is already a product state. We then define the *Boltzmann map with local independence* to be the map

$$\rho \mapsto \otimes_{x\in\Lambda} (\mathcal{M}_x \tau^* \rho) \ , \tag{10.6}$$

which obeys both laws of thermodynamics. A simpler description arises if we use the Boltzmann map with LTE: after the map (10.6) we replace each factor with the canonical state on $\mathcal{A}_x$ with the same mean energy as $\mathcal{M}_x\tau^*\rho$. This increases entropy even more, and obviously conserves mean energy. The state of the system is described by a beta field, which obeys self-contained equations of motion determined by τ. This would be a natural model if τ is ergodic relative to H. If there is a conserved quantity, such as the number of atoms, then we would use the grand canonical state with the same energy and particle number instead of the canonical state. Then the state is, at any time, described by a beta field $\beta(x)$ and a potential field $\mu(x)$, or equivalently, by the extensive fields $E(x)$ and $N(x)$. In either case the states follow an orbit inside a compact set (we assume that Λ is finite).

We can now formulate conditions under which the iterated map, applied to any initial state, converges to equilibrium. Suppose that on each algebra $\mathcal{A}_E = P_E \mathcal{A} P_E$ the map τ is ergodic, and is a mixture of the identity map and another automorphism. Then the entropy is a strict Lyapunov function and any fixed point of τ^*_E is the uniform state on $\mathcal{A}_E$. Because of strictness, to be a fixed point of the Boltzmann map with LTE, the state must be invariant under both stages, and so must be a product of equilibrium states, as well as being the uniform state on each $\mathcal{A}_E$. We can now show that there is a unique LTE product state with these properties under a further assumption, which is very natural, namely, that the spectrum of each H_x is the same; this is a consequence of invariance under translation, which we have not yet imposed. We may suppose that each spectrum begins at zero, and is $\{0 < E_1 < \ldots\}$. The H_x all commute, since they are operators on different factors in a tensor product. They may therefore be simultaneously diagonalised. If the density matrix of the state is a constant on the energy shell $\mathcal{A}_{E_1}$, and at the same time is a product of canonical states at possibly different betas at different points x then all the betas must be equal. This follows immediately from the comparison of the coefficients of the projection onto the simultaneous eigenvectors of total energy E_1. It is even true that the only product state that is constant on each energy shell is a canonical state (Exercise (10.61)). Thus there is a unique LTE-state with a given mean energy. We can therefore prove that in models whose dynamics is given by an ergodic $T^\dagger$, followed by the LTE-map, the system converges to equilibrium from any initial state, by Lyapunov's theorem. Although the state-space is not compact, the set of product *LTE*-states with fixed mean energy is.

In Krylov's thesis [73] it is pointed out that more should be demanded of the theory than the convergence for large times; we should also get a prediction of the relaxation time. This is the time needed for the system, whatever its initial state, to get very close to equilibrium. Otherwise, we may simply be talking about the astronomical times of ergodic theory. It is an experimental fact that for many systems the convergence to equilibrium is exponentially fast, and the rate is independent of the initial state. But the notion is a bit vague, since when we are very close to equilibrium, further convergence is also exponential, namely, exponentially slow. The rate is controlled by the gap in the spectrum of $TT^\dagger$; if we have a gap γ_E on each energy-shell, then the convergence is in the sense of the Hilbert-Schmidt norm. In the classical isothermal case, in Theorem (5.33) we introduced a stronger norm (called the KMS norm) and showed that convergence could be expressed in terms of this. In the next section we obtain a related result in the quantum case.

10.2 The Quantum Heat-Particle

The quantum heat-particle is a quantum oscillator, and the effect of introducing it into a model is very similar to what happens in the classical case. In particular, it is fair to describe noise as thermalised heat-particles. The heat-particles themselves can represent the excitations of a mode of the electromagnetic field, or of the phonon field. The Hilbert space of the oscillator is ℓ^2, the space of sequences c of complex numbers $\{c_j\}$, $j = 0, 1, 2 \ldots$ such that $\sum |c_j|^2 < \infty$. The energy is the multiplication operator H_γ:

$$(H_\gamma c)_j = j\varepsilon c_j \text{ where } \varepsilon \text{ is the energy of one quantum} \tag{10.7}$$

so that H_γ is diagonal in the natural basis in ℓ^2. The suffix γ indicates that it is the energy of the heat-particle. The algebra of observables for the heat-particle is $\mathcal{A}_\gamma = B(\ell^2)$, the C^*-algebra of all bounded operators on ℓ^2. We could introduce any number of heat-particles, each with its own energy per quantum; $\mathcal{A}_\gamma$ is then the tensor product of the algebras, one of each type. We avoid the ambiguity in the tensor product of general C^*-algebras by choosing the spatial tensor product. That is, given two Hilbert spaces $\mathcal{H}_1$ and $\mathcal{H}_2$, we define $B(\mathcal{H}_1) \otimes B(\mathcal{H}_2)$ to be $B(\mathcal{H}_1 \otimes \mathcal{H}_2)$.

Consider the case with one heat-particle. The thermal state with beta β is easily computed, in the basis where H_γ is diagonal, to be the diagonal matrix $Z^{-1}e^{-\beta E_i}$. By Theorem (8.16) this is the state of maximum entropy, given the mean value of the energy. There is no grand canonical state for the heat-particle, as the number of quanta is simply proportional to the energy. In interactions, the number of heat-particles is not conserved; the total energy of the system is conserved, but the energy of the heat-particles is not.

The heat-particles are Bosons, designed to carry energy from one atom of material to another. The Hilbert space of a single oscillator is separable, and so is isomorphic to $L^2(\mathbf{R})$. While there are many isomorphisms between separable Hilbert spaces, in this case there are natural ones, which identify the basis vectors $\psi_0 = (1, 0, 0, \ldots)$, $\psi_1 = (0, 1, 0, \ldots) \ldots$ (up to a phase) with the eigenvectors of the number operator of the harmonic oscillator,

$$\mathcal{N} = \frac{1}{2}(p^2 + q^2 - 1)$$

where p and q are the Schrödinger operators for momentum and position, and we have put $\hbar = 1$. These p, q are the generators of the Weyl operators, $U(a)$ and $V(b)$, defined for $a \in \mathbf{R}$ and $b \in \mathbf{R}$ as one-parameter unitary groups on $L^2(\mathbf{R})$ by

$$\left(U(a)\psi\right)(x) = \psi(x + a); \qquad \left(V(b)\psi\right)(x) = \exp\{ibx\}\psi(x). \tag{10.8}$$

The unitary operators $U(a) = \exp\{ipa\}$ and $V(b) = \exp\{iqb\}$ are called the Weyl operators, and they obey the Weyl relations

$$U(a)V(b) = \exp\{iab\}V(b)U(a) \tag{10.9}$$

which yields the Heisenberg commutation relations $[q, p] = i1$ on differentiation. Weyl suggested that his was the more rigorous way to formulate the commutation relations, which involve unbounded operators. We can express the number operator as $\mathcal{N} = a^* a$, where a and its adjoint a^* are the annihilation and creation operators, related to p, q by

$$q = 2^{-1/2}(a^* + a), \qquad p = i2^{-1/2}(a^* - a).$$

The Heisenberg relations are then equivalent to $[a, a^*] = 1$. The annihilation operator a kills the ground state, that is, it obeys the "Fock condition" $a\psi_0 = 0$. The eigenvalues of $\mathcal{N}$ are $\{0, 1, 2, \ldots\}$, and are simple. The map which takes the normalised eigenfunction of $\mathcal{N}$ with eigenvalue n to the vector $\psi_n \in \ell^2$ is well defined up to a phase, and is the natural isomorphism we were talking about above. Under this isomorphism $\mathcal{A}_\gamma$ becomes identified with $B\left(L^2(\mathbf{R})\right)$.

In a system with local structure, we may attach a heat-particle to each site $x \in \Lambda$, all with the same energy per quantum. The algebra is then taken to be

$$\mathcal{A} = \bigotimes_x \left(\mathcal{A}_{c,x} \otimes \mathcal{A}_{\gamma,x}\right).$$

The physical interpretation of $H_{\gamma,x}$ could then be the kinetic energy of all the material located at x, added to the energy of the electromagnetic field around x. The same arguments used in the classical case hold here too: the kinetic energy in liquids thermalises much faster than the rate of chemical reactions, and the dynamics can be modelled by the Boltzmann map which decouples the heat-particles from the chemicals. Thus the dynamics is determined by an energy-conserving bistochastic map τ, and one time-step is the map

$$\rho \mapsto \mathcal{M}_c(\tau^*\rho) \otimes \mathcal{M}_\gamma(\tau^*\rho).$$

The map τ can cause a molecular reaction at x, by changing the number of heat-particles at x, producing them in an exothermic reaction and absorbing them in an endothermic reaction. It can also contain a hopping term, either for the molecules or the heat-particles.

If we want a model with a simpler description than this and with even more randomness, we can follow the Boltzmann map by the stoss map

$$\rho \mapsto \otimes_x \mathcal{M}_x \rho.$$

In these models the microstate of the heat energy at x at any time is described by the state $\mathcal{M}_{x\gamma}\rho$. This contains much more detail than is needed for chemistry, and we now discuss ways to simplify the treatment without spoiling the two laws of thermodynamics. If the state at any time, restricted to the algebra $\mathcal{A}_{\gamma,x}$, is a thermal state, we can define a local, non-equilibrium value for beta at x. The hypothesis of local thermodynamic equilibrium can be expressed by saying that the local kinetic energy at x thermalises so fast that in one time step, on the scale of the chemical reactions, the local state of the heat-particle evolves into the thermal state with the same energy as it had after the map τ. We can implement this mathematically by following the Boltzmann map and the stoss map by the LTE map

$$\rho_{\gamma,x} \mapsto \rho_{\beta(x)},$$

where $\rho_{\beta(x)}$ is the thermal state with the same mean energy as $\rho_{\gamma,x}$. Models of this sort describe the heat content by a field $\beta(x)$, and do not keep tabs on the details of which molecules possess the heat. It is usually adequate for chemical reactions not involving the specific excitation of certain individual atoms, e.g by lasers.

It is also possible to localise the heat-particle on the bond, or link, between neighbouring sites, as we did in the classical case. This is natural in modern quantum field theory, where gauge fields are thought to reside on the links between the particles.

We now develop the theory of Bosons further. We can identify ℓ^2 with the Fock space over the one-dimensional Hilbert space $\mathbf{C}$. To include the case with several heat-particles, let us give the construction of Fock space for any Hilbert space $\mathcal{H}$, which is called the "one-particle space". In our case, a vector in $\mathcal{H}$ describes a single heat-particle with a wave-function spread over the points of Λ. If several particles are present, we must remember that Bosons, spoken of as particles, are indistinguishable, and their configurations require the field description. This is achieved in quantum mechanics by the quantised field. A pure state of several particles is described by a vector in the symmetric tensor product

$$\mathcal{H}_n = (\mathcal{H} \otimes \mathcal{H} \otimes \ldots \otimes \mathcal{H})_s \qquad n \text{ factors.}$$

The suffix s indicates that only vectors in the tensor product that are invariant under the permutation of the factors are to be included. In doing Exercise (10.62) you will prove that the space of symmetric tensors is a vector subspace of the full tensor product, and is spanned by symmetric product vectors, that is, vectors of the form $\otimes^n \psi = \psi \otimes \psi \ldots \otimes \psi$, with $\psi \in \mathcal{H}$. Then the *Fock space over* $\mathcal{H}$ is the direct sum

$$F(\mathcal{H}) = \mathbf{C} \oplus \sum_1^\infty {}^\oplus \mathcal{H}_n. \tag{10.10}$$

In particular, if $\mathcal{H} = \mathbf{C}$, then $F(\mathcal{H}) = F(\mathbf{C}) = \ell^2$, since all tensor powers of $\mathbf{C}$ give back just $\mathbf{C}$ itself. Thus the algebra of a single heat-particle acts on the Fock space over $\mathbf{C}$. The first term in the direct sum in Eqn. (10.10) is $\mathbf{C} = \mathcal{H}^0$, and the vector $1 \in \mathbf{C}$ is called the Fock vacuum. We shall denote this by Ψ_F.

Fock spaces have a "functorial" property,

$$F(\mathcal{H}_1 \oplus \mathcal{H}_2) = F(\mathcal{H}_1) \otimes F(\mathcal{H}_2). \tag{10.11}$$

This means that the Hilbert space of two Boson degrees of freedom, which is the *unsymmetrised* tensor product of two oscillator spaces, is the Fock space over $\mathbf{C}^2$; similar remarks hold for any number of degrees of freedom. This answers a question that has puzzled some students: why is the tensor product (not symmetrised) of two Bose systems correctly obtained by introducing an internal label to the particle, and then using the symmetric Fock construction? In the Fermion case, this was done instinctively by Heisenberg when he introduced isotopic spin, after which he treated the proton and the neutron as two states of the same particle. We shall see that Fermion space has a similar functorial property. The functorial property has induced some people to use the notation $e^{\mathcal{H}}$ for $F(\mathcal{H})$; Eqn. (10.11) is easily proved using the very useful concept of *coherent state*, which in Fock space has a simple description:

Definition 10.12 *For $\psi \in \mathcal{H}$, the coherent state with wave-function $\psi \in \mathcal{H}$ is the vector denoted e^ψ, in $F(\mathcal{H})$, given by*

$$e^\psi = 1 \oplus \psi \oplus (2!)^{-1/2} \psi \otimes \psi \oplus (3!)^{-1/2} \otimes^3 \psi \oplus \ldots. \tag{10.13}$$

Theorem 10.14 *The coherent states form a total set in $F(\mathcal{H})$.*

Proof (Araki). It suffices to show that the Hilbert space spanned by the coherent vectors contains each $\otimes^n\psi$, for any $\psi \in \mathcal{H}$, since such vectors span the symmetric tensor product (Exercise (10.62)). Proceed by induction. For $n = 0$, it is clear that $1 = e^0$ spans $\mathbf{C} = \mathcal{H}^0$. Now assume that $(\otimes^j\mathcal{H})_s$ is spanned by the vectors e^ψ. Consider the family of vectors

$$\Psi_n(\lambda) = e^{\lambda\psi} - \sum_{j=0}^{n-1}(j!)^{-1/2}\,(\otimes\lambda\psi)^j\,.$$

We have

$$\lim_{\lambda\to 0}\Psi_n/(\lambda^n) = (n!)^{-1/2}\,(\otimes\psi)^n$$

in the sense of strong convergence. Since the limit lies in the closure of the span of the coherent vectors, we have the induction hypothesis for n. □

The notation e^ψ is suggested by the easy relation

$$\langle e^\phi, e^\psi\rangle = \exp\{\langle\phi,\psi\rangle\}. \tag{10.15}$$

This is proved by summing the exponential series, noting that direct sums are orthogonal. Now the functorial property is easy:

Theorem 10.16 *There is an isomorphism U between $F(\mathcal{H}_1\oplus\mathcal{H}_2)$ and $F(\mathcal{H}_1)\otimes F(\mathcal{H}_2)$ such that*

$$Ue^{\phi_1\oplus\psi_2} = e^{\psi_1}\otimes e^{\psi_2}.$$

Proof.

$$\|e^{\psi_1\oplus\psi_2}\|^2 = e^{\|\psi_1\oplus\psi_2\|^2} = e^{\|\psi_1\|^2+\|\psi_2\|^2} = e^{\|\psi_1\|^2}.e^{\|\psi_2\|^2} = \|e^{\psi_1}\otimes e^{\psi_2}\|^2.$$

Hence U is an isometric and hence bijective map between total sets, and so can be extended to a unitary map. □

We note that the Fock vacuum is the coherent state corresponding to zero wave-function: $\Psi_F = e^0$. To any unitary operator U on $\mathcal{H}$, we define its "second quantisation" ΓU on $F(\mathcal{H})$ by its action on the coherent vectors

$$\Gamma U e^\psi = e^{U\psi}.$$

We can check that this is unitary on each pair of coherent states, and thus on the linear span. It can therefore be extended by continuity to the whole Fock space, since the coherent states are total. This can be done for each element of a one-parameter group, say $U(t)$, acting on $\mathcal{H}$. This might be the time-evolution of a single particle, for example, with infinitesimal generator given by the one-particle Hamiltonian, h say. Then $\Gamma U(t)$ is a one-parameter unitary group acting on $F(\mathcal{H})$, whose generator is called the second quantisation $d\Gamma h$ of h. This notation and most of the theory of this is due to Segal [57, 74]. By differentiating $\Gamma U(t)$ in time we find that $d\Gamma h$ is the correct free Hamiltonian for a many-particle system:

$$d\Gamma h(\psi \otimes \psi \otimes \ldots \otimes \psi) = \sum_j \psi \otimes \psi \ldots \otimes h\psi \otimes \psi \ldots \otimes \psi \tag{10.17}$$

where the one-particle operator h acts on the jth factor. As an example, consider the simplest one-parameter group on $\mathcal{H}$, namely $u(t) = e^{it}1$, where 1 is the unit operator on $\mathcal{H}$. Then the generator of $\Gamma u(t)$ is the number operator on $F(\mathcal{H})$. The coherent states e^{ψ} we have defined are not normalised. This is easily remedied by dividing by $e^{-\|\psi\|^2/2}$. The Weyl operators $U(a)$, $V(b)$ have a neat version in terms of their action on coherent states. To each $\psi \in \mathcal{H}$, define the "Segal-Weyl" operator $W(\psi)$ by

$$W(\psi)e^{\phi} = \exp\{\frac{1}{2}(\|\phi\|^2 - \|\phi + \psi\|^2) + i\, Im\langle\phi, \psi\rangle\}e^{\phi+\psi}. \tag{10.18}$$

In Exercise (10.63), you will check that this is a unitary map when acting on the coherent states. Since these are total, we can extend $W(\psi)$ uniquely to a unitary operator on the whole of Fock space. The operators $W(\psi)$ obey the Segal-Weyl relations

$$W(\psi)W(\phi) = \exp\{iIm\langle\phi, \psi\rangle\}W(\psi + \phi). \tag{10.19}$$

To prove this, just verify it on a general coherent state e^{χ} (Exercise (10.64)). The secret behind this is that the Segal-Weyl operators $W(\psi)$ are unitary, obey Eqn. (10.19), and that $W(\psi)$ creates the coherent state e^{ψ} when acting on the vacuum, except for normalisation. If we choose the phase so that

$$W(\psi)c^0 = \exp\{-\|\psi\|^2/2\}e^{\psi},$$

this is enough to determine the operator $W(\psi)$. We see that the product of two Segal-Weyl operators is a complex multiple of another. Since obviously $W(\psi)^* = W(-\psi)$, we conclude that the linear span of Segal-Weyl operators, as ψ runs over the one-particle space, is a $*$-algebra. We can obtain the Weyl operators $U(a)$ and $V(b)$, if $\dim \mathcal{H} = 1$, by choosing a particular complex conjugation on $\mathcal{H}$. For example, if we identify $\mathcal{H}$ with $\mathbf{C}$, we could take complex conjugation on $\mathbf{C}$. Then we can speak about real vectors in $\mathcal{H}$. This is the quantum analogue of choosing a Lagrangian subspace in classical Hamiltonian theory. The point is that Weyl's (or Schrödinger's) form of the commutation relations depends on a selection of one canonical variable as the co-ordinate q, and its conjugate, p, as momentum. This is needed in Lagrange's equations of classical mechanics, and in the Lagrangian approach to quantum mechanics. Hamilton's form is invariant under canonical transformations; in quantum mechanics, linear canonical transformations are exactly those preserving the bilinear form $Im\langle\phi,\psi\rangle$. Thus if $\mathcal{H} = \mathbf{C}$, we can identify $F(\mathcal{H})$ with ℓ^2 and for $a + ib \in \mathbf{C}$ put $U(a) = W(-a/\sqrt{2})$, $V(b) = W(ib/\sqrt{2})$. Then these obey the Weyl relations. This provides a link between Segal's canonically covariant formalism and the Schrödinger theory. By the isomorphism Theorem (10.16), we can identify the Hilbert space $L^2(\mathbf{R}^n)$ of n oscillators ($n < \infty$) with $F(\mathbf{C}^n)$.

It is known that the Schrödinger operators x and $-id/dx$ form an irreducible representation of the Heisenberg commutation relations on $L^2(\mathbf{R})$. It follows that the Segal-Weyl operators form an irreducible family on Fock space (by the isomorphism between the spaces). Thus every bounded operator on Fock space is in the strong closure of the algebra generated by the Segal-Weyl operators. In this book we limit discussion to a finite number of degrees of freedom. Then we can take the C^*-algebra $\mathcal{A}_\gamma$ of the Bose field to be $B(F(\mathcal{H}))$, the set of all bounded operators on Fock space, just as we did above. The Stone-von Neumann theorem tells us that if $\dim\mathcal{H} < \infty$, every continuous irreducible representation of the Segal-Weyl relations is unitarily equivalent to the Fock representation. We need to clarify this; all Fock spaces are isomorphic, and so we need to distinguish between models with different numbers of degrees of freedom and oscillator frequencies, but in an algebraic way valid for all representations of $\mathcal{A}_\gamma$. Segal does this by defining a Weyl system not simply as $\mathcal{A}_\gamma$, but to be a regular map W from the one-particle space $\mathcal{H}$ into $\mathcal{A}_\gamma$ obeying the Segal-Weyl relations Eqn. (10.19). Two representations π_1 and π_2 of the Weyl system, on Hilbert spaces $\mathcal{H}_1$ and $\mathcal{H}_2$, are unitarily equivalent if there exists a unitary map, U, from $\mathcal{H}_1$ to $\mathcal{H}_2$, such that

$$\pi_1(W(\psi)) = U\pi_2(W(\psi))U^{-1} \qquad \text{for all } \psi \in \mathcal{H}.$$

A state ρ on $\mathcal{A}_\gamma$ is said to be *regular* if in the corresponding cyclic representation π_ρ the map $\psi \to W(\psi)$ is continuous. In this case by Stone's theorem its generators, related to the operators p and q, are self-adjoint, and could be taken to represent observables. Because of the Segal-Weyl relations, any operator in $\mathcal{A}_\gamma$ is the limit of *sums* of the form $\sum c_i W(\psi_i)$ with complex constants c_i. Thus any regular state on $\mathcal{A}_\gamma = B(F(\mathcal{H}))$ is determined by the expectation $\mathcal{F}$ it gives for the Weyl operators:

$$\mathcal{F}(\psi) = \rho(W(\psi)) \qquad \text{(the generating function given by } \rho). \tag{10.20}$$

This is the analogue of the "characteristic function" of a probability measure $P(x)$ on the line, as in Eqn. (2.22),

$$\mathcal{F}(t) = \int \exp\{itx\}dP(x)$$

and it has similar properties. Thus $\mathcal{F}(\psi)$ is a sort of non-abelian Fourier transform. The following is a version of the Araki-Segal twisted Bochner theorem [75, 74]:

Theorem 10.21 *Let M be a finite-dimensional real vector space with a non-degenerate symplectic form $B(\bullet, \bullet)$, and let $\psi \mapsto \pi_\rho(W(\psi))$ be a representation of the Weyl-Segal relations, given by the regular state ρ. Then $\mathcal{F}(\psi) = \rho(W(\psi)) = \langle \psi_\rho, \pi_\rho(W(\psi))\psi_\rho \rangle$ obeys*

1. *F(0)=1;*

2. *for each finite set $\{c_j\} \subseteq \mathbf{C}$, $\{\psi_j\} \subseteq M$,*

$$\sum_{j,k} \mathcal{F}(\psi_j - \psi_k) \exp\{iB(\psi_j, \psi_k)\} \bar{c}_k c_j \geq 0;$$

3. *the map $\psi \mapsto \mathcal{F}(\psi)$ is continuous.*

Conversely, a functional $\mathcal{F}$ obeying these properties comes from a unique regular state ρ.

The necessity of the conditions is not hard; (2) follows from the positivity of the norm of $\sum c_j W(\psi_j)\psi_\rho$; for the sufficiency, we note that $\mathcal{F}$ supplies us with a positive linear functional on the linear span of the symbols $W(\psi)$, which we have noted form an algebra when they obey the Segal-Weyl relations. This defines a putative norm which gives $\|1\| = 1$, and so the linear functional is bounded, and therefore continuous. The extension by continuity to the completion then gives us the claimed regular state on the Weyl system. If the state ρ is a bit more than regular (being an analytic vector for the number operator is enough, for example) then by differentiating the generating function along one-parameter subgroups we can generate the "Wightman" functions; for one degree of freedom these are the moments of the operators p and q in the cyclic vector ψ_ρ:

$$\langle \psi_\rho, p^n q^m \psi_\rho \rangle = (-i)^{m+n} \frac{\partial^m}{\partial a^m} \frac{\partial^n}{\partial b^n} \rho(V(b)U(a)) \tag{10.22}$$

and $\rho(V(b)U(a)) = \rho(W(ib/(\sqrt{2})W(-a/\sqrt{2}))$ can be expressed in terms of $\mathcal{F}$. Moments with the operators in any other order can be obtained similarly. It is known that the logarithm of $\mathcal{F}$ generates the correlation functions (also known as the cumulants, or the truncated functions, ρ_T). For any state ρ, these are defined recursively in terms of the moments, or Wightman functions, of any collection of operators $A_1, \ldots, A_m$ by

$$\rho(A_1 \ldots A_n) = \rho_T(A_1 \ldots A_n) + \sum_I \rho_T(I_1)\rho_T(I_2) \ldots \rho_T(I_r). \tag{10.23}$$

Here, I runs over all partitions $I = I_1 \cup I_2 \cup \ldots \cup I_r$ of the set $\{A_1, \ldots, A_n\}$ into r parts, and then r runs from 2 to n. In each partition, I_k stands for the product of $A_j \in I_k$ in the order of increasing j. We start the induction with the one-point function $\rho_T(A_k) = \rho(A_k)$. The correlation functions do not satisfy the positivity conditions of a state; they correspond in perturbation theory to the sum of the connected Feynman diagrams.

In classical probability, a Gaussian measure leads to a characteristic function which is the exponential of a quadratic form. Its logarithm is therefore a quadratic polynomial, and all correlations beyond the second order vanish. In a Segal-Weyl system, a similar class of states is the *quasi-free* class [76], with generating function

$$\mathcal{F}(\psi) = \exp\{-\frac{1}{2}\|\psi\|^2 - A(\psi,\psi) + iIm\langle f,\psi\rangle\} \tag{10.24}$$

where A is a positive quadratic form on M, and $f \in M$. Since for a quasifree state, $\log \mathcal{F}$ is a quadratic polynomial, we see that in a quasifree state, all correlation functions of the fields of order greater than 2 vanish. It was shown in [77] that if we have a state of the CCR-algebra, and all correlation functions vanish beyond some order, then the state is necessarily quasifree. This is the quantum analogue of the theorem of Marcinkiewicz [13].

Let ρ be a quasifree state of the CCR with $f = 0$, and let $(\pi_\rho, \mathcal{H}_\rho, \psi_\rho)$ be the Gelfand-Naimark-Segal representation determined by ρ. There is the following explicit realisation: let $F(\mathcal{H})$ be the Bosonic Fock space over $\mathcal{H}$, ψ_0 the Fock vacuum and $W_0(\psi)$ the Fock representation of the Segal-Weyl algebra on $F(\mathcal{H})$. Let J be a conjugation on $\mathcal{H}$ commuting with A. Then we may put

$$\begin{aligned}
\mathcal{H}_\rho &= F(\mathcal{H}) \otimes F(\mathcal{H}); \\
\psi_\rho &= \psi_0 \otimes \psi_0; \\
\pi_\rho(W(\phi)) &= W_0\left(-A^{1/2}J\phi\right) \otimes W_0\left((1+A)^{1/2}\phi\right).
\end{aligned}$$

This tensor structure is typical of the KMS theory, and was first made clear by Araki and Woods [78]. A coherent state e^ϕ is quasifree, with $A = 0$ and $f = 2\phi$, as can be seen by computing its generating function

$$\mathcal{F}(\psi) = \exp\{-\|\phi\|^2\}\langle e^\phi, W(\psi)e^\phi\rangle = \exp\{-\frac{1}{2}\|\psi\|^2 - 2iIm\langle\psi, \phi\rangle\}. \tag{10.25}$$

It is a pure state, and in fact, all pure quasifree states are of this form [79]. The equilibrium state of a single harmonic oscillator is also quasifree; its generating function can be found explicitly. The main part of the calculation is to find the expectation of the anti-Wick ordered Weyl operator (in a Wick-ordered operator, all creators lie to the left, and all annihilators to the right; we want just the opposite):

Theorem 10.26 *Let $H = \varepsilon a^* a$ acting on $L^2(\mathbf{R})$. Then $\Delta\rho = Z^{-1}e^{-\beta H}$ is of trace class, and defines a quasifree state ρ such that*

$$\rho\left(\exp\{az\}\exp\{a^*\overline{z}\}\right) = \exp\{z\overline{z}/(1-\zeta)\} \quad z,\overline{z} \in \mathbf{C} \text{ where } \zeta = \exp\{-\beta\varepsilon\}. \tag{10.27}$$

Proof. Let $\psi_n \in F(\mathbf{C})$ be the normalised n-particle state

$$\psi_n = 1 \otimes 1 \ldots \otimes 1 = (n)^{-1/2} a^{*n}\psi_0 = \frac{\partial^n}{\partial \overline{z}^n}\exp\{a^*\overline{z}\}\psi_0.$$

Then

$$\begin{aligned}
\rho\left(e^{az}e^{a^*\overline{z}}\right) &= Z^{-1}Tr\,\exp\{-\beta\varepsilon a^*a\} \\
&= Z^{-1}\sum_n n\exp\{-\beta\varepsilon\}\langle\psi_n, \exp\{az\}\exp\{a^*\overline{z}\}\psi_n\rangle \\
&= Z^{-1}\sum_n \zeta^n (n!)^{-1/2}(n!)^{-1/2}\frac{\partial^n}{\partial z^n}\frac{\partial^n}{\partial \overline{z}^n}\langle\psi_0, \exp\{az\}\exp\{a^*\overline{z}\}\psi_0\rangle \\
&= Z^{-1}\sum_n \frac{\zeta^n}{n!}\frac{\partial^n}{\partial z^n}\frac{\partial^n}{\partial \overline{z}^n}\exp\{z\overline{z}\} \\
&= Z^{-1}\sum_n \frac{\zeta^n}{n!}\frac{\partial^n}{\partial z^n}\left(z^n\exp\{z\overline{z}\}\right) \\
&= \frac{\exp\{z\overline{z}\}}{Z}\sum_n \frac{\zeta^n}{n!}\left(n! + n.\frac{n!}{1!}z\overline{z} + \frac{n(n-1)}{2!}\frac{n!}{2!}(z\overline{z})^2 + \cdots\right) \\
&= \frac{e^{z\overline{z}}}{Z}\left(\sum_n \zeta^n + z\overline{z}\zeta\sum_n n\zeta^{n-1} + \frac{(z\overline{z}\zeta)^2}{(2!)^2}\sum_n n(n-1)\zeta^{n-2} + \cdots\right) \\
&= \frac{e^{z\overline{z}}}{Z}\left(\frac{1}{1-\zeta} + z\overline{z}\zeta\frac{\partial}{\partial\zeta}\frac{1}{1-\zeta} + \frac{1}{(2!)^2}(z\overline{z}\zeta)^2\frac{\partial^2}{\partial\zeta^2}\frac{1}{(1-\zeta)^2} + \cdots\right) \\
&= \frac{\exp\{z\overline{z}\}}{Z(1-\zeta)}\left(1 + \frac{z\overline{z}\zeta}{1-\zeta} + \frac{(z\overline{z}\zeta)^2}{2!(1-\zeta)^2} + \cdots\right) \\
&= \exp\{z\overline{z}\}\exp\{z\overline{z}\zeta/(1-\zeta)\}
\end{aligned}$$

since $Z(1-\zeta) = 1$. Collecting the exponents together gives the result. □
This is the hard part. Note that we have treated z and $\overline{z}$ as independent variables. Putting iz for z, $i\overline{z}$ for $\overline{z}$ and using the Segal-Weyl relations Eqn. (10.19) gives us the generating function. Its logarithm is quadratic in z, so the state is quasifree: all truncated functions beyond the second vanish. We can read off the two-point function, $\rho(aa^*) = (1-\zeta)^{-1}$, which leads to Planck's distribution

$$\rho(a^*a) = \rho(aa^*) - 1 = \zeta/(1-\zeta).$$

More generally, if we have N non-interacting harmonic oscillators then the Hamiltonian can be written as a sum of modes, and the trace in the definition of the generating function of the canonical state is a product of traces. That is, in the equilibrium state, the degrees of freedom are statistically independent. The generating function is thus a product of exponentials, each quadratic, and so the canonical state is quasifree. We now consider the general quadratic Hamiltonian in N degrees of freedom, with bilinear couplings between the modes. This can be reduced to diagonal form by a canonical transformation, in which we replace the a_i^*, a_i, $1 \leq i \leq N$, with the creators and annihilators of the normal modes, called quasiparticles. If the Hamiltonian is not positive, this is not an elementary result [80]. We note that the one-particle space $\mathcal{H}$ is a real vector space of dimension $2N$, and is furnished with the non-degenerate symplectic form $B(\psi, \phi) = 2Im\langle\psi, \phi\rangle$. This is non-degenerate in the sense that $B(\psi, \phi) = 0$ for all ψ implies that $\phi = 0$. It is *symplectic* because it is $\mathbf{R}$-bilinear and antisymmetric: $B(\psi, \phi) = -B(\phi, \psi)$. A canonical transformation is then a real linear map $S : \mathcal{H} \to \mathcal{H}$ that preserves this form:

$$Im\langle S\psi, S\phi\rangle = Im\langle\psi, \phi\rangle.$$

Such a map S is also called a symplectic transformation. A special case is a unitary map, which preserves the real as well as the imaginary part of the scalar product. The property of being unitary needs the choice of a complex structure J on the symplectic space. A *complex structure* is an $\mathbf{R}$-linear map J on a real symplectic space $(\mathcal{H}, B)$ obeying

1. $J^2 = -I$;
2. $B(J\phi, \psi) = -B(\phi, J\psi)$;
3. $(\phi, \psi)_J = B(\phi, J\psi)$ is a positive definite bilinear form.

Then we can form a complex Hilbert space by defining multiplication by i to be $\psi \mapsto J\psi$, and a complex scalar product to be

$$\langle\psi, \phi\rangle = (\psi, \phi)_J + iB(\psi, \phi).$$

Each choice of J leads to the construction of a Fock space, which is based on a complex Hilbert space, not a symplectic space. Thus the definitions of a and a^* depend on J. A symplectic transformation is complex-linear if it commutes with J. That S is not obliged to be complex-linear is what permits canonical transformations to mix up p and q, or a^* and a. The set of all symplectic maps is a real Lie group, denoted $Sp(2N, \mathbf{R})$, and called the real symplectic group. The maximal compact subgroup of $Sp(2N, \mathbf{R})$ is isomorphic to $SU(N)$. How this subgroup sits inside the symplectic group depends on which choice of complex structure we have made. For a choice of J allows us to define a real symmetric form which can act as the real part of a Hilbert scalar product.

If $N = 1$ a symplectic map is a real linear map of $\mathbf{R}^2$ that preserves the area. From this we see that the dimension of the group $Sp(2, \mathbf{R})$ is three, as it is given by the four parameters of a matrix in $GL(2, \mathbf{R})$ subject to one condition, that the area is preserved. We can think of $Sp(2N, \mathbf{R})$ as the set of $2N \times 2N$ real matrices S preserving the "symplectic metric" G:

$$\begin{pmatrix} S_{11}^t & S_{21}^t \\ S_{12}^t & S_{22}^t \end{pmatrix} \begin{pmatrix} 0 & I \\ -I & 0 \end{pmatrix} \begin{pmatrix} S_{11} & S_{12} \\ S_{21} & S_{22} \end{pmatrix} = \begin{pmatrix} 0 & I \\ -I & 0 \end{pmatrix} = G.$$

Here, S_{ij} are $N \times N$ real matrices, and I is the unit $N \times N$ matrix. The metric G arises from the Heisenberg commutation relations; write $z = \{q_1, q_2, \ldots, q_N\}$. The CCR can be written

$$[z_i, z_j] = i\hbar G_{ij};$$

then the new coordinates $z' = Sz$ obey the commutation relations if and only if $SGS^t = G$, as in Exercise (10.67), and if we write z as a complex coordinate $\zeta = q + ip$, $p \in \mathbf{R}^N$, $q \in \mathbf{R}^N$, then $Im\langle \zeta', \zeta\rangle = z'Gz$. So the symplectic maps are canonical, and preserve the Segal-Weyl relations. We can think of a symplectic map as a change in the complex structure from one J to another. We immediately see that S induces an automorphism of the C^*-algebra $\mathcal{A} = B(F(\mathcal{H}))$, where $\mathcal{H} = \mathbf{C}^N$. Thus if $\psi \mapsto W(\psi)$ obeys the Segal-Weyl relations, then so does $\psi \mapsto W(S\psi)$, and therefore gives a representation of $\mathcal{A}$. By the Stone-von Neumann theorem, this representation is unitarily equivalent to the original one, so there exists a unitary operator $U(S)$ in $\mathcal{A}$ such that $W(S\psi) = UW(\psi)U^{-1}$ for all $\psi \in \mathcal{H}$. This is true of any representation, as the automorphism is inner. We can throw the automorphism on the states, by duality; it is clear that the states transform by the inverse unitary map

$$\Delta\rho \mapsto U^{-1}\Delta\rho U;$$

This converts density operators to density operators, so we do not leave the class of normal states. A unitary conjugation does not change the spectrum of any operator; nor does it change the trace, and so a canonical transformation does not change the entropy of any state. The implementing operator U coincides with ΓS, the second-quantised operator determined by S, when S is unitary. A quadratic Hamiltonian will mean an operator of the form

$$H = z^t hz, \qquad \text{where } z^t = (q_1 \ldots, q_N, p_1, \ldots, q_N)$$

and h is a matrix which in the (Lagrangian) splitting above is of the form

$$h = \begin{pmatrix} C & A^t \\ A & B \end{pmatrix} \qquad B = B^t, \qquad C = C^t.$$

If we can find S which diagonalises the matrix h then the corresponding U will diagonalise the operator H. It can be proved that any quadratic Hamiltonian whose second-quantised form (by which we mean that the p and q making it up are operators obeying the Heisenberg relations) is a positive operator can be diagonalised to a sum of harmonic oscillators by a canonical transformation. In Exercise (10.68) you are asked to try it out for a single degree of freedom. The case of two degrees of freedom is solved in detail in [80]. If a canonical state $\rho = Z^{-1}e^{-\beta H}$ is to be of trace class, clearly H must be bounded below. When written in diagonal form, we apply our result above to show that ρ is quasifree. The symplectic transformation does not change the quadratic nature of the exponent in the generating function, so we conclude that all density operators with quadratic Hamiltonians are quasifree. The converse is also true; the general quasifree state has generating function of the form Eqn. (10.24), and we can diagonalise the operator A by a unitary transformation, and remove the f by a change in origin, $z \mapsto z - \zeta$, where ζ is a suitable c-number. Then the generating function has the form of a product over quasiparticle modes, and is therfore a canonical state for some diagonal oscillator Hamiltonian. This state has the maximum entropy of any state with the same mean energy.

We can now find the entropy of a general quasifree state for N degrees of freedom. As usual, we first treat the case $N = 1$. From $\rho = Z^{-1}e^{\beta H}$ we find that for $H = \varepsilon a^* a$ we get

$$S = -Tr\,\rho\log\rho = \log Z + \beta\langle\rho, H\rangle = -\log(1 - e^{-\beta\varepsilon}) + \beta\varepsilon e^{-\beta\varepsilon}/(1 - e^{-\beta\varepsilon}).$$

We can rewrite this in terms of the mean number

$$\nu = \rho(a^* a) = \zeta/(1 - \zeta),$$

where $\zeta = e^{-\beta\varepsilon}$. We find

$$S(\rho) = -\nu\log\nu + (1 + \nu)\log(1 + \nu). \tag{10.28}$$

We see that the energy of the mode does not enter directly. It follows that if we have N independent modes, the canonical state is a product of independent modes, and the entropy is additive:

$$S(\rho) = \sum_{k=1}^{N} \left(-\nu_k \log\nu_k + (\nu_k + 1)\log(\nu_k + 1)\right).$$

The classical theory uses just the first term, which is correct in the low-density limit, but which is not by itself positive if ν is not small. For the general quasifree state, we first diagonalise it into modes, and then use Eqn. (10.28).

In quantum field theory the momentum of a particle can be moderately sharp only if the state of the field (of which the particle is the quantum) shows phase correlations over a fair distance. The trouble with the stoss map is that it destroys the correlations even over very short distances; one result of this is that there is no flow term $\mathbf{v}.\boldsymbol{\nabla}$ in the resulting dynamics. We therefore are unable to reproduce the hydrodynamic equations if we randomise by the stoss map. The models studied so far exhibit a diffusion term but no convection, which needs collective flow. In classical mechanics the momentum of a small element of fluid is one of the labels of phase-space. In our treatment so far, these elements lose their individual velocity, because the system is thermalised by the LTE map, which assigns all kinetic energy to a communal pot. This is too drastic a simplification if we want to describe convection rather than diffusion of heat. The quasifree states of a Boson field offer us a way to simplify the description, and to retain at least the one-body (i.e. the quadratic) correlations over some distance, while retaining the second law of thermodynamics. Instead of the LTE map, or the stoss map, as a way to mix the energy-shells, we have available the *quasifree map*. Let us denote by Q the (non-linear) map on the set of regular states of the Segal-Weyl algebra that we get by replacing a state ρ by the quasi-free state with the same moments up to second order. In the derivation of the general form of the quasifree state Eqn. (10.24), we use the positivity condition on the moments, but only up to second order. In other words, any state ρ with finite one- and two-point functions does indeed define a unique quasifree state $Q\rho$ having the same first and second moments. The importance of this is in the crucial remark, made by Wichmann [81], that the entropy of $Q\rho$ is not less than that of ρ. This follows from Theorem (8.16), which also shows that the increase in entropy under Q is strict. Another proof of strictness is to be found in [82]. The argument goes as follows. Let ρ be any state with finite two-point functions, and let $\rho_\beta = Q\rho$. Then there exists a self-adjoint operator H, quadratic in the creators and annihilators, such that ρ_β is the canonical state of H. Since $\langle\rho, H\rangle = \langle\rho_\beta, H\rangle$ we can apply the theorem. The proof only applies to systems with a finite number of degrees of freedom, because otherwise we must consider H to be an infinite sum of quadratic expressions, and we must discuss convergence.

A similar result was proved for some systems with an infinite number of degrees of freedom in [83]. Lanford and Robinson show that the state of a free, infinite system of Bosons (or Fermions), initially in any clustering state, converges weakly to the quasifree state with the same two-point function (the one-point function is zero). The entropy is constant throughout the motion, and is discontinuous, and increases, at the limit. Both an infinite space and an infinite time are needed to get the quasifree state. There is no local production of entropy, but information escapes off to infinity in spatial directions. Krylov [73] has criticised the philosophy behind this type of ergodic theorem, saying that we need a theory in which entropy increases with time, and reaches near equilibrium in a finite time, the relaxation time. He attributed the production of entropy to mixing, and would not have agreed that free systems produce entropy. One might argue that an element of fluid, small when measured on the time-scale of every-day objects is well approximated, on the scale of the thermalising forces, by an infinite volume of fluid observed in an infinite time. This, together with [83] is a possible justification of the quasifree map. Wichmann [81] adopts a Bayesian view (previously advocated by Jaynes [7], and also by Ingarden [4]); he argues that if we have information only about the mean values of certain observables, then we should assign the state which maximises the entropy subject to these mean values. If the observables are all quadratic expressions in the creation and annihilation operators, such as densities and correlations, then the state of maximum entropy with these mean values is the quasifree one. This type of argument was anticipated by Krylov, who derides it as not being physics: the physicist's job, he says, is to devise good theories for deriving the state at time t, given the state at time 0; his job is not to guess what the state at time zero is. We may take it that Krylov would not have been happy with the book by Katz [84] (apart from the fact that it has no references whatever); Katz bases his book on the ideas of Jaynes. The principle of maximum information is too subjective; our knowledge or lack of it does not make any difference to the way a system evolves, says Krylov. In considering the various ways to mix the energy shell (the stoss map, the LTE-map, the quasifree map) we must think of these as giving us different models of the system. As always in theoretical physics, we must try to test the predictions against experiment, and decide which is the best. All are worthy models in that they obey the laws of thermodynamics. We have seen that entropy is a strict Lyapunov function for the quasifree map. In a model where there are Boson fields representing electromagnetic radiation, say from a laser, on phonons, and we wish to retain some of the coherence of this energy in one time-step of the model, then the quasifree map can be used instead of the LTE map. We can also describe atoms and molecules as Bosons, if the effects of the hard core can be neglected. Thus, if $x \in \Lambda$ represents a region that can contain an enormous number of atoms, it is convenient to choose the Hilbert space of the chemical to be $F(\mathbf{C}^{|\Lambda|})$. Then the map Q in place of the LTE-map will retain the coherence of the one-particle wave-functions over the whole space Λ, and will allow collective motion to persist, in spite of the thermalising nature of the dynamics.

10.3 Fermions and Hard-core Ions

Not all particles are Bosons; particles with half-integer spin satisfy Fermi-Dirac statistics, which is the algebraic statement of the Pauli exclusion principle. The requirement that the wave-function of n indistinguishable particles should be antisymmetric under permutations of the position variables ensures that the wave-function is zero if any two particles sit at the same point. For electrons, which have two spin states, the wave-function is anti-symmetric under the combined exchange of the positions and the spin states of any pair of particles. This ensures that the Pauli principle takes the form "at most two electrons can occupy the same position in space". This anti-symmetry can be ensured if we use creation and annihilation operators obeying anti-commutation relations, instead of the commutation relations of the Boson oscillators. Thus suppose that the one-particle space is the complex Hilbert space $\mathcal{H}_1$; let $\psi \mapsto a^*(\psi)$ be a linear map from $\mathcal{H}_1$ to $B(\mathcal{K})$, the bounded operators on a Hilbert space $\mathcal{K}$, such that, putting $a(\psi) = (a^*(\psi))^*$, we have

$$a(\psi)a^*(\phi) + a^*(\phi)a(\psi) = \langle \psi, \phi \rangle_{\mathcal{H}} \tag{10.29}$$

$$a^*(\psi)a^*(\phi) + a^*(\phi)a^*(\psi) = 0 \tag{10.30}$$

for all wave-functions $\psi \in \mathcal{H}$ and $\phi \in \mathcal{H}$. These are called the CAR-relations. If $\dim \mathcal{H} < \infty$ then there exists only one irreducible representation of these relations [85], where it is also shown that the algebra generated by any representation of the CAR-relations is isomorphic to the tensor product of $\dim \mathcal{H}$ copies of $\mathbf{M}_2$. From the second relation we see that $a^*(\psi)^2 = 0$, which expresses that we cannot create two particles in the state ψ. In any representation, the operators $a^*(\psi)$ are all bounded, and indeed obey $\|a^*(\psi)\| \leq \|\psi\|$. See Exercise (10.69). If $\dim \mathcal{H} = \infty$, we must form the CAR-algebra $\mathcal{A}$ by taking the inductive limit of these finite-dimensional algebras, as we increase the dimension. In this case there are uncountably infinitely many inequivalent irreducible representations of $\mathcal{A}$. The "Fock" representation is the unique one containing a vector, Ψ_0, the Fock vacuum, which is annihilated by all the operators $a(\psi)$, thus: $a(\psi)\Psi_0 = 0$ for all $\psi \in \mathcal{H}$. The Fock representation can be given in an explicit manner on the antisymmetric Fock space over $\mathcal{H}$:

$$\bigwedge \mathcal{H} = \mathbf{C} \oplus \mathcal{H} \oplus \bigwedge^2 \mathcal{H} \oplus \bigwedge^3 \mathcal{H} \cdots \tag{10.31}$$

where $\bigwedge^n \mathcal{H}$ is the anti-symmetric part of the n-fold tensor product $\mathcal{H} \otimes \cdots \otimes \mathcal{H}$. This space is spanned by the vectors

$$\psi_1 \wedge \psi_2 \cdots \wedge \psi_n = (n!)^{-1/2} \sum_{\gamma \in \Gamma_n} P(\gamma) \psi_{i_1} \otimes \cdots \otimes \psi_{i_n}, \tag{10.32}$$

where Γ_n is the permutation group on n symbols and $P(\gamma)$ is the parity of the permutation

$$\gamma = \begin{pmatrix} 1, & 2, & \cdots, & n \\ i_1, & i_2, & \cdots, & i_n \end{pmatrix}.$$

Then define the creation operator $a^*(\psi)$ by linear extension of

$$a^*(\psi)\psi_1 \wedge \cdots \wedge \psi_n = \psi \wedge \psi_1 \cdots \wedge \psi_n. \qquad (10.33)$$

In words, $a^*(\psi)$ creates the particle with wave-function ψ. The adjoint is then computable, from the definition of the scalar product in $\bigwedge \mathcal{H}$ as the sum of the scalar products in the summands. See and do Exercise (10.70). If $\dim \mathcal{H} = n < \infty$, then the direct sum stops at the "top form" $\bigwedge^n \mathcal{H}$: there cannot be more than n particles in any state, which is Pauli's exclusion principle. The dimension of Fock space is then 2^n. The Fermionic Fock space has a functorial property similar to the Bosonic Fock space:

$$\bigwedge(\mathcal{H}_1 \oplus \mathcal{H}_2) = \bigwedge \mathcal{H}_1 \otimes \bigwedge \mathcal{H}_2.$$

This is the property implicitly used by Heisenberg when he introduced isotopic spin and thereafter treated the proton and neutron as two states of the same particle.

A state ρ on the CAR-algebra $\mathcal{A}$ is determined by the "n-point functions"

$$\rho\left(a^*(\psi_1)\ldots a^*(\psi_m)a(\psi_1)\ldots a(\psi_{n-m})\right). \qquad (10.34)$$

Because of the anti-commutation relations, we only need to give the expectation values of all the Wick ordered products (in which the creators are to the left, and the annihilators are to the right). Then the expectation value of a product in any order is determined. For example, the Fock vacuum is determined by the vanishing of all those in Eqn. (10.34).

There are two important subalgebras of the CAR-algebra $\mathcal{A}$, namely the even subalgebra and the gauge-invariant subalgebra. Each is defined in terms of invariance under an automorphism of the CAR-algebra. The CAR relations, Eqn. (10.29) and (10.30), are invariant under any unitary transformation of the one-particle space. That is, if $\psi \mapsto a(\psi)$ and $\psi \mapsto a^*(\psi)$ satisfy Eqn. (10.29) and (10.30), then so do $\psi \mapsto a(U\psi)$ and $\psi \mapsto a^*(U\psi)$, for any unitary operator U on the one-particle space $\mathcal{H}$. Now the CAR-algebra is unique up to isomorphism; the sets of operators $\{a(\psi) : \psi \in \mathcal{H}\}$ and $\{a(U\psi) : \psi \in \mathcal{H}\}$ are exactly the same, so the uniqueness means that there is an automorphism γU of the CAR-algebra such that $\gamma U(a(\psi)) = a(U\psi)$ for all ψ. If $\dim \mathcal{H} < \infty$ the γU are all spatial automorphisms, implemented by unitary operators ΓU, called the second quantisation of U. It has the explicit form on the n-particle state

$$\Gamma U(\psi_1 \wedge \psi_2 \wedge \ldots \wedge \psi_n) = U\psi_1 \wedge U\psi_2 \ldots U\psi_n.$$

The automorphisms $\gamma\left(e^{i\theta}\right)$ are called gauge transformations (of the first kind). One can argue that all observables are gauge invariant, and that the algebra of observables of a theory with Fermions should not be the whole CAR-algebra $\mathcal{A}$, but only its gauge-invariant part, $\mathcal{A}_g$; this is obviously a subalgebra of $\mathcal{A}$. A state is said to be gauge invariant if the dual action to $\gamma(e^{i\theta})$, acting on the states, leaves it invariant. It is often said that only gauge invariant states are physical. However, the modern point of view is dual to this: the only observables are gauge invariant, and two states are physically the same if they reduce to the same state on the gauge-invariant subalgebra. From this stand-point, a gauge-non-invariant state always defines a state on $\mathcal{A}_g$. By integrating the state over the circle group we get a more mixed state of $\mathcal{A}$ that is gauge invariant, and which is physically the same as the original one. Thus the non-gauge-invariance of a state does not mean that gauge invariance is broken in the corresponding cyclic representation. If a state is gauge invariant, then the n-point functions are zero unless the number of creators (m in Eqn. (10.34)) is the same as the number of annihilators, $n - m$.

The other important sub-algebra of $\mathcal{A}$ is the even subalgebra. This is the algebra invariant under the single gauge transformation $\psi \mapsto -\psi$. A state is even if it is invariant under the dual map induced on the states. An even state has the property that the expectation value of the product of an odd number of creators and annihilators vanishes. For even states, we can define the useful concept of the correlation functions (cumulants, truncated functions are two other names). Let b denote a or a^*; then let ρ_T denote the correlation function, defined inductively by

$$\begin{aligned} \rho\left(b(\psi_1)\dots b(\psi_n)\right) &= \rho_T\left(b(\psi_1)\dots b(\psi_n)\right) \\ &+ \sum_I P(I)\rho_T\left(I_1\right)\rho_T\left(I_2\right)\dots\rho_T\left(I_r\right). \end{aligned} \tag{10.35}$$

Here, I runs over all the partitions $I = I_1 \cup I_2 \cup \dots \cup I_r$ of the set $\{b(\psi_1), \dots b(\psi_n)\}$ into r parts, and then r runs from 2 to n. Because the state is even, only parts with an even number of elements enter. In each partition, the symbol I_m stands for the product of the $b(\psi_j)$'s in increasing order in j. Finally, $P(I)$ is the parity of the permutation taking the order $I_1, I_2, \dots, I_m$ to the natural order. Because each part contains an even number of terms, this parity does not depend on the choice (arbitrary) of the order in which we write the parts. The two-point correlation functions are the same as the two-point functions, since the mean of a single power of b is zero. In general the $2n$-point correlation function measures the mutual correlation of the $2n$ operators in the state ρ.

Balslev and Verbeure [86] define a quasifree state of the CAR-algebra to be an even state for which the correlation functions vanish for $n \geq 3$. For gauge-invariant states this is equivalent to the earlier definition of Shale and Stinespring [87]. From the definition, Eqn. (10.35), we see that for a gauge-invariant quasi-free state, the n-point function is determined by the two-point functions by the formula

$$\rho\left(a^*(\psi_1)\dots a^*(\psi_n)a(\phi_1)\dots a(\phi_n)\right) = \det[\rho\left(a^*(\psi_i)a(\phi_j)\right)]. \qquad (10.36)$$

The general two-point function of a gauge-invariant state on the CAR-algebra is determined by the operator A such that

$$\rho\left(a^*(\psi)a(\phi)\right) = \langle\phi, A\psi\rangle_{\mathcal{H}}. \qquad (10.37)$$

Since this is positive definite, A must be a positive operator on the one-particle space $\mathcal{H}$. More, since $a(\phi)$ is bounded by $\|\phi\|$, the operator A must be bounded by the unit operator. Conversely, any A obeying $0 \leq A \leq 1$ defines the two-point function by Eqn. (10.37) of a quasifree state. It follows that we may define the (non-linear) quasifree map Q on any state ρ as taking ρ to the quasifree state with the same two-point function. It can be shown that the quasifree state defined by the operator A is pure (as a state on $\mathcal{A}$) if and only if A is a projection. If A is a projection and $\dim A < \infty$, then the subspace defined by A represents those states in $\mathcal{H}$ that are occupied, or "filled" by particles. The representation given by the state is then equivalent to the Fock representation. If $\dim A = \infty$ then the state is not a vector in Fock space, because infinitely many levels are occupied. The one-particle states in the subspace defined by A fill a "sea", and the representation is non-Fock.

We shall be interested in a generalisation of the concept of gauge group, because we need to describe a process in which a bound state splits into several Fermions, so the quantum number, total number of Fermions, is not conserved. This can occur in nuclear physics, atomic physics as well as chemistry and molecular physics. Of course, the total charge q is conserved, as is the total number N_B of baryons. The lepton number, ℓ_e, which is the number of electrons minus the number of positrons, adjusted for loss by neutrino conversion, is also conserved. Each of these quantum numbers is believed to generate a superselection rule. This means the following. Each type of particle, say the i-th particle, has a definite integral amount, q_i, N_{Bi}, ℓ_i of charge, baryon number and lepton number, and so should be an eigenstate of the operators representing charge, etc. To implement this, introduce a basis vector ψ_i, $(i = 1, 2, \dots, n)$ in the one-particle space $\mathcal{H}$; define on $\mathcal{H}$ the self-adjoint operators q, $\mathcal{N}_B$, ℓ_e to be the diagonal matrices, with the corresponding eigenvalues down the diagonal. For example, the one-particle charge operator is given by

$$q\psi_i = q_i\psi_i.$$

Then three one-particle gauge groups can be defined:

$$U_q(\theta) = \exp\{iq\theta\} \qquad U_B(\theta) = \exp\{i\mathcal{N}_B\theta\} \qquad U_e(\theta) = \exp\{i\ell_e\theta\}.$$

The second-quantisation of these unitary groups give us three automorphism groups of the CAR-algebra over $\mathcal{H}$. Let us denote these by $\alpha_q(\theta)$ etc., acting on the right. The observable algebra is the subalgebra invariant under these groups: $\mathcal{A}_0 = \{A \in \mathcal{A} : A\alpha_q(\theta) = A \text{ for all } \theta\}$. This replaces the gauge-invariant subalgebra $\mathcal{A}_g$ in models where the particles have not all got the same charge, baryon number and lepton number. This means that the physical states can be chosen to be invariant under the automorphism groups, since given any state ρ, gauge-invariant or not, we can integrate it over the three gauge groups to get a gauge-invariant state giving the same expectation values for all observables. For example, the group-average of ρ, $\int \alpha_q^*(\theta)\rho\, d\theta$ gives the same expectation values for all gauge-invariant operators as does ρ. This is what is meant by saying that when there is a superselection rule, there can be no coherent superposition of states with different values of the charge (or other quantum number defining the gauge group). It should be said that our gauge groups are of the first kind; gauge invariance of the second kind involves using a different element of the gauge group at each point in space or space-time, coupled with a compensating change in the quantum field of the gauge Boson. We have not been talking about this.

Even with this extension of the notion of gauge invariance, we might still be interested in non-gauge-invariant states, because it may be that non-gauge-invariant states, showing correlations between the superselection sectors, evolve differently from the corresponding gauge-invariant averages. This possibility arises because in statistical dynamics the equations of motion are not linear in the state. This question does not seem to have been fully analysed.

The CAR over $\mathcal{H}$ can be related to the Clifford algebra over $\mathcal{H}$, regarded as a *real* vector space with scalar product $(\psi,\phi) = Re\langle\psi,\,\phi\rangle_{\mathcal{H}}$. That is, we forget the complex structure on $\mathcal{H}$, leaving just the orthogonal structure defined by $(\bullet,\,\bullet)$. We then seek a real-linear map $A : \mathcal{H} \to B(\mathcal{K})$ such that

1. $A(\psi)A(\phi) + A(\phi)A(\psi) = 2(\psi,\,\phi)_{\mathcal{H}} 1_{\mathcal{K}}$;
2. $A(\psi)^* = A(\psi)$

for all ψ and ϕ in $\mathcal{H}$. Such a mapping A is said to be a representation of the Clifford algebra over the real Hilbert space $\mathcal{H}$. If A is a representation, then so is $A \circ T$, where T is an orthogonal operator on $\mathcal{H}$; see exercise (10.71). To get a representation of the CAR from A, we must choose a *complex structure* J on $\mathcal{H}$; that is, J is an orthogonal operator on $\mathcal{H}$ such that $J^2 = -1$. This gives a definition of "multiplication by $\sqrt{(-1)}$", making $\mathcal{H}$ into a complex Hilbert space. Thus using J, we define the multiplication of any vector $\psi \in \mathcal{H}$ by the complex number $\alpha + i\beta$ to be $\alpha\psi + \beta J\psi$. The scalar product needs an imaginary part; it is easy to show that

$$\langle\psi,\,\phi\rangle_J = (\psi,\,\phi) - i(\psi,\,J\phi) \tag{10.38}$$

makes $\mathcal{H}$ into a complex Hilbert space; see Exercise (10.72). Then define maps a and a^* from $\mathcal{H}$ to $B(\mathcal{K})$ by

$$a(\psi) = \frac{1}{2}\left(A(\psi) - iA(J\psi)\right); \qquad a^*(\psi) = \frac{1}{2}\left(A(\psi) + iA(J\psi)\right). \tag{10.39}$$

These then obey the CAR; (Exercise (10.73)). In addition to the choice of J we may specify a conjugation $\bar{\bullet}$ on $\mathcal{H}$. This is a J-antilinear map whose square is the identity. This enables us to define the *real* vectors in $\mathcal{H}$, namely, those for which $\overline{\psi} = \psi$. Choosing the conjugation is like the choice of a Lagrangian submanifold in classical mechanics; we already met this idea in the theory of Bosons. It enables us to define the Fermion field $\Psi(\psi) = 2^{-1/2}(a^*(\psi) + a(\overline{\psi}))$, which is **C**-linear in the vector ψ. This is what is needed in quantum field theory, where ψ is a test-function [88]. Physically, the conjugation is related to charge conjugation.

The orthogonal group of $\mathcal{H}$ is larger than the unitary group; the latter leaves both the real and imaginary parts of the complex scalar product of Eqn. (10.38) invariant, whereas the orthogonal group leaves only the real part invariant. Recall that for Bosons, it is exactly the opposite: the canonical transformations which mix p and q are the symplectic transformations, that leave the imaginary part of the scalar product invariant. A useful theorem states that any positive Hamiltonian, quadratic in the creation and annihilation operators, can be diagonalised by an orthogonal transformation. In general, canonical maps are called *Bogoliubov* transformations. They do not leave the Fock vacuum invariant (when the automorphism is thrown onto the states) unless the map is unitary; it can be shown ([89]) that any pure quasifree state of the CCR or the CAR is a Fock state, that is, it is obtained from the Fock vacuum by a Bogoliubov automorphism. In various models, an approximate ground state is constructed, which is pure and quasifree; it is not annihilated by the operators $a(\psi)$ of the original Hamiltonian, but, being a Fock state, there is a Bogoliubov transformation to new variables $a'(\psi)$ which do kill it. The states created by a'^* are called "quasi-particles". They are collective modes of the original quanta, and occur in several important models such as the BCS (Bardeen, Cooper, Schrieffer) model of superconductivity. The quasifree states can be completely characterised:

Theorem 10.40 *Let $\mathcal{H}$ be a real Hilbert space with scalar product $(\bullet, \bullet)$, and let $\psi \mapsto A(\psi) \in \mathcal{A}$ define the Clifford algebra over $\mathcal{H}$. Let ρ be a quasifree state. Then ρ determines a bounded operator T on $\mathcal{H}$ by the equation*

$$\rho\left(A(\psi)A(\phi)\right) = (\psi, \phi) + i(\psi, T\phi); \qquad \psi, \phi \in \mathcal{H} \tag{10.41}$$

and T satisfies $T^+ = -T$ and $\|T\| \le 1$. Moreover ρ is pure if and only if $T^2 = -1$, in which case T is a conjugation, and the representation is the Fock *representation with this T as the J of Eqn. (10.38). Conversely, any such T determines a quasifree state.*

Here, T^+ is the adjoint relative to the real scalar product $(\bullet, \bullet)$.
Proof. We shall prove the necessity of the conditions; you are referred to [89] for the sufficiency.

Suppose then that ρ is any state on the Clifford algebra. From the anticommutation relations, it is easy to see that

$$\rho\left(A(\psi)A\phi)\right) - (\psi,\, \phi) = i\sigma$$

defines a real, antisymmetric bilinear form σ. A necessary condition for positivity is

$$\rho\left([A(\psi) + iA(\phi)][A(\psi) - iA(\phi)]\right) \geq 0,$$

since by (2), $A(\psi)$ is Hermitian. This yields $\|\sigma\| \leq 1$, so by the Riesz duality theorem, σ defines a bounded operator T such that $\sigma = (\psi,\, T\phi)$. The antisymmetry of σ gives that of T. Finally, if $T^2 = -1$, it is a conjugation, and the corresponding creation and annihilation operators, Eqn.(10.39) obey $\rho(a^*a) = 0$, so the representation is the Fock representation, known to be irreducible, so ρ is pure. □

In the proof of the converse (that any operator T obeying the conditions of the theorem defines a quasifree state), Balslev, Manuceau and Verbeure give a useful formula for the representation obtained from ρ by the Gelfand-Naimark-Segal construction. The case when $\dim_{\mathbf{R}} \mathcal{H}$ is even or infinite is simpler, and is also more important, since only in that case can $\mathcal{H}$ be complexified. So let J be a complex structure on $\mathcal{H}$, and define $j = J \oplus -J$. This is a complex structure on $\mathcal{H}' = \mathcal{H} \oplus \mathcal{H}$. Let π_j be the Fock representation of the CAR over $\mathcal{H}'$. This takes place in the tensor product of the J-Fock and $-J$-Fock spaces, by the functorial property of these spaces. Let T' be the map from $\mathcal{H}$ into $\mathcal{H}'$ given by $\psi \mapsto 2^{-1/2}\left((1 + |T|)^{1/2}\psi \oplus (1 - |T|)^{1/2}\psi\right)$. Then the representation of the Clifford algebra defined by ρ is equivalent to $\pi_j \circ T'$, applied to the cyclic subspace generated from the vacuum. We can see a similarity with the tensor structure of the representation defined by a KMS state; in fact, this is exactly what we have, for the *even* subalgebra. All this was already understood by Araki and Wyss in [90]. For the Fermion operators $a(\psi)$, $a^*(\phi)$ it is not quite a tensor product, since the parts in the two factors anticommute rather than commute. The special case $T = 0$ corresponds the tracial state, (the microcanonical state, the state with $\beta = 0$) in the case where $\dim \mathcal{H} < \infty$. In the infinite case the vacuum defines a central state, which is one obeying $\rho(AB) = \rho(BA)$ for all elements A, B of the algebra. This shows that the representation generates a W^*-algebra of type II_1 in von Neumann's classification. In general, $0 < \|T\| < 1$ and the representation is of type III, which was a big surprise to mathematicians. It follows from the theorem that the quasifree map, which replaces a state by the quasifree state with the same two-point functions, is well defined: given any state, there always exists such a quasifree state.

While quasifree states have a simple description, they carry no correlations between three or more particles. It might be hoped that a more detailed description might be possible, by allowing for example the four-point correlations to be non-zero, but taking for simplicity all higher than four to be zero. Unfortunately, there are no such states; as with the classical case, and the Bosonic case, any state of the CAR with correlation functions zero beyond some n must be quasifree. Thus unless we include all the correlations, however remote, we may include none beyond those of second order, without violating the condition of positivity of the state. This has hampered the construction of non-quasifree theories. The idea of higher-order temperature [4] is a way out. This will be explained in Chapter 13.

Quasifree states are quite tractable: the entropy of a quasifree state has an explicit expression, and the canonical and grand canonical states of an oscillator are quasifree. Further, the quasifree map is entropy-increasing. To find the entropy, consider one Fermion mode, so that $\mathcal{H} = \mathbf{C}$. The Fermionic Fock space is then just $\mathcal{F} = \mathbf{C} \oplus \mathbf{C}$. A quasifree state given by the operator $A = \lambda$ where λ is just a real number, $0 \leq \lambda \leq 1$, which is the probability of occupation. The density matrix of the state is $\operatorname{diag}(1-\lambda,\, \lambda)$, and its entropy is $S = -\lambda \log \lambda - (1-\lambda)\log(1-\lambda)$. We now consider the case $\dim \mathcal{H} > 1$. Suppose now we have more than one mode. When A has finite rank, we may diagonalise it into modes and can take $\dim \mathcal{H}$ to be finite. Write $A = \sum \lambda_j P_j$. The relevant part of Fock space is, by the functorial property, the tensor product of copies of the above two-dimensional space. The state is also a product-state, so the entropy of the quasifree state determined by A is the sum of the entropy of the factors:

$$\begin{aligned} S &= -\sum_j \{\lambda_j \log \lambda_j + (1-\lambda_j)\log(1-\lambda_j)\} \\ &= -Tr\,\{A \log A + (1-A)\log(1-A)\}. \end{aligned} \tag{10.42}$$

Even if $\dim \mathcal{H} = \infty$ Eqn. (10.42) may make sense; we then say that the state has finite entropy. It is interesting to compare the Fermion entropy with the entropy of a quasifree Boson, Eqn. (10.28); we see that as there it is determined by the mean particle density, and that we have a similar formula except for some signs. Again the first term is the same as the Boltzmann entropy.

We shall find the canonical state if $\dim \mathcal{H} < \infty$ for any Hamiltonian which is the second quantisation of a one-particle operator. These are sometimes called quadratic Hamiltonians. They lead to linear equations of motion for the fields. We shall be interested in the case where the Hamiltonian H has discrete spectrum with finite multiplicity; we first diagonalise it by a canonical transformation. So suppose that $\{\psi_j\}$ are the normalised eigenfunctions of H. The second quantisation of H can then be written

$$d\Gamma H = \sum_j \mathcal{E}_j a^*(\psi_j)a(\psi_j).$$

Let us start with the simple case where $\dim_{\mathbf{C}} \mathcal{H} = 1$ with energy $\mathcal{E}$. The state ρ is determined by its generating function

$$\mathcal{F}(\lambda,\mu) = \rho\left(\exp\{\lambda a^*\}\exp\{\mu a\}\right) = Z^{-1}Tr\ \left(\exp\{-\beta\mathcal{E}a^*a\}\exp\{\lambda a^*\}\exp\{\mu a\}\right).$$

It will be quasifree if all the n-point functions are given by Eqn. (10.36). For one degree of freedom we can use the representation

$$a^* = \begin{pmatrix} 0 & 1 \\ 0 & 0 \end{pmatrix} \qquad a = \begin{pmatrix} 0 & 0 \\ 1 & 0 \end{pmatrix} \qquad a^*a = \begin{pmatrix} 1 & 0 \\ 0 & 0 \end{pmatrix}. \tag{10.43}$$

It is elementary to find the exponentials of these matrices, since $a^2 = 0$. We get (Exercise (10.74))

$$\mathcal{F}(\lambda,\mu) = 1 + \frac{e^{-\beta\mathcal{E}}}{1+e^{-\beta\mathcal{E}}}\lambda\mu.$$

We see that the partition function is $Z = 1 + e^{-\beta\mathcal{E}}$ and that the two-point function is

$$\rho(a^*a) = \frac{e^{-\beta\mathcal{E}}}{1+e^{-\beta\mathcal{E}}}.$$

We get the mean density with several modes by summing; the mean energy is then

$$\overline{E} = \sum_j \mathcal{E}_j \frac{e^{-\beta\mathcal{E}_j}}{1+e^{-\beta\mathcal{E}_j}}. \tag{10.44}$$

This is the famous Fermi-Dirac distribution. We note that the generating function contains only a constant term and the quadratic term proportional to $\lambda\mu$. The higher powers are zero, which comes from the fact that $a^2 = 0$. Similarly, the generating function for the case of several modes, defined to be

$$\mathcal{F}(\lambda_i,\mu_j) = \rho\left(\exp\{\sum_i \lambda_i a_i^*\}\exp\{\sum_j \mu_j a_j\}\right)$$

will contain a product $\prod_i \lambda_i\mu_i$ with no repetitions. So the n-point functions in Eqn. (10.36) are all zero unless all the creators refer to different modes, and the same for the annihilators. Then only one term survives, in which each creator is contracted with its own mode, and we arrive at the determinant of Eqn. (10.36) with only one non-zero term. So the state is quasifree.

Now we can easily see that the quasifree map is entropy increasing. For, to maximise the entropy among all states, subject to finitely many quadratic conditions, leads via Lagrange multipliers to a canonical state with a quadratic Hamiltonian. But this is quasifree. This proof is easier than the original one [83], which also applies to the infinite-dimensional case.

We can describe a gas of Fermions, moving on a lattice Λ, by the *CAR*-algebra over $\mathcal{H} = \oplus_x \mathcal{H}_x$, where we have a finite-dimensional Hilbert space $\mathcal{H}_x$ at each point. The states in $\mathcal{H}_x$ label the spin states of the Fermion, and also the different types of Fermion, such as neutron, proton, heavy hydrogen etc. To get a simple model, we choose the Hamiltonian to be that of the free particles; thus if the chemical energy per molecule, of the j-th type of particle is ε_j, then the Hamiltonian is

$$H_c = \sum_{x\in\Lambda}\sum_j \varepsilon_j a_j^* a_j.$$

This is the simplest choice, in which there is no kinetic energy assigned to the molecules. Usually in quantum mechanics the kinetic energy enters through gradient terms in the quadratic part; on a lattice the corresponding expression should be a linking of nearest neighbours. This is simplest if Λ is a cubic lattice, with periodic boundary conditions. Let us illustrate this in one dimension, and choose $\Lambda = \mathbf{Z}_N$, a ring of N atoms. The finite translation group $\mathbf{Z}_N$ acts on itself, and a translation-invariant kinetic energy can be chosen to be

$$H_1 = \frac{1}{2m}\sum_x \Big((a_{x+1}^* - a_x^*)(a_{x+1} - a_x)\Big)$$

for each atom, whose mass is m. In the term $x = N$, we interpret $x+1$ as 1, as we have a ring. Clearly, H_1 is invariant under translations of the lattice, which leads to the conservation of momentum. There are N eigenstates of momentum, labelled by the dual group (here also $\mathbf{Z}_N$). The total energy is taken to be $H = H_c + H_1$, with spectral projections P_j say. Let $\mathcal{A}$ be the *CAR*-algebra of all atoms at all sites. The energy-shells are the subalgebras $\mathcal{A}_{ij} = P_i\mathcal{A}P_j$. We get a model by choosing a bistochastic completely positive map T on the $\mathcal{A}$, which maps each $\mathcal{A}_{ij}$ into itself. This is designed to give a non-zero transition probability for each process we want to describe by the model, and at the same time, is to map to themselves each "number-shell" of any conserved quantity (like charge, baryon number). The number-shell is defined similarly to the energy-shell. Because the kinetic energy H_1 links different sites, it is no longer true that the stoss map conserves the mean energy. But we have available the quasi-free map Q, which conserves the mean (in the current state ρ) of all quadratic expressions, including each term in the sum for the kinetic energy. Since Q also increases the entropy, we have a viable theory if we take one time-step of the dynamics of $\mathcal{A}$ to be

$$\rho \mapsto QT^{\dagger}\rho.$$

This will obey both laws of thermodynamics. In the next section we show that the quantum Boltzmann equation (for discrete time) is obtained if the map T is a suitable unitary conjugation commuting with H.

We can add a heat-particle to the above model, to allow for the energy located in the elecromagnetic field. It is natural to localise the heat-particle on the bond between the sites, rather than at each site. If we omit the kinetic energy-term, we can use the heat-particle also to model the kinetic energy. In that case we can use the stoss map to add more mixing. This simpler model is like the classical models discussed in part I: there is no mechanism for collective flow, as the correlations between distant parts (and even neighbours) is destroyed. The advantage of the quasifree map over the stoss map is that it increases the entropy without destroying the correlations between distant particles. It is this correlation that gives rise to convection.

In a theory with a single type of Fermion the exclusion principle acts as a repulsion between the particles, which behave as if they had a hard core. When there are several Fermions, then each site can be occupied by more than one particle, and it is no longer true that the theory seems to describe particles with a hard core. Actual molecules very often exhibit a strong repulsion, and a large kinetic energy is then needed to bring them together, before a reaction can occur. This energy is called the threshold by physicists and the activation energy by chemists. We can model n atoms with a hard core by choosing at each site $x \in \Lambda$ the Hilbert space $\mathcal{H}_x = \mathbf{C}^{n+1}$. Let $\{\psi_j\}_{j=0,\ldots,n}$ be an orthonormal basis in $\mathcal{H}_x$; one vector, say ψ_0, will be taken to represent the empty state (no particles present) and the remaining n basis vectors will represent the presence of just one of the atoms. Then the algebra $\mathcal{A} = \bigotimes_x \mathcal{A}_x$ will be called the *hard-core* algebra. It generalises the Fermion model, in that if $N = 1$ the algebra reduces to the CAR-algebra of $|\Lambda|$ Fermions. The n excited states of an excitable atom are well described by the hard-core model. For some chemicals, the hard-core model might not be a good one; a large organic molecule might occupy more space than the parts out of which it is made. But for some gases and dilute solutions, Guy Lussac's law is nearly true. This states that the size of each molecule is approximately the same. Let us choose the lattice of points in space, Λ, so that one point can accommodate at most one molecule. When two molecules combine to form a third, this can occupy just one site, unlike the two molecules of which it is made. As a result, the total number of occupied sites in a given volume changes when a reaction takes place, and this means that the pressure changes as well. In this respect, the hard-core model gives predictions quite different from those of Fermions.

We can easily find the algebraic properties of the hard-core algebra. We first represent the operators of $\mathbf{M}_{n+1}$ by matrices with rows and columns labelled $\{0, 1, \ldots, n\}$. The one-dimensional projection operators onto the states ψ_j, $j = 1, \ldots, n$ are the number operators N_j for the n molecules, and the creation operator a_j^* for the j-th particle moves ψ_0 to ψ_j, and so is represented by the matrix with zeros everywhere except in the zeroth row, j-th column, where the entry is 1. Let us call this the *standard* representation of the hard-core algebra. In the standard representation the number operator $N_j = a_j^* a_j$ is the matrix of zeros except on the diagonal in the j-th place, and $a_j^* a_k = E_{j,k}$, the matrix unit with zeros except in the j-th row and the k-th column. In the standard representation N_0, the projection onto the vacuum, has a 1 in the zeroth row and column. We have

$$\begin{aligned} a_j a_j^* &= N_0 \text{ for } j = 1, 2, \ldots, n \ ; \qquad & a_j a_k &= 0 \ ; \\ a_j a_k^* &= 0 \text{ if } j \neq k \ ; \qquad & N_0 a_j &= a_j \text{ for all } j \ . \end{aligned} \tag{10.45}$$

For Fermions, $n = 1$ and the relations reduce to the CAR. A representation of the hard core algebra is then a mapping a from $\{1, 2, \ldots, n\}$ into a C^*-algebra obeying the relations (10.45). These relations characterise the algebra up to isomorphism. That is, we have the following theorem:

Theorem 10.46 *Let $\{a_1, \ldots, a_n\}$ be an irreducible set of operators obeying (10.45) such that $N_0 \neq 0$. Then $a_1, \ldots, a_n$ generate $\mathbf{M}_{n+1}$ and are equivalent to the standard representation.*

Proof. We see that $N_0^2 = N_0 a_j a_j^* = a_j a_j^* = N_0$, so N_0 has eigenvalues 0 and 1. If $N_0 \neq 0$ then there is a vector ψ_0 in the representation space such that $\|\psi_0\| = 1$ and for all j, $a_j a_j^* \psi_0 = \psi_0$. Consider $\psi_j = a_j^* \psi_0$; we find that these are normalised and mutually orthogonal. They are also orthogonal to ψ_0, since $\langle \psi_0, a_k^* \psi_0 \rangle = \langle a_j a_j^* \psi_0, a_k^* \psi_0 \rangle = \langle a_j^*, a_j^* a_k^* \psi_0 \rangle = 0$. By direct calculation we show that Span $\{\psi_0, \psi_1, \ldots, \psi_n\}$ is invariant under the action of a_j, a_k^*, and as the algebra is irreducible, this is the whole space. We find the matrix element $\langle \psi_j, a_\ell \psi_k \rangle = 0$ unless $j = 0$ and $\ell = k \neq 0$, so the representation in this basis is standard. □

From the defining relations (10.45) we can prove that $N_0 + \sum_j a_j^* a_j = 1$. This reduces to the CAR when $n = 1$. We note that the operator $\theta = 1 - 2N_0$ anticommutes with all the a_j and a_j^*.

Let us define a quasifree state with given two-point functions

$$\rho(a_j a_k), \ \rho(a_j^* a_k), \ \rho(a_j^* a_k), \ \rho(a_j^* a_k^*)$$

to be the state with greatest entropy having these values. Such a state always exists; it is unique by the strict concavity of the von Neumann entropy on $\mathbf{M}_{n+1}$.

By the method of Lagrange multipliers, the density matrices of the quasifree states will be exponentials of quadratic operators in a_j, a_k^*.

On a set Λ, we can introduce one copy of $\mathbf{M}_{n+1}$ at each site, and take the tensor product. In that way the creators at different sites commute. This is isomorphic to the usual algebra of spins. In the Fermion algebra (the case with n=1) the creators at different sites anticommute instead of commuting as in the algebra of spins. This is achieved by putting a twist in the tensor product, using the operator θ. We get the analogue of the Fermion algebra inductively as follows. Let Λ be given, and let $\mathcal{A}(\Lambda) = \mathbf{M}_{n+1}^{\otimes|\Lambda|}$. Suppose that π is a representation of the hard-core algebra over Λ, such that the creators at different sites anticommute. Let $\theta_x = 1-2N_0(x)$, $x \in \Lambda$; this anticommutes with all the creators in $\mathcal{A}_x$, and commutes with all the others in Λ. Let $y \notin \Lambda$, and let $a_j(y)$ be the standard representation of the hard-core algebra in $\mathbf{M}_{n+1}(y)$. Then the hard-core algebra over $\Lambda \cup \{y\}$ is represented in $\mathcal{A}(\Lambda) \bigotimes \mathbf{M}_{n+1}$ as follows:

- $a_j(x)$, with $x \in \Lambda$, is represented by the ampliation $\pi\left(a_j(x)\right) \otimes 1$;
- $a_j(y)$ is represented by $\prod_{x\in\Lambda} \theta(x) \otimes a_j(y)$.

The creators at different points anticommute. Starting at one site, we construct the anticommuting hard-core algebra for two sites, and so on, getting an isomorphism between the resulting algebra and the usual algebra of spins. The actual isomorphism depends on the order in which points of Λ are included. For $n = 1$ this construction reduces to that for the CAR, due to Jordan and Wigner [85].

A typical Hamiltonian for a system of chemicals moving in Λ will consist of the chemical energy $\sum_x \sum_j \mathcal{E}_j a_j^* a_j$ and some kinetic energy

$$(2m)^{-1} \sum_{x,y} \sum_j (a_j^*(x) - a_j^*(y))(a_j(x) - a_j(y)).$$

Here, y runs over the neighbours of x. We should also add the heat-particle, as a balancing item. This could be located on the sites, or on the bonds between neighbours. The Hamiltonian is then the sum of the chemical, kinetic and heat energy. To get one time-step in the dynamics, we apply a bistochastic completely positive map T on the algebra, which leaves invariant the spectral projections of the Hamiltonian. T is conveniently constructed in terms of creators and annihilators, which are used to make up the operators X_j in Kraus's form for a completely positive map. That is, X_j is a product of creators and annihilators that conserve energy, in that the sum of the energy of the creators balances that of the annihilators. These operators will not be localised at one point $x \in \Lambda$, because of the hard-core property. Several points will be needed to arrive at a non-zero polynomial. For example, we could choose one X, a unitary operator

e^{iK}, where K is a typical Hermitian interaction Hamiltonian $\lambda a_j(x)a_k(y)a_\ell^*(z)$ (plus Hermitian conjugate) which represents the combination of atoms j at x and k at y to produce atom ℓ at z, and the inverse. To ensure that K commutes with the kinetic energy, we sum over all sites. Let τ^* be the induced action on the states.

The next step in statistical dynamics is to choose a suitable randomising map that conserves means of all the quantities we want to include in the first law. These will already have been conserved by τ^*. It is clear that we need a generalisation of quasifree map; if the atoms have different quantum numbers such as baryon number most of the two-point functions will be zero, and we lose a lot of quantum coherence that we might want to keep. We shall return to this in Chapter 7 in the context of generalised temperatures, where a more general method will be introduced.

10.4 The Quantum Boltzmann Equation

The quantum Boltzmann equation was written down by Uehling and Uhlenbeck in 1933 [91]. For the Bosonic process

$$A \rightleftharpoons B + \gamma$$

it takes the form

$$\frac{dn_A}{dt} = -\frac{dn_B}{dt} = -\frac{dn_\gamma}{dt} = \kappa\left(-n_A(1+n_B)(1+n_\gamma) + n_B n_\gamma(1+n_A)\right). \quad (10.47)$$

The equation for the same process, if A and B are Fermions, is

$$\frac{dn_A}{dt} = -\frac{dn_B}{dt} = \frac{dn_\gamma}{dt} = \kappa\left(-n_A(1-n_B)(1+n_\gamma) + n_B n_\gamma(1-n_A)\right). \quad (10.48)$$

In these equations, n is the mean number of the particle labelled in the suffix. The parameter κ determines the rate at which the process occurs. If there are several modes, the rate for each is a sum over all processes; the original version [91] had a continuum of modes, and the righthand side was an integral over these modes; then κ was related to the cross-section of the process. We must decide whether we wish to regard the Boltzmann equation as giving an approximation to a Hamiltonian dynamics, or to regard it as defining a theory in its own right. Let us first adopt the Hamiltonian point of view, and introduce creators a_A^*, a_B^* and a_γ for the A, B and heat-particle respectively. Then a justification in the simplest terms of Eqn. (10.47) is to choose an interaction Hamiltonian

$$H_I = e\left(a_A^* a_B a_\gamma + a_B^* a_\gamma^* a_A\right). \quad (10.49)$$

This is to be added to the free Hamiltonian

$$H = H_0 + H_I \quad \text{where } H_0 = \mathcal{E}_A a_A^* a_A + \mathcal{E}_B a_B^* a_B + \mathcal{E}_\gamma a_\gamma^* a_\gamma. \tag{10.50}$$

We see that H_I commutes with H_0 if and only if we have a balance of energy:

$$\mathcal{E}_A = \mathcal{E}_B + \mathcal{E}_\gamma. \tag{10.51}$$

For, by the commutation relations for creators and annihilators,

$$[aa^*, a^*] = a^* \text{ and } [aa^*, a] = -a \text{ for each species.}$$

It follows that $[\mathcal{N}_A,\, a_A a_B^* a_\gamma^*] = -a_A a_B^* a_\gamma^*$, and so

$$[H_0, a_A a_B^* a_\gamma^*] = (-\mathcal{E}_A + \mathcal{E}_B + \mathcal{E}_\gamma)\, a_A a_B^* a_\gamma^* = 0\ .$$

Taking Hermitian conjugates and adding, gives $[H_0,\, H_I] = 0$ if and only if the energy balances. In the interaction picture, the unitary time-evolution is determined by

$$U_I(t) = e^{-iH_0 t} e^{iHt}$$

which causes the transitions of Eqn. (10.47) to occur. The usual derivation of the Boltzmann equation starts with a computation of the matrix element $\langle \psi_1,\, H_I \psi_2 \rangle = M$. Then according to the Fermi golden rule, $|M|^2$ is proportional to the transition rate from ψ_1 to ψ_2. Let a^* be a Bosonic creator, and denote the normalised vacuum by ψ_0. Then $\|a^{*n}\psi_0\| = \sqrt{(n!)}$; this is elementary and is exercise (10.75). Let us denote the state $a_A^{*n}\psi_0$ by ψ_A^n etc., and let us omit the tensor product symbol from the product of such states. Then the transition matrix for the forward process is

$$\langle \psi_A^{n_A} \psi_B^{n_B} \psi_\gamma^{n_\gamma},\, H_I \psi_A^{n_A - 1} \psi_B^{n_B + 1} \psi_\gamma^{n_\gamma + 1} \rangle.$$

Here, $n_A = \langle \rho, \mathcal{N}_A \rangle$ etc., where ρ is the current state. Only the term $e a_A a_B^* a_\gamma^*$ contributes, and we find

$$|M|^2 = e^2 n_A (n_B + 1)(n_\gamma + 1).$$

There are some extra factors arising from the golden rule, which is obscure in this simple model. We shall explain these below. The transition probability for the inverse is computed in the same way. We see that the rate of the forward process contains the factors $(1 + n_B)$ and $(1 + n_\gamma)$. This means that the more B particles or heat-particles that the final state contains the faster the initial state decays to it. The part proportional to n is the celebrated "stimulated emission" predicted with uncanny insight by Einstein in his seminal paper [92]. The remaining part, proportional to 1, he called the spontaneous part. Quantum field theory gives both parts together in a natural way. Since the right-hand side of the equation is the transition rate, we put it equal to the rate of *loss* of n_A. At this point we have ensured that the dynamics is irreversible. By using the transition probability $|M|^2$ instead of a complex amplitude, we increase randomness. It can be pictured as a measurement process, in which we measure n_A, n_B and n_γ. This step, in modern terms, should be implemented by a completely positive map.

This whole argument has been criticised by Krylov in his brilliant polemic [73]. In the first place, the eigenvalues and eigenvectors of H_0 are not the same as those of H. If we take H as being the true energy, then the time-evolution e^{iHt} cannot cause any transitions between the eigenstates. Why should we be interested in the transitions between the eigenstates of H_0, which is not the true energy? The answer we get cannot be a good approximation to the rate of transition between the eigenstates of H, since the latter is zero. The dynamics using the full Hamiltonian cannot lead to Eqn. (10.47), since the latter converges to a mixed state when we start from a pure state. In addition to these points made by Krylov, the use of Fermi's golden rule for this model is dubious, as the rule is derived in a theory with space-translation invariance in which an asymptotic condition holds. In fact, in the special case when H_0 and H_I commute we have $U_I(t) = e^{iH_I t} = 1 + iH_I t + \ldots$ and we get an expansion for the transition amplitude:

$$\langle\psi_1,\, U_I(t)\psi_2\rangle = \langle\psi_1,\, \psi_2\rangle + it\langle\psi_1,\, H_I\psi_2\rangle + \ldots.$$

Taking ψ_1 and ψ_2 to be orthogonal, the first term is zero, the term of lowest order is proportional to t, so the transition probability has a factor t^2 multiplying $|M|^2$. The transition *rate* is then proportional to t, which goes to zero as $t \to 0$. This is what is behind the "Zeno paradox" of continuous measurement, sometimes called "the paradox of the watched pot" (which never boils). This has often been cited as one of the puzzles of quantum theory. It seems that if we compute the instantaneous rate of the process, then this is zero. We therefore need a more careful study of the problem. A serious attempt to arrive at the Boltzmann equation from a purely Hamiltonian theory must involve some sort of limit. The most successful idea is that of van Hove [93]: the parameter t must go to infinity, to represent the fact that the macroscopic time we are using for one time-step in the Boltzmann equation is very large on the microscopic scale. We then use one of the time-factors in $|M|^2$ to provide the factor $e^2 t = \tau$, which is held fixed by taking the coupling e to zero. Then τ is taken to be one unit of macroscopic time. The result is the weak coupling limit. It has been implemented rigorously for a certain class of models [94, 95]. The other factor t multiplying $|M^2|$ is removed in a more involved way, which we now explain.

The equation of motion in the interaction picture is

$$\frac{dU_I(t)}{dt} = iH_I(t)U_I(t)$$

where $H_I(t) = e^{-iH_0t}H_I e^{iH_0t}$. We impose the boundary value $U_I(t_1) = 1$. This is equivalent to the integral equation

$$U_I(t) = 1 + \int_{t_1}^{t_2} H_I(s)U_I(s)\,ds.$$

Scattering takes place from time $t_1 = -\infty$ to time $t_2 = +\infty$, and so we are interested in the transition amplitude from ψ_1 at time $-t$ to ψ_2 at time t, as t becomes large. The Heisenberg dynamics must be compared with the free dynamics in the same period of $2t$; that is, the free dynamics is to be cancelled out by using $U_I(t)$ instead of the full dynamics $U(t)$. Using perturbation theory, we can solve by the Picard iteration method, to get

$$U_I(2t) = 1 + \int_{-t}^{t} H_I(s)\,ds + \ldots.$$

Then if ψ_1 and ψ_2 are orthogonal, we get

$$\begin{aligned}\langle\psi_1, U_I(2t)\psi_2\rangle &= \langle\psi_1, \int_{-t}^{t} \exp\{-iH_0s\}H_I \exp\{iH_0s\}\psi_2\rangle \\ &= M\int_{-t}^{t} \exp\{is(\mathcal{E}_2 - \mathcal{E}_1)\}ds.\end{aligned}$$

Here, $\mathcal{E}_j$ is the energy of ψ_j, taken to be an eigenstate of H_0. For this to make sense, we must regard H_0 as the energy of the free ingoing and outgoing states, as in scattering theory. In the model we have here, the asymptotic condition does not hold. But in a more realistic theory, with quantised fields in an infinite space and a local interaction H_I, we can expect that the particles become asymptotically free, at least for a dilute gas. The transition probability is obtained by squaring the amplitude, so we get for this

$$\begin{aligned}|M|^2\int_{-t}^{t}\int_{-t}^{t} \exp\{i(s-s')(\mathcal{E}_2 - \mathcal{E}_1)\}\,ds\,ds' &= |M|^2(2t)\int_{-t}^{t} \exp\{is(\mathcal{E}_1 - \mathcal{E}_2)\}\,ds \\ &\to 2t|M|^2 2\pi\delta(\mathcal{E}_2 - \mathcal{E}_1)\end{aligned}$$

as $t \to \infty$. This has the factor $2t$ over the period of time $2t$. The rate is therefore $2\pi|M|^2$, which is Fermi's golden rule. The other factor t in the rate has gone out, as promised, to provide us with an overall energy-conserving δ-function. In our theory with a discrete set of energy-levels, this factor is infinite. We must imagine that our discrete levels are obtained by starting with a continuum theory in an infinite volume, for which scattering theory holds, and then forming the discrete model by integrating over energy-bands. The δ-function might then be replaced by a Kronecker δ. Something like this has been achieved by Kiegerl and Schürrer [96] by starting with the classical Boltzmann equation and discretising it. One of the difficulties with the van Hove limit is that each of the terms of higher order in the perturbation series become infinite. Thus, for example if energy balance Eqn. (10.51) holds, the term of second order in the Dyson expansion is

$$e^2t^2/(2!)\langle\psi_1,\, H_I^2\psi_I\rangle$$

whose square has the factor $e^4t^4 = \tau^2t^2$. So the rate has the factor $\tau^2 t$, which has no limit as $t \to \infty$.

Our point of view is not to derive the Boltzmann equation from a purely Hamiltonian theory, but to show that it comes from a bistochastic map T followed by a randomising procedure which increases entropy. This will be enough to show that it leads to dynamics obeying both laws of thermodynamics. For comparison with the above "derivation", our total energy is analogous to the asymptotic energy H_0, and the bistochastic map $T^\dagger$ plays the role of the scattering operator (or a random mixture of such), which in the limit $t \to \infty$ commutes with the energy operator of the free ingoing and outgoing particles. In a theory with discrete energy levels, the energy-conserving δ-function requires that T maps the energy-shells to themselves, which has been our requirement all along.

The fixed points of Eqn.(10.47) obviously obey

$$\frac{n_A}{1+n_A} = \frac{n_B}{1+n_B}\frac{n_\gamma}{1+n_\gamma}\,. \tag{10.52}$$

Suppose that each particle is in the grand canonical state at the same temperature; that is, the density matrix of the system is the tensor product of the states

$$\begin{aligned}
\rho_A &= Z_A^{-1}\exp\{-\beta(\mathcal{E}_A-\mu_A)a_A^*a_A\}\,, &(10.53)\\
\rho_B &= Z_B^{-1}\exp\{-\beta(\mathcal{E}_B-\mu_B)a_B^*a_B\}\,, &(10.54)\\
\rho_\gamma &= Z_\gamma^{-1}\exp\{-\beta\mathcal{E}_\gamma a_\gamma^*a_\gamma\}\,. &(10.55)
\end{aligned}$$

Then the mean values of the number operators are related to the intensive parameters β, μ_A and μ_B by

$$n_A = \frac{\lambda_A}{1-\lambda_A}; \qquad n_B = \frac{\lambda_B}{1-\lambda_B}; \qquad n_\gamma = \frac{\lambda_\gamma}{1-\lambda_\gamma}.$$

Here, λ_A etc. are the activities:

$$\lambda_A = \exp\{-\beta(\mathcal{E}_A-\mu_A)\}; \qquad \lambda_B = \exp\{-\beta(\mathcal{E}_B-\mu_B)\}; \qquad \lambda_\gamma = \exp\{-\beta\mathcal{E}_\gamma\}.$$

Then the condition for equilibrium is simply $\lambda_A = \lambda_B\lambda_\gamma$, which we have met as the condition for equilibrium in chemistry (because a is used as an annihilation operator, we use λ for the activity in this section). In view of $\mathcal{E}_A + \mathcal{E}_B + \mathcal{E}_\gamma$, this condition becomes simply $\mu_A = \mu_B$, and the equilibrium state is the grand canonical state of the whole system with a certain temperature and a common chemical potential for the chemicals. The dynamics can be regarded as a flow through the space of the extensive parameters n_A, n_B, n_γ, or alternatively through the space of intensive parameters μ_A, μ_B, β. The ratio of the backward to the forward rate contains the factor $n_\gamma/(n_\gamma + 1) = \lambda_\gamma$, which we recognise as the Arrhenius factor, or detailed balance. It is very likely that in getting his kinetic law for stimulated emission, Einstein worked backwards from the equilibrium, known to be the Planck distribution, using detailed balance. It is not possible to arrive at the law of stimulated emission uniquely by this method, since we have seen that the "activity-led" chemical laws

$$\frac{dn_A}{dt} = \kappa(-\lambda_A + \lambda_B\lambda_\gamma)$$

have the same fixed point and obey detailed balance. These seem to be good laws for reactions in dense liquids; they are not good for photons. Einstein may have been guided to the right law by the following argument (or was he simply trying to avoid fractions?). The rate of the inverse process $B + \gamma \to A$ must be proportional to the number of B present. It must also be proportional to the number of γ present. So the direct process $A \to B + \gamma$, which ends up with $(n_B + 1)$ B's and $(n_\gamma + 1)$ heat particles, must be proportional to $(n_B+1)(n_\gamma+1)$, by microscopic reversibility. It is also proportional to n_A. This assumes that the particles do not get in each other's way. We must not confuse the probability rate from an occupied level with exactly n particles in it with the rate of change of the mean number of particles. The Boltzmann equation uses the means. I shall now concoct a *classical* model which uses this intuition to choose a bistochastic map T leading to the quantum Boltzmann equation with discrete time. This is adapted from [17] and appeared in [32].

Example 10.56 Consider the model consisting of two chemicals, A and B, and one heat particle, γ. The sample space is $\Omega = \Omega_A \times \Omega_B \times \Omega_\gamma$, where each factor is $\ell^2 = \{0, 1, 2, \ldots\}$, the classical Fock space. A typical sample point is $\omega = (i, j, k)$ where $i, j, k = 0, 1, 2, \ldots$. The algebra of observables is $\mathcal{A} = \ell^\infty(\Omega)$, the abelian algebra of all bounded random variables on Ω. The dynamics will be given by a bistochastic energy-conserving map T on $\mathcal{A}$, which is linear, followed by a non-linear randomising map to mix the energy-shells. For this, we are going to use the LTE map, which was called Q, to provide the increase in entropy. Then the state p at time t will be in LTE, and so can be written $p = p_A \otimes p_B \otimes p_\gamma$

where

$$p_A(i) = Z_A^{-1}\lambda_A^i; \qquad p_B(j) = Z_B^{-1}\lambda_B^j; \qquad p_\gamma(k) = Z_\gamma^{-1}\lambda_\gamma^k.$$

Any random variable on Ω is a function $f(\omega) = f(i,j,k)$; it can be written as a sum of the projections $P_{i,j,k}$:

$$P_{i',j',k'}(\omega) = \begin{cases} 1 & \text{if } i = i', \;\; j = j', \;\; k = k'; \\ 0 & \text{otherwise.} \end{cases} \tag{10.57}$$

In Eqn. (10.57), ω is the point (i,j,k). Then we can write

$$f(\omega) = \sum_{\omega'} f(\omega')P_{\omega'}(\omega)\ .$$

In particular, $\mathcal{N}_A$ is given by

$$\mathcal{N}_A = \sum_{\omega'} i' P_{\omega'}.$$

A state is also a function on Ω. Thus the state p in LTE has the form

$$p(\omega) = Z_A^{-1}Z_B^{-1}Z_\gamma^{-1} \sum_{i',j',k'=0}^{\infty} \lambda_A^{i'}\lambda_B^{j'}\lambda_\gamma^{k'} P_{i',j',k'}(\omega).$$

Because of the map Q, in one time-step we move from one LTE-state, say $p = (n_A, n_B, n_\gamma)$ to another, say $p' = (n'_A, n'_B, n'_\gamma)$. Since Q conserves $\mathcal{N}_A$, $\mathcal{N}_B$ and $\mathcal{N}_\gamma$, any change in n_A in the time-step is caused by bistochastic map T:

$$\begin{aligned} \langle p', \mathcal{N}_A\rangle &= \sum_\omega QT^\dagger p(\omega)\mathcal{N}_A(\omega) \\ &= \langle T^\dagger p,\, \mathcal{N}_A\rangle. \end{aligned} \tag{10.58}$$

The action of $T^\dagger$ on p, or of T on a random variable, is given by its action on all the $P_{i,j,k}$. We shall choose T to be symmetric in the Hilbert-Schmidt scalar product, so that $T = T^\dagger$. This is the simplest expression of microscopic reversibility. Following Einstein's intuition, T can be split into a forward part, T_+, associated with $A \to B + \gamma$, and a backward part, T_-, associated with the reverse. We shall define these, and then put $T = T_+ = T_-$. We now define

$$\begin{aligned} T_+P_{i,j,k} &= (1 - i(j+1)(k+1)\kappa)P_{i,j,k} + i(j+1)(k+1)\kappa P_{i-1,j+1,k+1}\ , \\ T_-P_{i,j,k} &= (1 - (i+1)jk\kappa)P_{i,j,k} + (i+1)jk\kappa P_{i+1,j-1,k-1}. \end{aligned} \tag{10.59}$$

In order for the discrete map T to be stochastic, for a given i,j,k we must have

$$(i(j+1)(k+1) + (i+1)jk)\kappa \leq 2.$$

Thus κ must be taken smaller, the greater the number of particles there are in the state. As κ is proportional to the time-step, we must reduce the time-step when we apply T to states of more and more particles. Thus, there is no stochastic map on the algebra which works for all elements. This is one of the features of Bosonic quantum fields – the rate of reactions gets large when there are many particles. The differential equation with continuous time, however, seems to show no pathology. Taking a small enough κ, the change of n_A is then got by inserting T in Eqn. (10.58):

$$\begin{aligned}\langle T_+^\dagger p, \mathcal{N}_A\rangle &= \sum_{\omega,\omega'} i'p(i,j,k)\langle\{P_{i,j,k}(1-\kappa i(j+1)(k+1)) \\ &+ P_{i-1,j+1,k+1}\kappa i(j+1)(k+1)\}, P_{i',j',k'}\rangle\end{aligned}$$

which reduces to

$$n'_A = n_A - \kappa n_A(n_B+1)(n_\gamma+1)$$

giving the forward part of the discrete version of Eqn. (10.47) except for a factor 2 in the definition of κ. Similarly, T_- gives the backward part. The limit as the time step goes to zero gives us Eqn. (10.47).

To derive Eqn. (10.47) in the spirit of *quantum* statistical dynamics, let us assume energy balance, Eqn. (10.51), and put $S = \exp\{iH_I\}$. This involves questions of rigour, since to exponentiate an operator it must be essentially self-adjoint. One way this might be achieved in the Bosonic case is to add a high power of a conserved positive operator such as the number operator, and to show that H_I is Kato-small [97]. This problem does not arise for Fermions in a discrete space. The operator S, once defined, is thus expressed as a function of the free creators and annihilators, in the spirit of the Haag expansion [98]. Then conjugation of the operators with S is a completely positive bistochastic map leaving the spectral projections of H_0 invariant. We regard this as the scattering operator in one large time-step. We imagine that the scattering is over and done with, even during a small fraction t of this time-step. Thus we can consider e^{iH_It} as the scattering that takes place in t steps, the ingoing states for one time step being the outgoing states of the previous step. The time-evolution of the number operator under this unitary group is

$$\mathcal{N}_A(t) = \mathcal{N}_A(0) + it[\mathcal{N}_A, H_I] + (it)^2/2![[\mathcal{N}_A, H_I], H_I] + \dots \qquad (10.60)$$

Let us follow the scattering over one time-step of size t not by another scattering automorphism, but by the quasifree map. This acknowledges that after the time t, which is large on the microscopic scale, the particles are far from their starting points, and meet new particles. The higher correlations among them are lost. This is the modern form of the Fermi golden rule. We start with an even state,

ρ, which is quasifree, and take the expectation of Eqn. (10.60). Since H_I is an odd power of the fields (it is cubic) the term of first degree in t has zero expectation value. The second term gives

$$-t^2/2[[\mathcal{N}_A, H_I], H_I] = e^2t^2\{(\mathcal{N}_A+1)\mathcal{N}_B n_\gamma - \mathcal{N}_A(\mathcal{N}_B+1)(\mathcal{N}_\gamma+1)\}.$$

Suppose that ρ is gauge invariant under the separate $U(1)$ groups generated by the number operators, and put $\rho(N_A) = n_A$ etc.; then the expectation of the second-order term is

$$e^2t^2\{(n_A+1)n_B n_\gamma - n_A(n_B+1)(n_\gamma+1)\}.$$

If we put $e^2t = \kappa$, and divide by t, and take the limit as $t \to 0$, we get the Boltzmann equation Eqn. (10.47). The higher-order terms vanish in this limit. After one time-step, we construct the quasifree state with the same two-point functions and start again with the scattering automorphism.

We can get the Fermi version of the Boltzmann equation, Eqn. (10.48) in the same way by choosing the A and B to be Fermions, and taking the same form for H_I.

We have mentioned that one of the limitations of using the CCR and CAR 9is that there are no states in which only a finite number of the many-body correlations are not zero except the quasifree states. Such states might omit correlations that might have physical importance. Even worse, if the interaction energy is a quartic expression in the creators and annihilators then the quasifree map will not conserve the mean energy. In Chapter (13) we shall use the idea of higher order temperatures to suggest a wider class of states and an entropy-increasing map which is more convenient than the quasifree map for interacting systems.

10.5 Exercises

Exercise 10.61 Let $\mathcal{A}_x = \mathcal{M}_2$ for all $x \in \Lambda$, where $|\Lambda| < \infty$. Let $H_x = \sigma_3$ for all x. Let $\mathcal{A} = \bigotimes \mathcal{A}_x$, and $H = \sum H_x$. Show that the only state that is both uniform on the energy-shell and a product state is a canonical state.

Exercise 10.62 Show that the permutation group S_n acts in a unitary way on $\otimes^n\mathcal{H}$ by linear extension of the map

$$\pi\left(\psi_1 \otimes \psi_1 \ldots \psi_n\right) = \psi_{\pi(1)} \otimes \psi_{\pi(2)} \otimes \psi_{\pi(n)} \qquad \text{where } j \mapsto \pi(j) \text{ is in } S_n.$$

Show that the space of symmetric tensors is spanned by product vectors $\psi \otimes \psi \otimes \ldots \otimes \psi$, with $\psi \in \mathcal{H}$.

Exercise 10.63 Show that the Segal-Weyl operators obey the relation

$$\langle W(\psi)e^{\chi}, W(\psi)e^{\phi}\rangle = \langle e^{\chi}, e^{\phi}\rangle$$

for all coherent states e^{χ} and e^{ϕ}.

Exercise 10.64 Show that $W(\psi)$ obeys Eqn. (10.19).

Exercise 10.65 Show that

$$\langle e^{\phi}, W(\chi)e^{\psi}\rangle = \exp\left(-\|\chi\|^2/2 - \langle\chi, \psi\rangle + \langle\phi, \chi\rangle + \langle\phi, \psi\rangle\right).$$

Hence obtain Eqn. (10.25).

Exercise 10.66 Show that the canonical state of 2 oscillators, with given mean values for all the first and second moments, is quasifree.

Exercise 10.67 Suppose that $z = \{q_1, q_2, \ldots, q_N, p_1, p_2, \ldots, p_N\}$ and $z' = Sz$, where q_i and p_j obey Heisenberg's commutation relations, and $S \in GL(2N, \mathbf{R})$. Show that $[z'_j, z'_k] = i\hbar G_{jk}$ if and only if $SGS^t = G$, where G is the symplectic metric.

Exercise 10.68 Put $z = q + ip \in \mathbf{C}^N$, and the same with dashed symbols. Show that

$$Im\langle z', z\rangle = z'Gz.$$

Exercise 10.69 Show that in any representation π of the CAR, we have

$$\|\pi(a(\psi))\| = \|\pi(a^*(\psi))\| = \|\psi\|_{\mathcal{H}}\,.$$

Exercise 10.70 Show that the Fermion creation and annihilation operators defined on the antisymmetric Fock space by Eqn. (10.33) obey the CAR relations Eqn. (10.29), (10.30).

Exercise 10.71 Suppose that A is a representation of the Clifford algebra over the real Hilbert space $\mathcal{H}$. Let T be an orthogonal operator on $\mathcal{H}$, that is, an $\mathbf{R}$-linear map obeying $(T\psi, T\phi) = (\psi, \phi)$ for all ψ and ϕ in $\mathcal{H}$. Show that $A \circ T$ is a representation of the Clifford algebra.

Exercise 10.72 Show that $\langle\bullet, \bullet\rangle_J$ defined in Eqn. (10.38) is a scalar product; that is, it is positive definite, $\mathbf{C}$-linear in the second variable, and Hermitian: $\langle\psi, \phi\rangle_J^* = \langle\phi, \psi\rangle_J$.

Exercise 10.73 Let A be a representation of the Clifford algebra over the real Hilbert space $\mathcal{H}$, and let J be a complex structure on $\mathcal{H}$. Let a and a^* be given by Eqn. (10.39). Show that they provide a representation of the CAR over the complex Hilbert space $(\mathcal{H}, J)$.

Exercise 10.74 Find the generating function for a single Fermion mode

$$\rho\left(\exp\{\lambda a^*\}\exp\{\mu a\}\right)$$

where $\rho = Z^{-1}\exp(-\beta\mathcal{E}a^*a)$.

Exercise 10.75 Let ψ_0 be the Fock vacuum of a Bosonic oscillator; show that the norm of $a^{*n}\psi_0$ is $\sqrt{n}!$

Chapter 11

Isothermal and Driven Systems

11.1 Isothermal Quantum Dynamics

Consider a quantum system, described by a C^*-algebra $\mathcal{A}_c$, coupled to a heat-bath at beta β, described by the algebra $\mathcal{A}_c$. The combined system is described by $\mathcal{A} = \mathcal{A}_c \otimes \mathcal{A}_\gamma$. In the simplest models, the energy is taken to be the sum $H = H_c \otimes 1 + 1 \otimes H_\gamma$. The isolated dynamics of the combined system is given by an energy-conserving bistochastic map τ, followed by the stoss map. Thus one time step is the map

$$\rho \mapsto \tau^*\rho \mapsto \mathcal{M}_c(\tau^*\rho) \otimes \mathcal{M}_\gamma(\tau^*\rho) .$$

Even if the state at time t is the product of a state ρ_c with a thermal state $\rho_{\gamma\beta}$, the interaction τ^* will spoil this form. In the isothermal dynamics we assume that the chemical dynamics is so slow that within the time-step it is possible to restore the state of the heat bath to the same state $\rho_{\gamma\beta}$ as before the action of τ^*. Thus the isothermal dynamics is given by the map

$$T^\dagger \rho_c = \mathcal{M}_c(\tau^*(\rho_c \otimes \rho_{\gamma\beta})) \tag{11.1}$$

with $\rho_{\gamma\beta}$ being constant in time. Unlike the isolated dynamics with stoss map, this is a linear map in the state ρ_c. Further, this dynamics does not conserve the mean value of the energy, either of the chemical system or the combined system; physically, this is due to the assumed heat flows needed to restore the state $\mathcal{M}_\gamma(\tau^*(\rho_c \otimes \rho_{\gamma\beta}))$ back to $\rho_{\gamma\beta}$. In the case where τ is unitary conjugation (non-random dynamics), this map has been called the "reduced Heisenberg dynamics" in [99].

The map τ is assumed to be energy-conserving; thus, if $H = \sum_j \mathcal{E}_j P_j$ is the spectral resolution of the total energy then we assume that for each j, we have $P_j\tau = P_j$. Moreover, we require that τ^*, the adjoint of τ on the Hilbert space of Hilbert-Schmidt operators, also leaves each P_j invariant. As a consequence, the canonical state $\rho_{c\beta}$ of H_c is a fixed point of $T^\dagger$. Thus

$$T^\dagger \rho_{c\beta} = \mathcal{M}_c\left(\tau^*(\rho_{c\beta} \otimes \rho_{\gamma\beta})\right) = \mathcal{M}_c\left(\rho_{c\beta} \otimes \rho_{\gamma\beta}\right) = \rho_{c\beta}.$$

Here, we have used the fact that each P_j projects to a finite dimensional space and so is of Hilbert-Schmidt class.

Just as in the classical theory, in isothermal dynamics the second law is expressed as the free-energy theorem. This states that the thermodynamic function $\Psi(\rho) = S(\rho) - \beta\langle\rho, H_c\rangle$ does not decrease under a completely positive stochastic map. This is effectively contained in a result of Lindblad [100] of which a simple form is

Theorem 11.2 *Let T be a completely positive stochastic map on $\mathbf{C}^n$ and $T^\dagger$ the induced map on the states; denote the density matrix of a state ρ by $\Delta\rho$. Then*

$$Tr\,\left(\Delta\rho(\log\Delta\rho - \log\Delta\sigma)\right) \geq Tr\,\left(\Delta T^\dagger\rho(\log\Delta T^\dagger\rho - \log\Delta T^\dagger\sigma)\right). \quad (11.3)$$

The point is that if σ is the canonical state for the Hamiltonian H_c and is a fixed point of $T^\dagger$, then $\log\Delta\sigma = -\beta H_c - Z_\beta$ and the inequality becomes

$$S(T^\dagger\rho) - \beta\langle T^\dagger\rho, H_c\rangle \geq S(\rho) - \beta\langle\rho, H_c\rangle \quad (11.4)$$

as the terms in $\log Z$ cancel. In the classical case all positive maps are completely positive, and the result reduces to a theorem on relative entropy contained in [19].

We shall prove a version of this result for isothermal dynamics; it does not use that the map is completely positive, and this leads one to ponder whether complete positivity automatically holds for any isothermal dynamics. This is true, as was first proved by Kraus [69] for the reduced Heisenberg dynamics, in which τ is a unitary conjugation, and so by convexity, is true for random reversible dynamics. For the general bistochastic case, it is not known. Our more general result [31] is

Theorem 11.5 (The Quantum Free-energy Theorem) *Any isothermal dynamics $T^\dagger$ obeys*

$$S\left(T^\dagger\rho_c\right) - \beta\langle T^\dagger\rho_c, H_c\rangle \geq S(\rho_c) - \beta\langle\rho, H_c\rangle. \quad (11.6)$$

Proof. Let τ be the energy-conserving bistochastic map which defines the isolated dynamics with Boltzmann map by

$$\rho_c \otimes \rho_\gamma \mapsto \rho_c' \otimes \rho_\gamma' = \mathcal{M}_c\left(\tau_*(\rho_c \otimes \rho_\gamma)\right) \otimes \mathcal{M}_\gamma\left(\tau_*(\rho_c \otimes \rho_\gamma)\right).$$

The quantum H-theorem for isolated systems says that

$$S(\rho_c) + S(\rho_\gamma) \le S(\rho_c') + S(\rho_\gamma').$$

Apply this when $\rho_\gamma = Z^{-1}\exp -\beta H_\gamma$, in which case $\log(\rho_\gamma) = -\beta H_\gamma - \log Z$. We get, using Klein's inequality on the way

$$\begin{aligned} S(\rho_c) + S(\rho_\gamma) &= S(\rho_c) - Tr\,(\rho_\gamma(-\beta H_\gamma - \log Z)) \le S(\rho_c') - Tr\,(\rho_\gamma' \log \rho_\gamma') \\ &= S(\rho') - Tr\,(\rho_\gamma' \log \rho_\gamma) + Tr\,\left(\rho_\gamma'(\log \rho_\gamma - \log \rho_\gamma')\right) \\ &\le S(\rho') - Tr\,(\rho_\gamma' \log \rho_\gamma) = S(\rho') - Tr\,\left(\rho_\gamma'(\beta H_\gamma - \log Z)\right). \end{aligned}$$

As in the classical case and Lindblad's theorem, the term in $\log Z$ cancels, and we are left with

$$S(\rho_c) + \beta Tr\,(\rho_\gamma H_\gamma) \le S(\rho') + \beta Tr\,(\rho_\gamma' H_\gamma). \tag{11.7}$$

But the isolated dynamics preserves the mean total energy, say E, so

$$E = Tr\,(\rho_c H_c) + Tr\,(\rho_\gamma H_\gamma) = Tr\,(\rho_c' H_c) + Tr\,(\rho_\gamma' H_\gamma)$$

which substituted in Eqn. (11.7) gives

$$S(\rho_c) - \beta\langle \rho_c, H_c\rangle \le S(\rho') - \beta\langle \rho_c', H_c\rangle$$

which is the F-theorem. □

If, in the proof, we use the quantum Kullback inequality instead of the Klein inequality, we see that there is an increase of at least $\varepsilon = 1/2\|\rho_\gamma' - \rho_\gamma\|_2^2$ which is positive unless $\rho_\gamma' = \rho_\beta$; that is, there is a strict increase in the thermodynamic function Ψ unless the state of the heat-particle is not disturbed by the isolated dynamics. As in the classical theory, the gain in entropy in one time-step is at least $\beta(\langle \rho', H_c\rangle - \langle \rho, H_c\rangle) = dQ/T$ when the heat flow (needed for the readjustment of the state of the heat particle) is dQ, in addition to ε coming from the sharp form of the Kullback lemma.

Since the isothermal dynamics $T^\dagger$ is an affine map state space $\Sigma(\mathcal{A}_c)$, it can be extended to a linear map on $\operatorname{Span}\Sigma_c$ and so is the dual of a linear map on $\mathcal{A}_c$. Let us give some of its properties.

Theorem 11.8 *Any isothermal dynamics $T^\dagger$, coming from a bistochastic map τ, obeys*

1. *$T^\dagger$ is the dual of a stochastic map T and has $\rho_{c\beta}$ as fixed point. The relation*
$$\langle \rho_c, T(A)\rangle = \langle \rho_c \otimes \rho_{\gamma\beta}, (A \otimes 1_\gamma)\tau\rangle, \tag{11.9}$$
holds, where 1_γ is the unit operator in $\mathcal{A}_\gamma$.

2. If τ is completely positive, then T is completely positive.

Proof.
(1) Since $T^\dagger$ is a bistochastic map followed by tracing over the heat-particle, it maps $\Sigma(\mathcal{A}_c)$ to itself, and so is normalised and positive. It is therefore the dual to a stochastic map T. We have already shown that $\rho_{c\beta}$ is a fixed point. For the last part,

$$\begin{aligned}\langle \rho_c, T(A)\rangle &= \langle T^\dagger \rho_c, A\rangle = Tr_1(AT^\dagger \rho_c)\\ &= Tr_1 A Tr_2 \tau^*(\rho_c \otimes \rho_{\gamma\beta}) = Tr_1 Tr_2 (A\otimes 1_\gamma)\tau^*(\rho_c \otimes \rho_{\gamma\beta})\\ &= Tr_1 Tr_2 (A\otimes 1_\gamma)\tau(\rho_c\otimes\rho_{\gamma\beta}) = \langle \rho_c \otimes \rho_{\gamma\beta}, (A\otimes 1_\gamma)\tau\rangle.\end{aligned}$$

[Recall that τ act on the right on elements of the algebra.] □
(2) You are referred to [69] for the case of reduced Heisenberg dynamics. The general case is similar. □
Note that [99], Appendix, has an error, since it assumes that the convex hull of product states is dense in the state-space of a tensor product.

Recall that a random reversible dynamics is a convex mixture of unitary conjugations, each of which leaves the spectral projections of the total Hamiltonian invariant. This is a special class of bistochastic energy-conserving maps, since not every bistochastic map is a mixture of automorphisms. If τ is a random reversible dynamics, then the corresponding isothermal dynamics has some further important properties. One of these is *detailed balance*; this is a quantum version of Def. (5.22). We give it in the general form suggested in [43].

Definition 11.10 *Let $\mathcal{A}_c$ be a C^*-algebra with identity and let T be a stochastic map on $\mathcal{A}_c$. Let ρ be a faithful state on $\mathcal{A}_c$, invariant under T, and let $\{\pi_\rho, \mathcal{H}_\rho, \psi_\rho\}$ be the representation, Hilbert space and cyclic vector given by the Gelfand-Naimark-Segal construction. Let $\pi_\rho(T)$ denote the operator on $\mathcal{H}_\rho$*

$$\pi_\rho(T)\pi_\rho(A)\psi_\rho = \pi_\rho(TA)\psi_\rho, \qquad (A\in\mathcal{A}) \tag{11.11}$$

defined on the dense domain $D = \pi_\rho(\mathcal{A}_c)\psi_\rho$. We say that T obeys detailed balance *relative to ρ if there exists a stochastic map $T^{(\rho)}$ such that $\pi_\rho(T)^* = \pi_\rho(T^{(\rho)})$ on D.*

Note that $T^{(\rho)}(A) = T^\dagger(A\rho)\rho^{-1}$ (Exercise (11.22)). In the following theorem, the *modular automorphism* is defined to be conjugation by $e^{iH_c t}$, that is, it is the reversible Heisenberg dynamics of the chemical algebra.

Theorem 11.12 *Let T be the isothermal dynamics dual to $\rho_c \mapsto T^\dagger\rho_c = \mathcal{M}_c\tau^*(\rho_c\otimes\rho_{\gamma\beta})$, where τ is a random reversible dynamics. Then*

1. *T obeys detailed balance relative to $\rho_{c\beta}$ in the general form of definition (11.10).*

2. *T commutes with the modular automorphism.*

3. *T is a contraction on $\mathcal{A}_c$ in the norm $\|A\|_\beta = |\langle\rho_{c\beta}, A^*A\rangle|^{1/2}$.*

4. *$T^\dagger$ is a contraction on $\Sigma(\mathcal{A}_c)$ in the norm (on density matrices) $\|\bullet\|_{-\beta}$.*

Proof.
(1) Let $A\tau = \sum\lambda_i U_i A U_i^*$ be the random dynamics, with each U_i commuting with H. Here, $A \in \mathcal{A} = \mathcal{A}_c \bigotimes \mathcal{A}_\gamma$, $U_i \in \mathcal{A}$ and H is the total Hamiltonian $H_c \otimes 1 + 1 \otimes H_\gamma$. Let $\rho_\beta = \rho_{c\beta} \otimes \rho_{\gamma\beta}$ be the canonical state and denote $T^{(\rho_{c\beta})}$ by $T^{(\beta)}$. Then

$$\begin{aligned}\tau^*(A\rho_\beta) = \sum_i \lambda_i U_i^* A\rho_\beta U_i &= \sum_i \lambda_i U_i^* A U_i U_i^* \rho_\beta U_i \\ &= \tau^*(A)\rho_\beta \end{aligned} \tag{11.13}$$

since each U_i commutes with ρ_β. It follows that

$$\begin{aligned}\langle\rho_c, T^{(\beta)}(A)\rangle &= \langle\rho_c, T^\dagger(A(\rho_{c\beta})\rho_{c\beta}^{-1}\rangle = \langle\rho_c\rho_{c\beta}^{-1}, T^\dagger(A\rho_{c\beta})\rangle \\ &= \langle T\left(\rho_c\rho_{c\beta}^{-1}\right), A\rho_{c\beta}\rangle = \langle\left(\rho_c\rho_{c\beta}^{-1}\otimes 1\right)\tau, A\rho_{c\beta}\otimes\rho_{\gamma\beta}\rangle \\ &= \langle\rho_c\rho_{c\beta}^{-1}\otimes 1, \tau^*\left(A\rho_{c\beta}\otimes\rho_{\gamma\beta}\right)\rangle \\ &= \langle\rho_c\rho_{c\beta}^{-1}, \tau^*(A\otimes 1)(\rho_{c\beta}\otimes\rho_{\gamma\beta})\rangle \quad \text{by Eqn. (11.13)} \\ &= \langle\rho_c\otimes\rho_{\gamma\beta}, \tau^*(A\otimes 1)\rangle.\end{aligned}$$

It follows that $T^{(\beta)}$ is the isothermal dynamics determined by τ^*, (which is bistochastic since τ is bistochastic), by Eqn. (11.9). It follows from theorem (11.8) that $T^{(\beta)}$ is stochastic. Note that it is the positivity that is the hard part, since $T^{(\rho_c)}1 = 1$ is easy (Exercise (11.22)). Note further that the condition of detailed balance follows from bistochasticity and the one condition Eqn. (11.13), and does not require that the dynamics be random conservative. Thus the theorem is rather more general than stated. □
(2) To show that T commutes with the modular automorphism, take any ρ_c; then we have

$$\begin{aligned}\langle\rho_c, \rho_{c\beta}T(A)\rho_{c\beta}^{-1}\rangle &= Tr\left(\rho_c\rho_{c\beta}T(A)\rho_{c\beta}^{-1}\right) = \langle\rho_{c\beta}\rho_c\rho_{c\beta}^{-1}, T(A)\rangle \\ &= \langle\rho_{c\beta}\rho_c\rho_{c\beta}^{-1}\otimes\rho_{\gamma\beta}, (A\otimes 1)\tau\rangle \\ &= Tr\left(\rho_{c\beta}^{-1}\rho_c\rho_{c\beta}\otimes\rho_{\gamma\beta}(A\otimes 1)\tau\right) \\ &= Tr\left(\rho_\beta^{-1}(\rho_c\otimes\rho_{\gamma\beta})\rho_\beta(A\otimes 1)\tau\right)\end{aligned}$$

$$
\begin{aligned}
&= Tr\left((\rho_c \otimes \rho_{\gamma\beta})[\rho_\beta(A \otimes 1)\rho_\beta^{-1}]_T\right) \\
&= \langle \rho_c \otimes \rho_{\gamma\beta}, [\rho_{c\beta} A \rho_{c\beta}^{-1} \otimes 1]_T \rangle = \langle \rho_c, T(\rho_{c\beta} A \rho_{c\beta}^{-1}) \rangle.
\end{aligned}
$$

Since ρ_c was an arbitrary state, we conclude that $T(A)$ commutes with conjugation by $\rho_{c\beta}$. Hence it commutes with all powers of $\rho_{c\beta}$ and hence with $\Delta = \log \rho_{c\beta}$ and so with the modular automorphism group, namely, conjugation with Δ^{it}. □
(3) I follow [101]. In fact, we show that the contractivity of T is a consequence of detailed balance (which by (1) is a consequence of the construction of T from a random reversible dynamics). So, choose the faithful state ρ mentioned in Def. (11.10) to be $\rho = \rho_{c\beta}$. We then have that $\pi_\rho(T)^*$ is induced by the stochastic map $T^{(\rho)}$. By Størmer's theorem, [65] both T and $T^{(\rho)}$ are norm contractions on $\mathcal{A}_J$. Then

$$
\begin{aligned}
\|\pi_\rho(T)\pi_\rho(A)\psi_\rho\|^2 &= \rho(T(A^*)T(A)) = \rho\left(A^* T^{(\rho)}(T(A))\right) \\
&\le (\rho(A^*A))^{1/2}[\rho\left(T^{(\rho)}(TA)^* T^{(\rho)}(TA)\right)]^{1/2} \\
&= [\rho(A^*A)]^{1/2}[\rho\left(A^* T^{(\rho)}(T(T^{(\rho)}(TA)))\right)]^{1/2} \\
&\le [\rho(A^*A)]^{3/4}[\rho\left(T^{(\rho)}(T(T^{(\rho)}(TA^*)))T^{(\rho)}(T(T^{(\rho)}(TA)))\right)]^{1/4}\ldots \\
&\le [\rho(A^*A)]^{(1-2^{-n})}[\rho\left(T^{(\rho)}(T\ldots(TA^*)T^{(\rho)}(T\ldots T(A)\ldots)\right)]^{2^{-n}} \\
&\to C\rho(A^*A)
\end{aligned}
$$

as $n \to \infty$, where $C \le 1$. In the last step we used that T and $T^{(\rho)}$ are contractions in norm, and so any iteration of the map $T^{(\rho)}T(\bullet)$ is a contraction.
(4) This follows from (3) by duality. □

The first results along the lines of (2) and (3) were obtained by Majewski [102] who used a version of the condition of detailed balance based on the physical idea of time-reversal invariance of an underlying dynamics. This was a generalisation of the property of microscopic reversibility. Majewski's version implies the version given here, but the converse has only been established in special cases [101]. We see that detailed balance is a strong and subtle condition, whose full implications have not yet been fully explored.

It is easy to show that $\rho_{c\beta}$ is the state with the smallest norm $\| \bullet \|_{-\beta}$. See Exercise (11.23). So the dynamics, if it converges to equilibrium, is a contraction from the initial state to the state of smallest norm. In the next section we give sufficient conditions which ensure this, at least if $\dim \mathcal{A}_c < \infty$.

11.2 Convergence to Equilibrium

The long-time behaviour of a dynamical system can be studied by Lyapunov's direct method, which starts with the construction of a Lyapunov function. In the case of isothermal dynamics, we have such a function, the free energy or its relation Ψ, both in the classical and the quantum versions. The general theorem of Lyapunov can be applied if three conditions hold: the state space is compact, the Lyapunov function is strict, and there is a unique fixed point. We are not free to choose the sense in which the state-space is compact, since the topology must be such that both the Lyapunov function and the dynamics are continuous, at least in the elementary version given in Theorem (3.23). There is no problem in this connection, if $\dim \mathcal{A}_c < \infty$. However, the theorem leaves something to be desired, and in the spirit of Krylov [73] we should ask for a stronger result: we need a strong (=large) norm in which the convergence takes place, and also we should get an estimate of the relaxation time. In good cases too there should be exponential approach to equilibrium. Finally, the theory should have something to say if $\dim \mathcal{A}_c = \infty$. We can achieve some of these objectives if we postulate strong enough ergodic conditions on the dynamics τ. Here we are guided by what happens in the case of random reversible dynamics.

Let $\mathcal{A} = \mathcal{A}_c \otimes \mathcal{A}_\gamma$ be the algebra of all bounded operators on $\mathcal{H} = \mathcal{H}_c \otimes \mathcal{H}_\gamma$, with Hamiltonian $H = H_c \otimes 1 + 1 \otimes H_\gamma$, and let $\tau = \sum \lambda_i U_i$ be a mixture of unitary conjugations of $\mathcal{A}$, each U_i commuting with H. Let $H = \sum \mathcal{E}_j P_j$ be the spectral resolution of H, with $\dim P_j < \infty$ for all j. The state space $\Sigma(\mathcal{A})$ spans a complex vector space, known as $\ell^1(\mathcal{A})$ by analogy with the commutative case; this, and the analogues of all the ℓ^p spaces, were introduced and studied by von Neumann and Schatten. The span of the normal states can be identified with a subalgebra $\mathcal{A}_1$ of $\mathcal{A}$ by the map Δ, which assigns the density operator $\Delta\rho$ to the normal state ρ. Then $\mathcal{A}_1$ is the direct sum of algebras

$$\mathcal{A}_1 = \overline{\bigoplus}_{ij} P_i \mathcal{A}_1 P_j. \tag{11.14}$$

The meaning of the bar is that the algebraic direct sum must be closed in the trace-norm. The diagonal algebras $P_i \mathcal{A} P_i$ contain the microcanonical states P_i, which act as units, and the others, $P_i \mathcal{A} P_j$, with $i \neq j$, are nilpotent. Each of these algebras is invariant under τ^*. It is therefore reasonable to say that a bistochastic map, τ, is energy-conserving if each $\mathcal{A}_{ij}$ is invariant under τ^*. Let us say that τ is ergodic relative to H if for each j, multiples of P_j are the only invariant elements of $P_j \mathcal{A} P_j$, and zero is the only invariant element of $P_j \mathcal{A} P_k$, $k \neq j$. We shall also need a further property, which ensures that the energy-shells link all the energy levels of H_c. For example if the energies were incommensurate, then only the identity map $\tau = Id$ would conserve energy. Let us say that two energy-levels $\mathcal{E}_{cj}$ and $\mathcal{E}_{ck}$ are linked if there exist two energy-levels $\mathcal{E}_{\gamma 1}$ and $\mathcal{E}_{\gamma 2}$ such that

$$\mathcal{E}_{cj} + \mathcal{E}_{\gamma 1} = \mathcal{E}_{ck} + \mathcal{E}_{\gamma 2}.$$

We can regard being linked as a relation like being nearest neighbours in a graph. To get mixing, we shall need to assume that each energy-level $\mathcal{E}_{cj}$ is linked to any other by a finite chain. A similar assumption was needed in the classical case (the irreducibility of the Markov chain). We express this by saying that the energy-shells are connected. Then we have

Theorem 11.15 *Let τ be energy-conserving and ergodic relative to H. Suppose that* $\dim \mathcal{A}_c < \infty$ *and that the energy shells of $H = H_c \otimes 1 + 1 \otimes H_\gamma$ are connected. Then the iterated isothermal dynamics converges to equilibrium, for any initial state ρ_c.*

Proof. Consider the action of τ. Any invariant state $\rho = \rho_c \otimes \rho_{\gamma\beta}$ can be written as the orthogonal sum (in the Hilbert-Schmidt sense) of its parts in the spectral subspaces of $H_c = \sum \mathcal{E}_c P_{cj}$ and $H_\gamma = \sum \mathcal{E}_\ell P_{\gamma\ell}$:

$$\rho = \sum_{jk\ell} (P_{cj} \otimes P_{\gamma\ell})(\rho_c \otimes Z_{\gamma\beta}^{-1} e^{-\beta\mathcal{E}_{\gamma\ell}} 1)(P_{ck} \otimes P_{\gamma\ell}). \tag{11.16}$$

Let us choose energy-shells (of H) labelled by $\mathcal{E}_1$ and $\mathcal{E}_2$, and consider the partial sum over all k, ℓ such that $\mathcal{E}_{cj} + \mathcal{E}_{\gamma\ell} = \mathcal{E}_1$ and $\mathcal{E}_{ck} + \mathcal{E}_{\gamma\ell} = \mathcal{E}_2$. This is an element of $P_1 \mathcal{A} P_2$, which is mapped to itself by τ^*. By the uniqueness of the Hilbert-Schmidt orthogonal sum, this element must be invariant, since ρ is invariant, and so by ergodicity is zero if $\mathcal{E}_1 \neq \mathcal{E}_2$, and is a multiple of P_1 if $\mathcal{E}_2 = \mathcal{E}_2$. We note that

$$\begin{aligned} H &= H_c \otimes 1 + 1 \otimes H_\gamma = \sum_{j\ell} (P_{cj}\mathcal{E}_{cj} \otimes P_{\gamma\ell} + P_{cj} \otimes P_{\gamma\ell}\mathcal{E}_{\gamma\ell}) \\ &= \sum_{\mathcal{E}} \sum_{j\ell:\mathcal{E}_{cj}+\mathcal{E}_{\gamma\ell}=\mathcal{E}} \mathcal{E}\,(P_{cj} \otimes P_{\gamma\ell}). \end{aligned}$$

It follows that

$$P_1 = \sum_{j\ell:\mathcal{E}_{cj}+\mathcal{E}_{\gamma\ell}} P_{cj} \otimes P_{\gamma\ell}$$

holds, and that the sum is an orthogonal one. Since we have just proved that the part of ρ, Eqn. (11.16) on the energy shell $\mathcal{E}_1$ is a multiple of P_1, we can equate the orthogonal components, to get

$$(P_{cj} \otimes P_{\gamma\ell})(\rho_c \otimes Z_{\gamma\beta}^{-1} e^{-\beta\mathcal{E}_{\gamma\ell}} 1)(P_{cj} \otimes P_{\gamma\ell}) = \mu P_{cj} \otimes P_{\gamma\ell}$$

for all j and ℓ leading to the same energy-shell $\mathcal{E}_1$. The factor μ depends on $\mathcal{E}_1$ but not on j and ℓ separately. So the part of ρ_c in the algebra $P_{cj}\mathcal{A}_c P_{cj}$ must be proportional to P_{cj}, with a factor $e^{-\beta\mathcal{E}_{cj}}$ to balance the factor $e^{-\beta\mathcal{E}_{\gamma\ell}}$. The overall factor is not determined by looking at one energy-shell, but we do get the Boltzmann factors

$$P_{cj}\rho_c P_{cj} = \lambda_j P_{cj} \text{ with } \lambda_j = e^{-\beta(\mathcal{E}_{cj}-\mathcal{E}_{ck})}\lambda_k$$

for all j participating in the energy-shell $\mathcal{E}_1$. But since we have assumed that the energy-shells are connected, we get a family of such relations including all levels, and we conclude that ρ_c is the canonical state with beta β. Thus τ^* has a unique fixed point. From the free-energy theorem, Ψ is a strict Lyapunov function, and by the compactness of $\Sigma(\mathcal{A}_c)$, we can apply Lyapunov's theorem to show that the isothermal dynamics converges. □

If the isothermal dynamics satisfies detailed balance then it is a contraction in the norm $\|\bullet\|_{-\beta}$, and we get convergence by contraction from the original state ρ_c to the state of smallest norm. This is because in finite dimensions all norms are equivalent. If $\dim\mathcal{A}_c = \infty$ we have seen that detailed balance is enough to ensure that the isothermal dynamics is a contraction in this norm, but we must make some assumptions about the growth in multiplicity of the energy-eigenvalues in order to get convergence [60]; it remains an open question whether we get convergence in the norm $\|\bullet\|_{-\beta}$ in the infinite case.

11.3 Driven Quantum Systems

Lebowitz and Spohn [103] formulate quantum dynamics in which various points are kept at different temperatures, by being coupled to heat-baths. They work in the limit of weak coupling, and impose the condition of detailed balance (at the local beta) on the local generator of a completely positive semi-group. The full dynamics is a convex sum of all these. We now develop a general scheme for driven systems, which includes a version of this construction as a special case. In the description of isothermal systems we start with one particle, the heat-particle, in a canonical state, and after one time-step, we inject heat, or remove it, to restore the heat-particle to its previous state. In a similar way, we can construct isopotential dynamics, for any chemical potential. Thus, suppose that the full system is described by the algebra $\mathcal{A} = \mathcal{A}_1 \bigotimes \mathcal{A}_2$, where $\mathcal{A}_2$ concerns the chemicals whose potentials are to be kept fixed. Let τ be a bistochastic map on $\mathcal{A}$. We assume that the initial state is $\rho_1 \otimes \rho_2$, where ρ_2 is a grand canonical state of all the chemicals described by $\mathcal{A}_2$. Then the isopotential dynamics is the map

$$\rho_1 \otimes \rho_2 \mapsto \mathcal{M}_1\left(\tau^*(\rho_1 \otimes \rho_2)\right) \otimes \rho_2.$$

The only difference between an isopotential map and a driven dynamics is that in the isopotential dynamics all the chemicals in $\mathcal{A}_2$ are conserved quantities, whereas in a driven system they are not. In the first case, we can use the conservation laws to establish a free-energy theorem, so that isopotential systems converge to equilibrium. Driven systems, in which some non-conserved quantities are fed from outside, might or might not have a Lyapunov function, and in general do not. In the models with several temperatures [103], the local energy is not conserved, so they are driven, not isopotential systems. We illustrate the general theory with the Fröhlich pumped phonon model [104, 105]. The model is "formulated at the kinetic level", which means that the correct Boltzmann equation is written down by instinct. We shall show that it fits in to statistical dynamics.

Example 11.17 The model consists of a system, which has n Bosons, called here A-particles, which are modes of a phonon field. So the Hilbert space of this is $\mathcal{F}(\mathbf{C}^n)$. Let a_j, a_j^* be the annihilators and creators, and let the energy of the system be $H_c = \sum_j \mathcal{E}_j a_j^* a_j$; we assume that $\mathcal{E}_k > \mathcal{E}_j$ if $k > j$. Thus $\mathcal{E}_1$ is the A-particle with the smallest energy. The A-particles can be absorbed directly by a heat-bath, and can be emitted by the bath. They can also be converted, one into the other, by absorbing or emitting a suitable quantum from the heat-bath. In addition, the population of phonons is fed by a source, which increases the number of phonons of type j at the rate s_j. Thus the system is clearly a driven one, but one in which the driving mechanism cannot modelled by an external chemical potential kept constant in time, as we have used earlier. The driving mechanism must be such that the actual input of quanta is at the predetermined rate s_j (which could depend on time in a more elaborate version). It is rather easy to model this; we just add s_j to the rate of change in the mean number of quanta j.

In statistical dynamics we first set up the theory of an isolated system obeying both laws of thermodynamics. To this end, we introduce a suitable collection of heat-particles, which act as the balancing item (so beloved of accountants). We then move to the isothermal system as usual by keeping the heat-particles at their canonical state, by feeding in heat or removing it. We then add the changes in the mean number due to the sources. Finally, we apply the quasifree map to find the quasifree state after one time-step.

The energy-levels $\mathcal{E}_j$ are not assumed to be rationally related to each other, yet they make transitions among themselves; so we need one heat-particle γ_{jk}, of energy $\mathcal{E}_{\gamma jk}$ for each pair of phonon modes, such that

$$\mathcal{E}_k - \mathcal{E}_j = \mathcal{E}_{\gamma jk}, \text{ where } k > j. \tag{11.18}$$

This allows the process

$$A_k \rightleftharpoons A_j + \gamma_{jk}$$

to proceed with the conservation of energy. Let b_{jk} denote the annihilator of this heat-particle. The model also allows a phonon. say A_j, to be absorbed by the heat-bath, or created by it; this means that it is converted directly into a heat-particle, so we need a heat-particle γ_j with energy exactly $\mathcal{E}_j$. Let b_j denote its annihilator. The absorption and its inverse is therefore the process

$$A_j \rightleftharpoons \gamma_j.$$

The energy of the heat-particles is thus

$$H_\gamma = \sum_j \mathcal{E}_j b_j^* b_j + \sum_{j<k} \mathcal{E}_{jk} b_{jk}^* b_{jk}.$$

A simple operator implementing the desired reactions is

$$H_I = \sum_j C_j(a_j^* b_j + b_j^* a_j) + \sum_{j<k} C_{jk}\left(a_j^* a_k b_{jk}^* + a_j a_k^* b_{jk}\right).$$

We only need to consider C_{jk} with $j < k$, since otherwise the reaction is forbidden by energy-conservation. We note that H_I has a quadratic term and a cubic term. Let $U_I = e^{iH_I}$. Then the Boltzmann equation for the isolated dynamics of any observable will be chosen to be conjugation by U_I; we throw this dynamics onto the states, by duality, and follow one time-step by the quasifree map. We shall also let the time-interval go to zero, and the couplings large, so that only the second-order term survives (the anti-van Hove limit). Suppose that at time $t = 0$ the state (which can be assumed to be quasifree, from the previous time-step) is a tensor product $\rho = \rho_c \otimes \rho_\gamma$. We then find the change in the state in the time-step to $t = 1$. The quasifree map does not change the mean numbers of any type of Boson, so the change of for example $n_j = \rho(a_j^* a_j)$ is given in one time step by the change due to the conjugation with U_I:

$$\begin{aligned} n_j' &= \rho(U_I a_j^* a_j U_I^{-1}) \\ &= \rho\left(a_j^* a_j + i[H_I, a_j^* a_j] + i^2/2!\left[H_I, [H_I, a_j^* a_j]\right]\right). \end{aligned} \tag{11.19}$$

The operator of first order is (Exercise (11.24))

$$\begin{aligned} i[H_I, \mathcal{N}_j] &= iC_j(b_j^* a_j - a_j^* b_j) \\ &+ i\sum_{k:k>j} C_{jk}\left(a_k^* b_{jk} a_j - a_k b_{jk}^* a_j^*\right) \\ &+ i\sum_{\ell:\ell<j} C_{\ell j}\left(a_j a_\ell^* b_{\ell j}^* - a_\ell a_j^* b_{\ell j}\right). \end{aligned} \tag{11.20}$$

This has a quadratic term, and a cubic term under the sum; each is linear in the heat-variables, and as we assume that the heat-particles are (in the isothermal dynamics) in the thermal state, the expectation value of (11.20) in ρ is zero. So the dynamics comes from the second-order term. If ρ_c is an even state, then the only parts which survive in the second-order term $i^2/2[H_I, [H_I, \mathcal{N}_j]]$ in Eqn. (11.19) are the commutator of the quadratic terms, and the commutator of the cubic terms. These are, respectively,

$$i^2/2 \sum_m C_m C_j \left[a_m^* b_m + a_m b_m^*, \, b_j^* a_j - b_j a_j^*\right]$$

and

$$i^2/2 \sum_{j'<k',k>j,\ell<j} C_{j'k'} \left[a_{j'}^* a_{k'} b_{j'k'}^* + a_{j'} a_{k'}^* b_{j'k'}, \, C_{jk}(a_k^* a_j b_{jk} - a_k a_j^* b_{jk}^*) \right.$$
$$\left. + C_{\ell j}(a_j a_\ell^* b_{\ell j}^* - a_\ell a_j^* b_{\ell j})\right].$$

These can be evaluated in the state ρ, which is quasifree. We get terms involving the two-point functions $\rho(a_j^* a_k)$, $\rho(a_j^* b_j)$ etc. These express correlations between the different modes of the phonon field, and between the phonons and the heat-particles. To get a closed system of equations we would need to find equations for all these two-point functions. In the spirit of statistical dynamics, we can use the stoss map to replace the state ρ' by the product state of the system and the heat-particles. This step is part of the change-over to isothermal dynamics from isolated dynamics. The heat-bath, by definition, always remains independent of the system. The stoss map is entropy increasing and conserves mean energy, so we remain with a viable theory if we assume that $\rho = \rho_c \otimes \rho_\gamma$, and maintain this independence after one time-step. As for the correlations between the different phonon modes, we get a tractable theory if the number of modes is small, say 2. The original model [104] did not derive the equations in this way, and ignored the phonon-phonon correlations. That is alright; we can apply a further randomising map, also entropy-increasing, putting these correlations zero after each step. Then ρ_c is a product state $\otimes \rho_{cj}$, one factor for each phonon, at each time. Evaluating the two expressions contributing to the second-order term, we find for the first,

$$n_j' = n_j - C_j(n_j - n_{\gamma j})$$

and defining C_{jk} for $j > k$ to be C_{kj} for convenience, the second is

$$\sum_{k \neq j} C_{jk}^2 \left(n_k(n_j+1)(n_{\gamma jk}+1) - n_j(n_k+1)(n_{\gamma kj}+1)\right).$$

We have assumed that the first moments of the phonon field are zero. If this is true at time zero, then it is true for all time. There is no freedom to put them zero as part of an entropy-increasing map, and if the initial state is a coherent state of a mode of the phonon, with non-zero one-point function, we must take account of the dynamics of this parameter. We would expect such a state to lose its coherence because of the use of entropy-increasing maps. Thus far, we have derived the isolated dynamics.

The move to isothermal dynamics consists in putting the Planck distribution at the same beta for all the n_γ occurring in the dynamics. We have already used the map which puts to zero the correlations between the heat-particles and the system. Finally, we drive the system by adding the source s_j to the rate of change of n_j, the mean number of quanta in the phonon mode j; we get the coupled system of equations

$$\begin{aligned} \frac{dn_j}{dt} &= s_j + C_j(n_{\gamma j} - n_j) \\ &+ \sum C_{jk}^2 \left(n_k(n_j+1)(n_{\gamma jk}+1) - n_j(n_k+1)n_{\gamma kj}\right). \end{aligned} \tag{11.21}$$

This seems to disagree with [105], p.210; but after correcting a misprint by the exchange of the Boltzmann factors in the second term in Eqn. (115) we can relate our constants C_j and C_{jk} to Sewell's constants a_j and b_{jk}, which become temperature-dependent.

If $s_j = 0$ then there is an equilibrium point when $n_j = n_{\gamma j}$, which is thermal equilibrium. The principle of detailed balance holds: both terms are separately zero at the fixed point; this holds by virtue of the relation (11.18). If $s_j > 0$, then there is a fixed point in which the state of the phonon A_j is a grand canonical state with some chemical potential related to the source strength [104, 105]. Duffield [106] has shown that independently of any derivation of the Fröhlich equations, they do define a one-parameter group of completely positive maps.

11.4 Exercises

Exercise 11.22 Show that $T^{(\rho)}(A) = T^\dagger(A\rho)\rho^{-1}$; hence show that $T^{(\rho_{c\beta})}1 = 1$.

Exercise 11.23 Let ρ be a faithful normal state on $B(\mathcal{H})$. Show that the unique normal state σ with the smallest value of $Tr\,(\rho^{-1}\sigma^2)$ is $\sigma = \rho$.

Exercise 11.24 As in the pumped phonon model, let

$$H_I = \sum_j C_j \left(a_j^* b_j + a_j b_j^*\right) + \sum_{j<k} C_{jk} \left(a_j^* a_k b_{jk}^* + a_j a_k^* b_{jk}\right)$$

and $\mathcal{N}_j = a_j^* a_j$. Prove Eqn. (11.20).

Chapter 12

Infinite Systems

In the theory of equilibrium states in statistical mechanics the important question as to whether a system undergoes a phase transition is best addressed in the thermodynamic limit. This is the limit in which the size of the system becomes infinite. Although any actual system is of finite size, this device is a very useful idealisation. Almost any sensible model of finite size shows no phase transition at all; all the thermodynamic variables are analytic functions of the intensive variables such as beta. A phase transition is recognised as a discontinuity in an interesting variable, usually as a function of beta and an external field. For example, a system of particles of spin 1/2 on a finite lattice Λ in an external magnetic field $\mathbf{B}$ is described by the algebra $\mathcal{A} = \bigotimes_{x\in\Lambda} \mathbf{M}_2(x)$, and the Hamiltonian $H = H_0 + \mathbf{B}.\boldsymbol{\sigma}$, where H_0 is the Hamiltonian in the absence of the magnetic field, and $\boldsymbol{\sigma}$ are the Pauli matrices. The magnetisation as a function of $\mathbf{B}$ is $\mathbf{m} = \rho_{\beta,\mathbf{B}}(\boldsymbol{\sigma})$. Here, $\rho_{\beta,\mathbf{B}}$ is the canonical state with Hamiltonian $H(\mathbf{B})$. We define the *spontaneous magnetisation* to be the limit of $\mathbf{m}$ as $\mathbf{B}$ converges to zero. It can be shown that if $|\Lambda| < \infty$, then the spontaneous magnetisation is zero at all beta. On the other hand, the celebrated work of Peierls [107, 108] shows that for the Ising model, the limit of the magnetisation per site converges, as $|\Lambda| \to \infty$ to a discontinuous function of beta, being zero for small beta, and non-zero above a critical beta, β_c, called the *Curie point*. The behaviour for a large but not infinite system is presumably near to that of the infinite system in some numerical sense, but there is no developed theory of functions that are continuous but nearly discontinuous. The infinite system not only shows the desired effect, but is simpler to describe. One reason for the simplicity is translation invariance, which holds for the infinite system. The strategy of studying the system in infinite volume is called "taking the thermodynamic limit". There are some advantages to adopting the same strategy in non-equilibrium statistical mechanics; this is particularly true in the fundamental problem of deriving irreversible behaviour from Hamiltonian dynamics. The weak limit of a sequence of unitary operators need not be unitary, but could be a contraction. So the dynamics of a limit of theories with reversible dynamics might be irreversible, just as in equilibrium theory the limit of functions continuous in the parameters $\beta, \mathbf{B}$ might not be continuous. The idea, then, is an attempt at a more fundamental theory, in which irreversibility is not put in, but is derived. This is only partly achieved, as the infinite-volume approximation has its own limitations. In this chapter we shall discuss some of the easier methods and results, though it is outside our self-imposed framework of dealing with only finite systems. We begin with the definition of the spin algebra for an infinite lattice. This uses the idea of an inductive limit. We get a definition of the hard-core algebra for an infinite system, which does not seem to have been defined before. In discussing time-evolution for infinite systems it is easier to choose the time to be continuous; then we are able to show that with an interaction of finite range, the time evolution is a strongly continuous group of automorphisms of the algebra, following [109]. In the next section, we analyse the possible one-parameter groups of completely positive semigroups of contractions. This is the heart of the subject, and was set up by Kraus, Gorini, Sudarshan, Kossakowski and Lindblad among others. I shall follow the elementary treatment described in [67], which assumes that the system is of finite dimension. But then, applying the same estimates as in the case of automorphisms, we can extend the semigroup to the inductive limit. In previous treatments it was thought that it was necessary to impose the condition of detailed balance on the dynamics. From our point of view, detailed balance (in the sense of Glauber) is a consequence of moving to the isothermal dynamics; we construct the isolated dynamics, in which the dissipative part is determined by a bistochastic map. We can then move to the isothermal version after this. This makes the construction of the theories easier than imposing detailed balance from the start.

12.1 The Algebra of an Infinite System

Let Λ be an infinite set, and let Λ_0 be a typical finite subset. As usual we define the algebra of observables $\mathcal{A}(\Lambda_0)$ to be the tensor product of $\mathbf{M}_n$ at each site $x \in \Lambda_0$. Each algebra $\mathcal{A}(x)$ is naturally embedded in each $\mathcal{A}(\Lambda_0)$ for which $x \in \Lambda_0$ and each operator in $\mathcal{A}(\Lambda_0)$ can be identified with a unique element (its ampliation) in $\mathcal{A}(\Lambda_1)$ for any finite Λ_1 containing Λ_0. This identification defines an injective C^*-homomorhpism,

$$\phi(\Lambda_0, \Lambda_1): \quad \mathcal{A}(\Lambda_0) \to \mathcal{A}(\Lambda_1)$$

which is transitive in the sense that $\phi(\Lambda_2, \Lambda_1)\phi(\Lambda_1, \Lambda_0) = \phi(\Lambda_2, \Lambda_0)$. This allows us to put an equivalence relation on the union of all $\mathcal{A}(\Lambda_0)$ over all finite subsets

Λ_0, namely, two elements $A_1 \in \Lambda_1$ and $A_2 \in \Lambda_2$ are equivalent if there exists Λ_0 containing both Λ_1 and Λ_2 such that $\phi(\Lambda_0, \Lambda_1)A_1 = \phi(\Lambda_0, \Lambda_2)A_2$. Since the ampliation is norm-preserving, we see that equivalent elements have the same norm. We can multiply equivalence classes by multiplying representatives; the result is independent of the choice of representative. We can add elements in the same way, and form the Hermitian conjugate. The C^*-identity $\|A^*A\| = \|A\|^2$ also holds. In this way, the set of equivalence classes becomes a *-algebra, which except for completeness, obeys all the axioms of a C^*-algebra. Let us call this algebra $\mathcal{A}_0$, the algebra of local observables. Its completion is therefore a C^*-algebra, $\mathcal{A}$, the algebra of quasilocal observables. As always when we complete a space, the original one is dense in the completion. So any quasilocal observable in the norm limit of a sequence of local observables. (We adopt the convention of calling the algebra the observable algebra even though it contains non-Hermitian elements.) Each $\mathcal{A}(\Lambda_0)$ can be identified with a unique subalgebra of $\mathcal{A}$. This construction of $\mathcal{A}$ from the local algebras is called the inductive limit. It is not hard to show that $\mathcal{A}$ is separable; it is one of the class of C^*-algebras introduced and studied by Glimm, the uniformly hyperfinite class. This algebra has a special state, the trace. We define the trace on the matrix algebra $\mathcal{A}(\Lambda_0)$ as

$$\rho_0(A) = d^{-1}\mathrm{Tr}\, A \qquad \text{where } d = n^{|\Lambda_0|}$$

is the dimension of $\mathcal{A}$. This is unchanged on ampliation, and so defines a positive linear functional on the space of equivalence classes. Since $\rho_0(1) = 1$, it is bounded, and so can be extended to the completion. This state obeys $\rho_0(AB) = \rho_0(BA)$. It corresponds to the state at infinite temperature, that is, zero beta. This state is a fixed point of all bistochastic maps on $\mathcal{A}$. We defined the hard-core algebra for any finite Λ_0. We could choose the operators so that different modes commute, by injecting the standard representation into $\mathcal{A}(\Lambda_0)$. Or, we could choose them so that different modes anticommute. Recall that this is achieved by ordering the modes and forming the tensor product of the algebras containing the modes one by one. Anticommutativity is achieved by inserting the factor θ which anticommutes with all modes used up to then. The hard-core algebra is defined as the mapping of the symbols obeying the hard-core relations into the C^*-algebra $\mathcal{A}$. The Fermion algebra is the special case where $n = 2$.

12.2 The Reversible Dynamics

If $\Lambda = \mathbf{Z}^s$, where s is the space-dimension, then the translation group $\{\sigma_x\}_{x\in\Lambda}$ acts naturally on $\mathcal{A}$ as automorphisms. Thus, if $A \in \mathcal{A}(x)$, then $A\sigma_y$ is the same element $A \in \mathbf{M}_n$ ampliated to be an element of $\mathcal{A}(x+y)$. The map σ_x

is bounded and can be extended to an automorphism of the quasilocal algebra. The map $\sigma: \quad x \mapsto \sigma_x$ is a homomorphism from the group Λ into the group $AUT\mathcal{A}$.

We now show how to define the time-evolution with continuous time in the simplest case, when the interaction provides a coupling between the operators at a point $x \in \Lambda$ and its nearest neighbours, and is translation invariant. We might as well assume that $s = 1$, the case of the spin chain, in which case there are two nearest neighbours, and the energy can be assigned to the bond between them. The proof for space of any dimension is very similar. This and other more general cases were treated in [109]. Thus let $H(x) \in \mathbf{M}_n(x) \otimes \mathbf{M}_n(x+1)$ represent the energy on the bond between x and $x+1$. The operator $H(x)$ is Hermitian, and is taken to be covariant under the translation group: $H(x+y) = H(x)\sigma_y$. The total energy in a finite interval $\Lambda_1 \subseteq \Lambda = \mathbf{Z}$ is taken to be

$$H(\Lambda_1) = \sum_{x\in\Lambda_1} H(x).$$

This is an operator in the interval Λ_1 extended by one point on each side. The sum for the energy of the whole system is divergent, and so the definition of the time-evolution needs attention. We denote by H the sum of indefinitely many terms.

We first define the time-evolution of a local operator, one in $\mathcal{A}_0$ and therefore in some region Λ_0: $A \in \mathcal{A}(\Lambda_0)$. We start with the series

$$A(t) = A + it[H, A] + (it)^2/2!\,[H, [H, A]] + \dots \; . \tag{12.1}$$

In each bracket of a multiple commutator we choose as many terms in H as is needed so that the inclusion of any more does not change the expression. We shall now prove that this converges in norm to an element of $\mathcal{A}$ if t is small enough. Taking the first commutator, we note that A commutes with $H(y)$ unless y is a neighbour of Λ_0, and there are $|\Lambda_0| + 2$ such y. So this is the number of terms needed. The commutator has two terms, and each is bounded in norm by $\|H(x)\|\,\|A\|$. So we get the estimate for the second term of the series

$$\|it[H, A]\| \leq 2\,(|\Lambda_0| + 2)\,|t|\,\|H(x)\|\,\|A\|.$$

Let Λ_1 denote Λ_0 together with its neighbours. So $|\Lambda_1| = |\Lambda_0| + 2$. Then each of the $2\,(|\Lambda_0| + 2)$ terms in $[H, A]$ lies in Λ_1. Now consider the number of terms in H in the outer bracket of third term, the double commutator. Since each of the $2\,(|\Lambda_0| + 2)$ terms in the inner commutator lies in Λ_1, we must keep $|\Lambda_1| + 2 = |\Lambda_0| + 4$ terms in H. There are four terms in each contribution to the double commutator, each of norm $\leq \|H(x)\|^2\,\|A\|$. Hence the third term is bounded by

$$\|(it)^2/2[H, [H, A]]\| \leq |t|^2/2.4.\,(|\Lambda_0| + 2)\,(|\Lambda_0| + 4)\,\|H(x)\|^2\|A\|.$$

The third term, the double commutator, lies in Λ_2 say, which is Λ_1 together with its nearest neighbours. This has $|\Lambda_0|+4$ points, so we must keep $|\Lambda_0|+6$ terms in the outermost H in the fourth term, the triple commutator, and it is similarly bounded in norm by

$$(|\Lambda_0|+2)\,(|\Lambda_0|+4)\,(|\Lambda_0|+6)\,8\|H(x)\|^3\,\|A\|\,|t|^3/3!.$$

By induction the ratio of the norms of the $n+1$-th to the n-th term in the series is bounded by

$$(|\Lambda_0|+2n)\,.2.\|H(x)\|.|t|/n \to r < 1$$

as $n\to\infty$ if $t<(4\|H(x)\|)^{-1}$. So for small t, the series converges in norm to an element of $\mathcal{A}$, as $\mathcal{A}$ is complete. Let $A\tau_t$ denote the limit. This result is the heart of the construction of the time-evolution. The rest is tidying up.

Now, for fixed Λ_1, $H(\Lambda_1)$ can be identified with a Hermitian matrix, and it is well known that

$$A+it[H(\Lambda_1),\,A]+(it)^2/2![H(\Lambda_1),\,[H(\Lambda_1),\,A]]+\dots$$

converges in norm to

$$A\tau_{1t}=e^{iH(\Lambda_1)t}Ae^{-iH(\Lambda_1)t}$$

and clearly, $\|A\tau_{1t}\|=\|A\|$. We see that for large Λ_1 the first terms of the series for $A\tau_{1t}$ coincide with the first terms of the series in Eqn. (12.1), and we get more terms coinciding as Λ_1 becomes large. Since the series converge in norm, we have for $A\in\mathcal{A}_0$ and t small,

$$A\tau_t=\lim_{\Lambda_1\to\mathbf{Z}}A\tau_{1t}.$$

It follows that $\|A\tau_t\|=\|A\|$, and that

$$(A\tau_t)^*=A^*\tau_t;\qquad (AB)\tau_t=A\tau_tB\tau_t;\qquad \text{and that } \tau_t \text{ is linear.}$$

The continuity of τ_t in time follows from the convergence of the series. Since τ_t is a bounded map, it can be uniquely extended to an automorphism of $\mathcal{A}$, the completion of $\mathcal{A}_0$. The extension is continuous in time, and obeys the group law for small time; see and do Exercise (12.13), and the next two as well. We can show that the resulting time-evolution commutes with space-translation; this is first done on $\mathcal{A}_0$ and small time simply by commuting τ_t and σ_j on the series in Eqn. (12.1) term by term, and using the space translation-invariance of the indefinite sums defining H. The commutativity can then be extended to the whole algebra.

We have constructed the time-evolution for a one-dimensional lattice, with an interaction of finite range, and with translation invariance. In [109] we deal with any dimension, and cover the case of interactions of any finite range (which must be independent of position). Further extensions to more general cases are described in [61, 58].

12.3 Return to Equilibrium

The tracial state ρ_0 is easily shown to be invariant under τ_t and σ_t. In fact, ρ_0 is *universally invariant*: it is invariant under all automorphisms and indeed all bistochastic maps. The representation obtained from it using the Gelfand-Naimark-Segal construction leads to a represented algebra which is of type II_1. It corresponds to the case $\beta = 0$. In a seminal paper [42] Haag, Hugenholtz and Winnink show how to define states at any beta. The starting point is the notable remark of Kubo [110], and of Martin and Schwinger [111] that the canonical state possesses some analytic properties, now dubbed the KMS condition. This goes as follows.

Definition 12.2 *Let $\mathcal{A}$ be a C^*-algebra, with continuous time-evolution τ_t. Let A_t denote the function of time $A\tau_t$. Then a state $\rho \in \Sigma(\mathcal{A})$ is said to obey the KMS condition relative to the time-evolution τ_t if*

1. *For any elements $A, B \in \mathcal{A}$, the function of time $\rho(A_t B)$ is the boundary value of an analytic function $F(t)$ of the complex variable t in the strip $-\beta < Im\, t < 0$, and is continuous on the boundaries. Also the function $G(t) = \rho(AB_t)$ is the boundary value of an analytic function in the strip $0 < Imt < \beta$, and continuous on the boundary.*

2. *$F(t) = G(t + i\beta)$ holds for all $t \in \mathbf{R}$.*

These certainly hold in finite systems, as can be seen by the cyclicity of the trace:

$$\begin{aligned} Tr\left(e^{-\beta H} e^{iHt} A e^{-iHt} B\right) &= Tr\left(e^{-\beta H} B e^{-\beta H} e^{iHt} A e^{-iHt} e^{\beta H}\right) \\ &= \rho\left(B e^{iH(t+i\beta)} A e^{-iH(t+i\beta)}\right). \end{aligned}$$

We expect the KMS condition to hold if we take the limit to an infinite system, since in the domain of complex t considered, the rising exponential e^{iHt} is beaten by the decaying exponential $e^{-\beta H}$; this relies on the positivity of the microscopic Hamiltonian. In fact, for our system of spins with finite-range interactions, and for a somewhat larger class too, Robinson [112] has proved that the canonical state for a region Λ_1, using the Hamiltonian $H(\Lambda_1)$, converges to a state obeying the KMS condition relative to the automorphism τ_t that we have constructed above.

Haag, Hugenholtz and Winnink [42] noticed that the KMS condition can be expressed in terms solely of the automorphism of the C^*-algebra, and does not depend on the Hilbert space in which we might have defined the operators. They are able to construct the equilibrium statistical mechanics of an infinite system entirely in terms of the Gelfand-Naimark-Segal representation defined by a KMS state. One of the basic results of the theory is that a KMS-state is always faithful. This reflects the property, obvious for finite classical systems, that the equilibrium measure is everywhere positive, and in finite quantum systems, the canonical density matrix has no zero eigenvalues. Another important theorem is that any KMS-state is invariant under the automorphism group $\{\tau_t\}$. This has the important consequence that the time-evolution is given by a unitary group with self-adjoint generator in the representation given by the KMS-state. The generator, called the *modular Hamiltonian*, is never bounded below (unless the algebras are of finite dimension). This is quite natural: an infinite system at a positive temperature contains an infinite amount of energy, and by removing some from a state we can get to another state of lower mean energy than we had before. This unboundedness below shows up in finite systems, in that the canonical state is conveniently taken to have zero energy; then states with some heat-particles removed will have negative energy. More recently, the theory has become based on the state itself, especially on its property of being faithful; the time-automorphism then become a derived concept, determined by the state. In the Tomita-Takesaki theory, we start with a C^*-algebra and a faithful state ρ. We then prove that there exists a unique automorphism group τ_t on the algebra such that the given faithful state obeys the KMS-condition relative to this automorphism. This is called the modular automorphism group defined by ρ. The state can be used in the Gelfand-Naimark-Segal construction to find a representation π_ρ on a Hilbert space $\mathcal{H}_\rho$. Surprisingly, the representations obtained are always of type III in von Neumann's classification (unless the system is finite). This has been a great stimulus to this branch of mathematics. See [58] for a complete account, with applications to equilibrium statistical mechanics. Of particular physical interest is the work of Woronowicz and Pusz [114]. They introduce the concept of *passive* states, which are very stable under small perturbations. To give a small flavour of the stability question, we now give a typical result in the subject. We shall also see the limitations.

Theorem 12.3 *Let $\mathcal{A}$ be a C^*-algebra and ρ a faithful state, such that the Gelfand-Naimark-Segal representation π_ρ on the Hilbert space $\mathcal{H}_\rho$ is of type III. Let ψ_ρ be the cyclic vector of the construction. Then there exists a dense set $\mathcal{D}$ of vectors in $\mathcal{H}_\rho$ such that for any $A \in \mathcal{A}$ and any normalised $\psi \in \mathcal{D}$, the mean of $A_t = A\tau_t$ in the state ψ converges to its equilibrium value:*

$$\langle \psi, A_t \psi \rangle \to \langle \psi_\rho, A \psi_\rho \rangle = \rho(A) \text{ as } t \to \infty.$$

Proof. Since ρ is faithful, the vector ψ_ρ is separating, that is, $\pi_\rho(A)\psi_\rho = 0$ implies that $A = 0$. Since ψ_ρ is separating for the algebra $\pi_\rho(\mathcal{A})$, it is cyclic for the commutant. Thus the vectors

$$\mathcal{D} = \{B\psi_\rho : \ B : \mathcal{H}_\rho \to \mathcal{H}_\rho \text{ and } [B, \pi_\rho(A)] = 0 \text{ for all } A \in \mathcal{A}\}$$

give us a dense set $\mathcal{D}$. We may scale the operator B so that $B\psi_\rho$ is normalised. Let H_m denote the modular Hamiltonian, and $U(t)$ the unitary time-evolution it generates. Then $\pi_\rho(A\tau_t) = U(t)\pi_\rho(A)U(-t)$, and $U(t)\psi_\rho = \psi_\rho$. We get

$$\begin{aligned}
\langle B\psi_\rho, \pi_\rho(A_t)B\psi_\rho\rangle &= \langle\psi_\rho, \pi_\rho(A_t)B^*B\psi_\rho\rangle \\
&= \langle\psi_\rho, U(t)\pi_\rho(A)U(-t)B^*B\psi_\rho\rangle \\
&= \langle\pi_\rho(A^*)\psi_\rho,\ U(-t)B^*B\psi_\rho\rangle \\
&\to \langle\pi_\rho(A^*)\psi_\rho,\ \psi_\rho\rangle\langle\psi_\rho,\ B^*B\psi_\rho\rangle \text{ as } t\to\infty \\
&= \rho(A) \text{ since } \|B\psi_\rho\| = 1.
\end{aligned}$$

by the Riemann-Lebesgue lemma. Here we use the assumption that the algebra is of type III only inasmuch as this implies that H_m has an absolutely continuous spectrum, except for the zero eigenvalue from the stationary state ψ_ρ. Our calculation shows that the vector state $B\psi_\rho$ converges weakly to the equilibrium state ρ for large times. □

The theorem gives no unique rate of convergence. Can we get the exponential convergence demanded by Krylov? By restricting B we can, but not with a uniform rate. For example, let $h(t)$ be a test-function such that its Fourier transform $\hat{h}$ has compact support. The replace B in the proof by $B(h) = \int U(t)BU(-t)h(t)\,dt$, which lies in the commutant of the algebra since the modular time-evolution $U(t)$ maps this to itself. Then the convergence given by the proof is exponential, but the rate is determined by the support of $\hat{h}$, not by an intrinsic physical property, as required by Krylov. For this type of model, this lack of relaxation time is quite natural; there are indeed some states which do converge extremely slowly to equilibrium. These might be very hard to find, and might correspond (in this model) to the states that exhibit the *spin-echo effect*. This is the name given to an apparent violation of the second law. In the experiment, energy was sent into a system of ordered spins in the form of spin-waves; after a long time on the atomic scale, the system was a jumble of spins, and it seemed at first sight that the spins had been randomised. However, on reversing the applied magnetic field, the spins retraced their steps, and nearly reproduced the ordered spins of the original sample. The true relaxation time for the system was very much longer than that seen in coarse-grained measurements. The state seen at the half-way stage was one of the complicated states that are deemed to be overwhelmingly improbable. This may be so, but experimentalists are very clever, and did make it, and the theorem we proved allows for it. Another way to put Krylov's requirement that a theory should predict a relaxation time is to ask, how can we make a choice of coarse-graining in a theory such as the infinite-volume spin-system, such that the chosen variables exhibit a relaxation to equilibrium exponentially at a minimum rate? We cannot do this in a self-contained Hamiltonian theory; the point is that it is the *fast* variables, the very variables that we do *not* use in our description, that cause the relaxation of the variables we do use.

12.4 Irreversible Linear Dynamics

If time is continuous, we can ask that the time-evolution should be a one-parameter group of completely positive maps; the general form for such systems was found by Lindblad [113], and by Gorini, Kossakowski and Sudarshan [99]. I follow the argument of [67]. This analysis does not include the conditions that each time-step be bistochastic and conserve energy. Recall the notation of Section (2.3.3): if ρ is a linear functional on the algebra $\mathbf{M}_n$, not necessarily positive, we denoted by $\Delta\rho$ the operator in $\mathbf{M}_n$ such that $\rho(A) = Tr\,\Delta\rho A$ for all $A \in \mathbf{M}_n$. If ρ is a state, then $\Delta\rho$ is the corresponding density operator; hence the symbol Δ. Recall too the correspondence $T \mapsto L_T$ between a linear map $T : \mathbf{M}_n \to \mathbf{M}_n$ and the linear functional $L_T : \mathbf{M}_n \times \mathbf{M}_n \to$; it is defined on product operators by

$$L_T(A \otimes B) = \rho_0\left(B^t T(A)\right) = n^{-1} Tr\left(B^t T(A)\right),$$

and by linear extension to the general operator. We shall use the same notation, ΔL, for the matrix in $\mathbf{M}_n \otimes \mathbf{M}_n$ which gives the linear functional L on $\mathbf{M}_n \otimes \mathbf{M}_n$, thus:

$$Tr_{12}\left(\Delta L(A \otimes B)\right) = L(A \otimes B).$$

Here, Tr_{12} means the trace over $\mathbf{M}_n \otimes \mathbf{M}_n$. We shall also use the realisation of $\mathbf{M}_n \otimes \mathbf{M}_n$ as the set of operators on the Hilbert space $\mathbf{M}_n$, which is furnished with the normalised Hilbert-Schmidt norm. This is the faithful representation of $\mathbf{M}_n \otimes \mathbf{M}_n$, given for product operators by

$$(A \otimes B)C = ACB^t \qquad \text{for } A,\, B,\, C \in \mathbf{M}_n \tag{12.4}$$

and by linearity for the general element, which is of the form $\sum A \otimes B$. In this realisation, ΔL denotes an operator on $\mathbf{M}_n$. This has sometimes been called a superoperator in the Physics literature, being an operator on matrices in $\mathbf{M}_n$. Naturally, the word is less used now, to avoid confusion with other uses of "super". Recall too the linear functional $L_{C,D}$ on $\mathbf{M}_n \otimes \mathbf{M}_n = B(\mathbf{M}_n)$ defined for each pair of elements C and D in $\mathbf{M}_n$ by $L_{C,D}(M) = \langle C, MD\rangle$. In this, $\langle \bullet, \bullet\rangle$ is the Hilbert-Schmidt scalar product

$$\langle A,\, B\rangle = Tr\,(A^* B)\,.$$

We shall need the relationship:

Lemma 12.5 *The linear functional $L_{C,D}$ corresponds to the operator $T_{C,D} = C^* \bullet D$ on $\mathbf{M}_n$.*

Proof. Let T denote $T_{C,D}$; then

$$\begin{aligned} L_T(A \otimes B) &= n^{-1} Tr\,(B^t T(A)) \\ &= n^{-1} Tr\,(B^t C^* A D) = n^{-1} Tr\,(C^* A D B^t) \\ &= \langle C, (A \otimes B) D \rangle = L_{C,D}(A \otimes B). \end{aligned}$$

Since this is true for all A and B, we have $L_T = L_{C,D}$. □

Let $P_1 \in B(\mathbf{M}_n)$ be the orthogonal projection onto the unit vector $1 \in \mathbf{M}_n$. This is the density matrix of the vector state on $B(\mathbf{M}_n)$ given by the unit vector $1 \in \mathbf{M}_n$, that is, the functional $M \mapsto \langle 1, M1 \rangle = L_{1,1}(M)$. In symbols, $\Delta L_{1,1} = P_1$. More, the linear map $T_{1,1}$ corresponding to $L_{1,1}$ given by Lemma (12.5) is the identity map of $\mathbf{M}_n$ onto itself. The following relations are elementary but very useful:

Lemma 12.6

$$P_1 \Delta L P_1 = \langle 1,\, L1 \rangle \Delta L_{1,1} \tag{12.7}$$

$$\Delta L P_1 = \Delta L_{1,\Delta L 1} \tag{12.8}$$

$$P_1 \Delta L = \Delta L_{\Delta L^* 1, 1}. \tag{12.9}$$

Proof. The first just says that if $P_1 = \psi \otimes \psi$ is a one-dimensional projection, and M is any operator, then $P_1 M P_1 = \langle \psi, M\psi \rangle P_1$. For the second,

$$\begin{aligned} Tr\,(\Delta L P_1 (A \otimes B)) &= Tr\,(P_1 (A \otimes B) \Delta L) \\ &= \langle 1,\, (A \otimes B) \Delta L 1 \rangle \\ &= L_{1,\Delta L 1}(A \otimes B). \end{aligned}$$

Since this is true for all $A \otimes B$, we get the result. The third equation, in which * denotes the adjoint on $B(\mathcal{M}_n)$, is similar. □

We can now define the tilde operation used in [67]. Let T be a linear map of $\mathbf{M}_n$ to itself, and L_T the corresponding linear functional on $\mathbf{M}_m \otimes \mathbf{M}_n$. Let $P^\perp = 1 - P_1$ be the projection onto the orthogonal complement of the vector $1 \in \mathbf{M}_n$. Then we have

$$\begin{aligned} \Delta L_T &= P_1^\perp \Delta L_T P_1^\perp + P_1 \Delta L_T P_1^\perp + P_1^\perp \Delta L_T + P_1 \Delta L_T P_1 \\ &= P_1^\perp \Delta L_T P_1^\perp + P_1 \Delta L_T + \Delta L_T P_1 - P_1 \Delta L_T P_1 \\ &= \widetilde{\Delta L_T} + \delta L_{\Delta L^* 1, 1} + \delta L_{1,\Delta L 1} - \langle 1,\, \Delta L_T 1 \rangle L_{1,1} \end{aligned}$$

by Eqn. (12.7), (12.8), (12.9). We have put

$$\widetilde{\Delta L_T} = P_1^\perp \Delta L_T P_1^\perp.$$

Now associate half of the negative term $-P_1 \Delta L_T P_1$ with each of the other two terms, and define

$$S = \Delta L_T 1 - \frac{1}{2}\langle 1, \Delta L_T 1\rangle 1 \in \mathbf{M}_n,$$
$$S' = \Delta L_T^* 1 - \frac{1}{2}\langle 1, \Delta L_T^* 1\rangle 1 \in \mathbf{M}_n.$$

Then we get

$$\Delta L_T = \widetilde{\Delta L}_T + \Delta L_{1,S} + \Delta L_{S',1}. \tag{12.10}$$

We correspondingly obtain, from Lemma (12.5)

$$T = \tilde{T} + S'^* \bullet + \bullet\, S. \tag{12.11}$$

The tilde operation is defined as the transformation $T \mapsto \tilde{T}$. We see that if T is Hermitian, then L_T is real and $\Delta L_T = \Delta L_T^*$. Hence $S = S'$. Moreover, if T is completely positive, then L_T and ΔL_T are positive, and so is $\Delta L_{\tilde{T}}$, as it is a positive operator sandwiched between projections $P_1^\perp$. Hence $\tilde{T}$ is completely positive.

Now suppose that $T(t)$ is a smooth semigroup of completely positive maps. By continuity, $\widetilde{T(t)}$ and $S(t)$, defined for each t by Eqn. (12.10), are smooth functions of t. Further, since the identity map corresponds to P_1 it follows that $\widetilde{Id} = 0$. Thus

$$\mathcal{J} = \frac{d}{dt}\widetilde{T(t)}|_{t=0} = \lim_{t\to 0} t^{-1}\widetilde{T(t)}$$

is the limit of completely positive maps, and so is completely positive. Then with the definition $K = \dot{S}_{t=0}$, we have proved the "only if" part of the structure theorem of [99] and [113]:

Theorem 12.12 *The map $\Gamma : \mathbf{M}_n \to \mathbf{M}_n$ generates a completely positive semigroup on $\mathbf{M}_n$ if and only if $\Gamma = \mathcal{J} + \mathcal{K}$, where $\mathcal{J}$ is completely positive, and $\mathcal{K} = K^* \bullet + \bullet\, K$ for some $K \in \mathbf{M}_n$.*

The converse follows easily by exponentiating, and using the Lie-product formula.

Suppose that the anti-hermitian part of K is H_1. This part of the generator gives a contribution to Γ of the form $i[H_1, \bullet]$. This generates a unitary dynamics, which does not itself contribute to dissipation. If the Hermitian part of K, or the completely positive part $\mathcal{J}$, does not commute with H_1, then H_1 does contribute indirectly to dissipation by helping to mix the system. It might also contribute to mixing if it does not commute with the non-linear part of the dynamics, like the quasifree map or stoss map. Some authors feel that H_1 should be identified with the Hamiltonian of the theory. This is not necessary in statistical dynamics. What we do need is that the semigroup as a whole leaves the spectral projections of the Hamiltonian invariant (the first law of thermodynamics). We shall also need to impose that the maps $T(t)$ are bistochastic (second law of thermodynamics). We are far from a general solution to either problem. We can find some interesting special cases as follows. Suppose that K is antihermitian, and that iK is the Heisenberg Hamiltonian H. For the semigroup to be stochastic, it is then sufficient that $\mathcal{J}1 = 0$. The rotation group acts on $\mathbf{M}_2$: to $R \in SO(3)$, conjugate with the unitary operator $U \in SU(2)$ corresponding to R. This representation decomposes into the scalar and vector parts. The projection onto the vector part (in the Hilbert-Schmidt space) is $A \mapsto A - 1/2\mathrm{Tr}\, A\, 1$. This can be taken to map $\mathcal{A}_x$ to itself, and by tensoring with unit maps, ampliated to an operator on the whole local algebra. Summing over $x \in \Lambda$ then gives $\mathcal{J}$, which commutes with H and annihilates 1.

12.5 A Model in Infinite Volume

The problem of finding a model of dissipative dynamics having all the desired properties is very hard if the time is continuous. For discrete time, however, we can find an elementary solution to the problem. Let us start with the Heisenberg ferromagnet chain. The limitation to one dimension is for simplicity only, and the idea is easily generalised to two and three dimensions. Again for simplicity, we study the spin-1/2 model with nearest-neighbour interactions

$$H = -\sum_{x \in \Lambda} J\mathbf{S}_x.\mathbf{S}_{x+1}.$$

Here, $\mathbf{S} = \boldsymbol{\sigma}/2$, where $\boldsymbol{\sigma}$ is the vector of Pauli matrices. You will prove in Exercise (12.13) that this gives rise to a group of automorphisms of the C^*-algebra of quasilocal observables, which in this case is $\mathcal{A} = \bigotimes_x \mathbf{M}_2$. We have also seen that the space translation group acts on $\mathcal{A}$ as automorphisms $\{\sigma_x\}$, and that in the case we have here, τ_t commutes with σ_x. In this, the original Heisenberg model, the Hamiltonian formally commutes with space-rotations. It is easy to adapt the proof of the existence of time evolution to show that rotations about the z-axis can be defined by the convergent series

$$A_\theta = A + i\theta\left[\sum_x S_z(x),\, A\right] + (i\theta)^2/2\left[\sum_x S_z(x),\, \left[\sum_x S_z(x),\, A\right]\right].$$

We take A to lie in some local algebra, and the sums, left indefinite in the series, are over enough sites in each terms so that all non-zero commutators are included. The sum converges, and defines an automorphism $r_3(\theta)$ giving rotation by θ about the third axis. Similarly rotation automorphisms about any axis can be defined. The resulting automorphisms commute with the time-evolution. The model can therefore be expected to show conservation of momentum and angular momentum. To get the first, we take Λ to be $\{0, 1, \ldots, N-1\}$ and identify the next point, N, with 0. Then the translation group is $\mathbf{Z}_N$ and the dynamics leaves the eigenstates of momentum fixed. For the infinite system we can introduce some mixing by forming a sum of the space-translation automorphisms, and this will give a completely positive bistochastic map. This is similar to the random dynamics of Chapter (2.3.2); thus, we form $T = \sum_i \lambda_i \sigma(x_i)$, with $\lambda_i \geq 0$ and $\sum \lambda_i = 1$. Then T commutes with the dynamics, and certainly causes some mixing of the energy-shell. It tends to average states over the sites related by the translations σ_i, and so spreads the disorder, causing diffusion. We could limit the sum to two terms, so that $T_1 = (1-\lambda)1 + \lambda\sigma$, where σ is translation by one time-step. This is bistochastic but not symmetric. The parameter λ determines the relaxation time, irrespective of the initial state. So it satisfies one of Krylov's criteria. We can also mix the shell of a given angular momentum by summing, or integrating the automorphisms $r(\theta)$ over the rotation group. This is a local operation, that is, maps each algebra $\mathcal{A}(x) = \mathbf{M}2$ to itself. The average

$$T_2 = (1-\lambda)1 + \lambda \int_{SO(3)} r(\theta)$$

using Haar measure gives a completely positive bistochastic map. If $\lambda = 1$ it completely thermalises the angular momentum in one time-step. If λ is small, it represents the rate at which angular momentum is thermalised. It commutes with space-translation as well as with time-translation. A simple model with discrete time can be constructed by taking $t = 1$ and taking the evolution in one time step to be given by the completely positive bistochastic map $T = \tau_1 T_1 T_2$. Alternatively, we can take a convex sum of τ_1, T_1 and T_2, and can include mixing in several time-steps 1, τ_1, $\tau_2 = \tau_1^2$.

What are the properties of this model? If Λ is finite, it certainly has enough mixing to ensure that the system converges to the uniform distribution on each energy-shell; that is, it will converge to a fixed point, which will be given by a density matrix that is a function of the Hamiltonian. For example, if the initial state is an energy-eigenstate, then the system will converge to the microcanonical state with that energy. There will be no mixing between energy-shells, and so we do not get the canonical state. For a large system, we can appeal to the theorem on the equivalence of ensembles; this says that the microcanonical state for the system in Λ, restricted to a finite region Λ_0, converges to the canonical state as Λ becomes large. This leads us to guess that the infinite system should relax to the canonical ensemble for large times. But this is a difficult question, There is a physical ambiguity about the infinite system at infinite time; which limit do we take first? We have seen that a small disturbance of the infinite system in equilibrium at beta β leads to a state that converges back to the original state at beta β, and not at to an equilibrium state near but not equal to the original one. It is not easy to arrive at convergence to a canonical state from an initial state that is not a local disturbance of a canonical state; the infinite system seems unable to determine its own temperature if it is not near equilibrium. The physical conditions at the boundary are lost in the thermodynamic limit. This is the main casualty of the idealisation of infinite volume.

To improve the model, so that it converges to the canonical state rather than the microcanonical state, we should stay with finite Λ and add more randomness; this might involve the stoss map, the Boltzmann map in a larger theory with heat-particles, or some form of quasi-free map. The stoss map removes all correlations between different sites in one step. That is, it replaces the state ρ on $\mathcal{A}$ with $\otimes_x \mathcal{M}_x \rho$. Recall that $\mathcal{M}_x$ denotes the marginal map onto $\Sigma(\mathcal{A}_x)$. The stoss map is not admissible in this model, since it does not conserve the mean energy-density at each pair of sites, and so cannot be proved to conserve mean energy. The Boltzmann map cannot be applied unless we enlarge the system, by introducing a heat-particle say on each bond between nearest neighbour sites. Then we must choose a bistochastic map T_3 on the product algebra, that conserves the total energy. This leads to a new class of models, whose relaxation time will involve the rate-constants in T_3, as well as the parameters of the model we already have. The quasifree map can be defined for Fermions, and the spin algebra is actually isomorphic to the Fermion algebra, but the Jordan-Wigner isomorphism involves high powers of the spin and there is no natural way to implement the quasifree map so that the energy-density is preserved. We need a map that conserves the mean energy, and increases entropy. The energy $\mathbf{S.S}$ involves products of number operators, which lead to four-point functions in the creators and annihilators. What we need is a simple state with non-zero four-point correlations, but whose equations of motion do not involve an infinite number of unknown higher correlations. In the next chapter we tackle the problem of the dearth of tractable states that are not quasifree, and we defer the next step in the model until then.

12.6 Exercises

Exercise 12.13 Let $\mathcal{A}$ be a C^*-algebra with dense sub*-algebra $\mathcal{A}_0$. Suppose that for small t, $\tau_t : \mathcal{A} \to \mathcal{A}$ is a *-algebra homomorphism preserving the norm. Show that τ_t can for small t be extended to an automorphism of $\mathcal{A}$.

Exercise 12.14 If in addition to the properties in Ex. (12.13), τ_t is norm continuous ($A\tau_t \to A\tau_s$ as $t \to s$) on $\mathcal{A}$, show that its extension is also norm continuous.

Chapter 13

Generalised Temperatures

In this Chapter we return to finite systems, and address the question of how to describe interacting systems in terms of quantum states with a simple description, but one more detailed than a quasifree state. We make use of a principle of maximum entropy, which is very close to the techniques of Jaynes and Ingarden; their theory is the subject of the next section. We recommend adopting their techniques but not all their philosophy.

13.1 The Jaynes-Ingarden Theory

The justification of the canonical state as that which most systems end up in has been a recurring theme of this book, and many before it. The practical failure of the ergodic programme, after the Kolmogorov-Arnol'd-Moser theorem (KAM), led physicists to think of other reasons why we should choose, sometimes the canonical state, sometimes the grand canonical state (or the great grand canonical state). The success of information theory, in the guise of the principle of minimum information, suggested a non-dynamical, subjective reason why nearly all states are canonical. Identifying entropy as negative information, Jaynes [7] proposed that in the absence of any information, we should assume that the state is that of maximum entropy. This makes sense for finite systems, and leads to the microcanonical state. As information becomes available, we modify the state accordingly. Thus if the mean energy is known, we maximise the entropy among all states with the observed mean energy. This gives us the canonical state. If, in addition, the mean particle-number is observed (and it was assumed that to be observable on the macroscopic scale, the particle-number should be conserved in time) then we get the grand canonical state. This idea is similar in spirit to the Bayesian method. There, we start with a prior distribution; this is often the uniform distribution (if the sample space is finite). We then feed in information by conditioning, using Bayes' theorem. The two procedures are slightly different, since in Bayes' method we use the result of a single sample, whereas in the Jaynes method we use the known mean of the population, and do not pick a subsample. Now, the Bayes method (quantum version) was the subject of Krylov's criticism [73] well before Jaynes' proposal. Krylov objected to what he called von Neumann's measurement theory, in which a state (taken to be the uniform, tracial state) is filtered by projections onto the subspaces of observed values of measurements, to get the conditioned state, purer than the original in general. What Krylov objected to was not the theory of filtering, but the presumption that the initial state is uniform. He was objecting to the choice of the state of maximum entropy, even if we have no information about the state; physics predicts the state at a later time, from the state at an earlier time, and does not make best guesses about the initial state; this is the job of a different discipline, prediction theory. The name, prediction theory, is a misnomer; it should really be "hedging theory", since its purpose is not to predict anything specific, but to devise a strategy so that one's wrong guesses are not likely to have serious consequences, on average. In spite of Krylov's objections, we shall use the principle of maximum entropy, as extended by Ingarden, to guide us in finding useful non-quasifree states, and a randomisation which can replace the quasifree map or the stoss map, for example in models with non-quadratic energy or energy not localised at a single site.

Ingarden's idea is that when we only know how to measure very crude things like the mean total energy, then we use the canonical state for the description of the state. But if we can measure more detail, such as the first n cumulants of the energy, then we should use n "higher-order" temperatures, being the Lagrange multipliers when we maximise the entropy subject to given values of all n cumulants of the energy, not just the mean, which is the first cumulant. Haag has described the increase in entropy as being due to "dust": it is too small to make any real changes in the means of physical observables. Let us see how one defines the second-order temperature for the harmonic oscillator. Here, the Hilbert space is ℓ^2 with basis $\psi_0 = (1, 0, 0, \ldots)$, $\psi_1 = (0, 1, 0, \ldots)$ and so on; the energy is $\varepsilon\mathcal{N}$, where $\mathcal{N}\psi_n = n\psi_n$. So we must maximise $-Tr\,\rho\log\rho$ subject to

- $Tr\,\rho = 1$
- $Tr\,\varepsilon\mathcal{N}\rho = \overline{E}$
- $Tr\,(\varepsilon\mathcal{N} - \overline{E})^2 = V$.

The answer (Exercise (13.20)) is of the form

$$\rho_{\beta,\beta'} = Z^{-1}e^{-\beta H - \beta'(H - \overline{E})^2} \tag{13.1}$$

Ingarden calls β', or its inverse, a higher-order temperature. Apart from the implicit relation between the intensive parameters β, β' and the given constraints, the state we get is quite tractable. We even have a formula for it. The state that maximises the entropy subject to given mean-values of all operators in a subspace $\mathcal{A}_o$ will be called the corresponding *Jaynes state.* We shall not need to make use of constraints non-linear in the state.

Note that we do not try to require the cumulants beyond 2 to be zero; this might lead to contradictions with positivity. The state maximising entropy, subject to such conditions, not only on the energy, and any other finite number of chosen random variables, certainly exists and is unique, by the concavity of the entropy. The choice of which variables we choose to fix seems subjective. One experimenter might be able to measure more than another. Ingarden answers this by saying that we always have to choose the level of description before constructing any model, and having done so the method gives us a unique procedure, with no further subjectivity. This is so: the level of description is part of the model, like the choice of sample space and Hamiltonian. However, I do not go all the way with Jaynes and Ingarden. First, our state is not intended to be stationary under the dynamics. Secondly, information theory has too little predictive power (like prediction theory in general). For example, suppose that a system is in equilibrium (so that the first objection cannot be made). Ingarden says, that if we can measure the second cumulant of the energy as well as the mean, then we should use two generalised temperatures in the description. In fact we can indeed often measure the second cumulant: it is the fluctuation of the energy. We might then think that we would often, or even sometimes, find that the equilibrium state is not the canonical state, but would need two parameters, as in Eqn (13.1). But this rarely happens: when we measure the fluctuations of a system in equilibrium, we nearly always find that they are thermal. This was the remarkable thing about the celebrated discovery of the thermal background radiation by the COBE satellite: the law was exactly Planckian. The simple explanation is that it is in a canonical state. When we are in equilibrium, and many high moments of the energy can be measured, Ingarden's theory does not say that the distribution cannot be Planckian; but it does not offer any explanation as to why it is Planckian. Ingarden argues that for a macroscopic system there is very little difference between the two states, and that we would need a mesoscopic or microscopic system to be able to detect the higher temperature.

In the following, we shall use Ingarden's methods to construct models. The reader might well find it hard to see any difference between our theory and Ingarden's. In our version, the states with generalised temperatures are not in equilibrium; the final state, at large times, will usually be the canonical or grand canonical state, (depending on the mixing properties), and we intend that this occur even for a mesoscopic system, such as a few atoms. There is little difference with Ingarden's theory if the relaxation time from the state with generalised temperatures to the final equilibrium is very long. Although the state at time t is described by a few parameters, we are still able to compute, and so predict, any statistical property of the system at time t, since we know the full state. We illustrate the idea by a detailed study of the Ising model.

13.2 Non-linear Ising Dynamics

We now apply the general method of Ingarden to define a dynamics for the Ising model, and in such a way that both laws of thermodynamics are true. The dynamics differs from the Glauber dynamics, in that the temperature does not appear in the dynamics of the isolated system. Our dynamics will also be non-linear, unlike the Glauber dynamics, which defines a stochastic process in the usual sense. For us, isothermal dynamics and detailed balance is a secondary construct, obtained from the isolated model by keeping the beta fixed using external heat-flows.

To illustrate the method, we describe the spin-1/2 Ising model on $\mathbf{Z}_N$. Recall that this is a classical model, with sample space $\Omega = \prod_x \Omega_x$, where $\Omega_x = \{-1, 1\}$. Thus a configuration, a point of Ω, is a sequence of N numbers $\omega(x)$, each either $+1$ or -1. The translation group $\mathbf{Z}_N$ acts on Ω by $(\omega\tau(j))(x) = \omega(x+j)$. Recall that the addition rule in $\mathbf{Z}_N$ is addition modulo N, and that $x = N+1$ is identified with 1. The energy is

$$\mathcal{E}(\omega) = -J\sum_x \omega(x)\omega(x+1).$$

There are two states of minimum energy; the lowest energy, $-NJ$, occurs when all the spins are $+1$, and when all the spins are -1. It is usual to introduce random variables to describe the spins, thus: $S_x(\omega) = \omega_x$.

The Ising model is a theory of equilibrium; there is no dynamics associated with the Hamiltonian. We cannot get anything from treating it as a quantum theory; since the Hamiltonian is a random variable, it commutes with all the observables. So the Heisenberg equations of motion lead to the trivial dynamics; each observable is constant in time. Dynamics has to be added, and there is no unique way to do this. In the model dynamics suggested by Glauber [6], transitions between configurations are assumed to take place with certain probabilities. These allow transitions between configurations of different energy, and therefore the model is not of the type we have called "isolated dynamics". If the energy-difference between two configurations is E, Glauber chooses the ratio of the forward and backward transition probabilities to be $e^{-\beta E}$. This has been called detailed balance in Eqn. (5.17); we preferred the chemists' term microscopic reversibility. It is clear that Glauber's model is an isothermal one, not an isolated one. The temperature appears in the dynamical law, and to this extent the model is rather far from a microscopic one. Our first task is to suggest another Markov chain, one that obeys the first law of thermodynamics. For simplicity, we shall assume that the stochastic matrix is symmetric (and therefore bistochastic). This is microscopic reversibility. We shall also assume that in one time-step at most one spin is flipped from $+1$ to -1 or *vice versa*. Naturally, we only allow flips that conserve energy. We first give a bistochastic map that flips only one point say $x = 2$ and affects its two neighbours. There are eight configurations of $\Omega_1 \times \Omega_2 \times \Omega_3$, namely

$$\{(+++), (++-), (+-+), (+--), (-++), (-+-), (--+), (---)\}.$$

Only four energy-conserving transitions are possible without involving other sites or flipping more than one spin. These are

$$(++-) \to (+--); \qquad (+--) \to (++-);$$
$$(-++) \to (--+); \qquad (--+) \to (-++).$$

In general the first two will not be on the same energy-shell as the last two; this depends on the next nearest neighbours. So the simplest possible (but not the only) choice of dynamics is to mix the first two with the matrix

$$T_{x=2} = \begin{pmatrix} 1-\lambda & \lambda \\ \lambda & 1-\lambda \end{pmatrix} \tag{13.2}$$

and to mix the last two with the same matrix. The suffix x denotes the point of Λ where the spin-flip is happening. The map is the identity on the other factors in Ω. We see that this map changes the numbers of spins up and down, and tends to equalise them. So the dynamics coming from this map will show a decay of magnetisation, the rate being proportional to λ. The map conserves energy, but moves it from one bond to the next. The dynamics will therefore give rise to a diffusion of heat, with rate related to the decay-rate of spin. We get a translation-invariant bistochastic map T by defining N similar maps T_x with any $x \in \Lambda$ replacing 2, and averaging them: $T = N^{-1} \sum_x T_x$.

We now have the linear part of the dynamics. It cannot mix the energy-shells, and we must add a further randomising map to lead us to the canonical state. Following our discussion, this should be a loss of information as in the stoss map or the Boltzmann map; we do not want to alter the mean energy. According to Ingarden we should do more; we should fix the level of description; in our model, we choose this to be the $2N$ variables given by the mean energy on each of the N bonds, and the mean spin at each of the N sites. We do not want the further randomising map to alter any of these means, as the flow of heat, and the decay of the spins, should be caused by T. That decides it; we define the map Q on the set of states to be the replacement of a state ρ by that state $Q\rho$ maximising entropy, subject to the constraints

$$Q\rho(S_x) = \rho(S_x); \qquad\qquad Q\rho(S_xS_{x+1}) = \rho(S_xS_{x+1}). \tag{13.3}$$

The map Q is by construction entropy-increasing, and conserves the local mean energy, and the particle number (if we interpret + as occupied, and − as unoccupied). It is very non-linear. We can nevertheless make some progress; introduce N Lagrange multipliers, β_x, to go with the constraint on the energy between x and $x+1$, and N Lagrange multipliers μ_x to go with the constraints on spin-density. The state $Q\rho$ therefore has the form

$$Q\rho = Z^{-1}\exp\{\sum_x (J\beta_x S_x S_{x+1} + \mu_x S_x)\}. \tag{13.4}$$

We can interpret β_x as the inverse temperature at x, or rather, on the bond $(x, x+1)$; it is the Legendre dual to the local energy it multiplies, but the word "intensive variable" would be a misnomer, as it is not uniform throughout the system. The variable μ_x is not exactly the chemical potential, as the latter usually appears in the state multiplied by β. Here, the betas at the two neighbouring bonds could be different; there is no beta at x itself; there is no external field in our model, and so there is no energy at x. We must get used to the fact that the vocabulary invented for the theory at equilibrium might need some changes for non-equilibrium dynamics.

One time-step in our dynamics is $p \mapsto QT^{\dagger}p$. We now show that this converges to the canonical state at some overall beta, whose value depends on the mean total energy of the initial state. The map $T^{\dagger}$, being a proper mixture of permutations, increases the entropy except at a fixed point. The constraints are linear in $p \in \Sigma$, so the concavity of the entropy ensures that Q is strictly entropy-increasing. Thus, entropy is a strict Lyapunov function for the dynamics, which is a motion through states of the form in Eqn. (13.4). By the principle of detailed balance, any fixed point must be a fixed point of each T_x. Since T_x alters the mean spin at x unless it is zero, any equilibrium state must have mean spin-density equal to zero. Since T_x exchanges energy between the bonds $(x-1, x)$ and $(x, x+1)$, it changes the state unless the betas are all equal. So there is a unique fixed point. The state-space is compact, so, applying Lyapunov's theorem, we see that starting at any state the system converges to equilibrium, which is a canonical state.

We now find the equations of motion. We write the rate of change in the local extensive variables S_x and $\mathcal{E}_x$, or rather their means, in terms of the state at the time. We write the state in terms of the local intensive variables β_x and μ_x. Thus the equations will be rather implicit. We saw a similar type of equation for chemical reactions, where the rate of change of density is given in terms of the activity. The form of the dynamics ensures that at time $t > 0$ the state is in the form Eqn. (13.4) and since Q does not change the mean values of the chosen extensive variables, any change must be due to $T^\dagger$. To find the change in $\langle S_x \rangle$ it is enough to work with the marginal probabilities on $\Omega_{x-1} \times \Omega_x \times \Omega_{x+1}$. We need only consider the effect of T_x alone, since the others do not flip at x. Let p' denote the state at time $t+1$, that is, $T^\dagger p$, with a similar notation for the marginal probability on the space $\Omega_{x-1} \times \Omega_x \times \Omega_{x+1}$. Let p_{+++} denote the marginal probability of p at the configuration $S_{x-1} = S_x = S_{x+1} = +1$ etc. The marginal maps are linear, and the map $T^\dagger$ is linear. So we can work entirely in the marginal variables, and our choice of the map T leads to the following change caused by the linear part of the dynamics:

$$\begin{aligned} p'_{+++} &= p_{+++}\,, & p'_{++-} &= p_{++-} - \lambda p_{++-} + \lambda p_{+--}\,, \\ p'_{+-+} &= p_{+-+}\,, & p'_{+--} &= p_{+--} - \lambda p_{+--} + \lambda p_{++-}\,, \\ p'_{-+-} &= p_{-+-}\,, & p'_{-++} &= p_{-++} - \lambda p_{-++} + \lambda p_{--+}\,, \\ p'_{---} &= p_{---}\,, & p'_{--+} &= p_{--+} - \lambda p_{--+} + \lambda p_{-++}\,. \end{aligned} \tag{13.5}$$

From this we can compute the change in

$$\langle S_x \rangle = p_{+++} + p_{++-} + p_{-++} + p_{-+-} - p_{+-+} - p_{+--} - p_{--+} - p_{---}$$

to be

$$\delta\langle S_x \rangle = \langle S_x \rangle' - \langle S_x \rangle = 2\lambda(p_{+--} - p_{++-} + p_{--+} - p_{-++}). \tag{13.6}$$

The right-hand side is expressed easily in terms of the variables β and μ at time t, but not easily in terms of S_x and $\mathcal{E}_x$. The left-hand side can be expressed in terms of the β, μ, β' and μ' by finding the mean of S_x in the state Eqn. (13.4), and again with β and μ replaced by β' and μ'. We get one such equation for each x, so we have N in all. We get N further equations by finding the changes in the N means $\langle \mathcal{E}_x \rangle$ due to T_x and T_{x+1}. For $\delta\langle S_x \rangle$ no other T_y for $y \in \Lambda$ contributes, but for $\delta\langle \mathcal{E}_x \rangle$ we must include the change due to T_{x+1} as well as the change due to T_x. We see that the change due to T_x leads to

$$\langle \mathcal{E}_x \rangle' = J\left(-p'_{+++} + p'_{++-} + p'_{+-+} - p'_{+--} - p'_{-++} + p'_{-+-} + p'_{--+} - p'_{---}\right).$$

Using Eqn. (13.5) again, we find that

$$\langle \mathcal{E}_x \rangle' = \langle \mathcal{E}_x \rangle - 2\lambda J(p_{+--} - p_{++-} + p_{-++} - p_{--+}). \tag{13.7}$$

T_x also causes a change in $\langle \mathcal{E}_{x-1} \rangle$, which is the opposite to the change in $\mathcal{E}_x$. Thus T_{x+1} causes a change in $\langle \mathcal{E}_x \rangle$ given by the same expression as Eqn. (13.7), but with opposite signs and where the $p_{\pm\pm\pm}$ are the marginal probabilities translated one step to the right. Both terms are to be added and divided by N to give the change due to $T = N^{-1}\sum_x T_x$. When we write $\langle \mathcal{E}_x \rangle$ in terms of β, μ and $\langle \mathcal{E}'_x \rangle$ in terms of β', μ' we get N equations. Together with the N equations for $\delta\langle S_x \rangle$ this gives us $2N$ equations for β', μ' in terms of β and μ; then we have made one time-step. These are very difficult equations, but they do give a well-defined procedure. They simplify a bit in continuous time, and give rise to a system of $2N$ first-order differential equations in t for $\beta(x,t)$ and $\mu(x,t)$.

In this model the transport of heat can be described as being by convection, where the carriers are the domain walls, that is, points where the direction of the spin changes. One process, the spin-flip, causes both the decay of spin-density and the transport of heat from one bond to the next. In the next section we study the rate of return to equilibrium, when near a canonical state.

13.3 Ising Model Close to Equilibrium

It is of interest to study the rate of return to equilibrium, when near a canonical state. For this we use the first method of Lyapunov. We illustrate this in the simplest non-trivial case, which is when $N = 3$. So the model is of a molecule composed of three spin 1/2 atoms in a ring. We maximise entropy subject to fixing the three mean spin-values $\langle p, S_x \rangle$ and the three mean energies $\langle p, \mathcal{E}_x \rangle = -J\sum_\omega S_x(\omega)S_{x+1}(\omega)$. The answer, by the usual method of Lagrange multipliers is described by six parameters $\boldsymbol{\beta}$ and $\boldsymbol{\mu}$. To keep symmetry, we denote by β_1 the Lagrange multiplier associated with the bond (23), and so on cyclically. The result is (Exercise (13.22))

$$\begin{aligned}
p_{+++} &= Z^{-1}\exp\{\mu_1+\mu_2+\mu_3+J(\beta_1+\beta_2+\beta_3)\} \\
p_{++-} &= Z^{-1}\exp\{\mu_1+\mu_2-\mu_3+J(-\beta_1-\beta_2+\beta_3)\} \\
p_{+-+} &= Z^{-1}\exp\{\mu_1-\mu_2+\mu_3+J(-\beta_1+\beta_2-\beta_3)\} \\
p_{+--} &= Z^{-1}\exp\{\mu_1-\mu_2-\mu_3+J(\beta_1-\beta_2-\beta_3)\} \\
p_{-++} &= Z^{-1}\exp\{-\mu_1+\mu_2+\mu_3+J(\beta_1-\beta_2-\beta_3)\} \\
p_{-+-} &= Z^{-1}\exp\{-\mu_1+\mu_2-\mu_3+J(-\beta_1+\beta_2-\beta_3)\} \\
p_{--+} &= Z^{-1}\exp\{-\mu_1-\mu_2+\mu_3+J(-\beta_1-\beta_2+\beta_3)\} \\
p_{---} &= Z^{-1}\exp\{-\mu_1-\mu_2-\mu_3+J(\beta_1+\beta_2+\beta_3)\}.
\end{aligned} \tag{13.8}$$

Here Z is not a seventh parameter, since it is fixed as the sum of the exponentials on the right. Since the general state for the three-site model needs seven parameters, and here we have six, there is not much reduction in the level of description, using Ingarden's terms. This is not altogether an accurate way to say what we have done, since the remaining parameter needed for a full description of the state is still in the model, and can be predicted.

The dynamics is given by $p \mapsto p' = Q \circ (T_1^\dagger + T_2^\dagger + T_3^\dagger)p/3$, where T_x is as in Eqn. (13.2). It is easy to see that the state, with $\boldsymbol{\mu} = \mathbf{o}$ and all betas equal, is a fixed point of the map. So the Lyapunov theory (first method) will set each μ_x to be a small quantity, and each β_x very close to β; then we study the linearised dynamics about the fixed point. One type of small perturbation, in which μ_x remains zero, and all β_x move together from β, and remain equal, will not return to the original state, since we have moved to a new canonical state with a slightly different β. This will show up in the analysis: one of the eigenvalues of the linearised equation of motion will be zero. The details will appear in [115].

Let us now find the equation of motion of $s_x = \rho(S_x)$. It is enough to do one, say s_2. We saw that

$$\delta s_2 = 2\lambda(p_{+--} - p_{++-} + p_{--+} - p_{-++}). \tag{13.9}$$

On substituting Eqn. (13.8) and keeping only first-order terms in μ_x amd $\beta - \beta_x$, we get (Exercise (13.23))

$$\delta s_2 = -8\lambda_0 Z^{-1}(\beta)e^{-J\beta}\mu_2 \,, \tag{13.10}$$

where $\lambda_0 = \lambda/3$ to take into account $T = \sum T_x/3$. Equation (13.10) shows the typical Onsager behaviour: the variable s_2 is driven by the dual variable μ_2; in this case, s_2 is driven back to zero by any non-zero μ_2, taking it for granted that the signs of the two are always the same. In fact, in our model the relation between the intensive and extensive variables is determined, and in the linear approximation, can be found. This is how. First, in this approximation,

$$Z = 2e^{3\beta J} + 6e^{-\beta J}; \qquad \text{see Exercise (13.24).}$$

Then, to first order

$$\begin{aligned} s_1 &= p_{+++} + p_{++-} + p_{+-+} + p_{+--} - p_{-++} - p_{-+-} - p_{--+} - p_{---} \\ &\approx 2Z^{-1}\{e^{3\beta J}(\mu_1 + \mu_2 + \mu_3) + e^{-\beta J}(3\mu_1 - \mu_3 - \mu_2)\} \end{aligned} \tag{13.11}$$

and similarly

$$s_2 \approx 2Z^{-1}\{e^{3\beta J}(\mu_1 + \mu_2 + \mu_3) + e^{-\beta J}(3\mu_2 - \mu_1 - \mu_3)\} \tag{13.12}$$

$$s_3 \approx 2Z^{-1}\{e^{3\beta J}(\mu_1 + \mu_2 + \mu_3) + e^{-\beta J}(3\mu_3 - \mu_2 - \mu_1)\}. \tag{13.13}$$

These can be solved for $\boldsymbol{\mu}$, and the solution, for μ_2 say, put in the form

$$\mu_2 = e^{\beta J}\kappa Z/8(-s_1 + 2s_2 - s_3) + \frac{Z}{2(3e^{3\beta J} + e^{-\beta J})}s_2, \tag{13.14}$$

where (Ex. (13.25))

$$\kappa = \frac{e^{4\beta J} - 1}{3e^{4\beta J} + 1} > 0.$$

This gives us the equation of motion for s_2, from Eqn. (13.10),

$$\delta s_2 = \lambda_0 \kappa \{(s_3 - s_2) - (s_2 - s_1)\} - \frac{4\lambda_0}{3e^{4\beta J} + 1} s_2. \tag{13.15}$$

This dynamics has two parts, which are both determined by the single rate parameter λ and the beta of the fixed point. The first represents diffusion of spins, with a beta-dependent rate determined by the theory in terms of the hopping parameter λ_0; the expression $s_1 - 2s_2 + s_3$ is the finite-difference approximation to the second spatial derivative of the spin field; in the continuum limit it gives the diffusion equation. This term conserves the total spin number. In addition, there is a decay term, with rate

$$4\lambda_0(3e^{4\beta J} + 1)^{-1}$$

converging to λ as $\beta \to 0$, and vanishing as $\beta \to \infty$. Physically, at $\beta = \infty$ all spins become aligned, and spin-flips become rare. The equation of motion for the energy-density, $\delta\mathcal{E}_1$, can be written as an equation for the rate of change of the beta; up to first order, this gives the heat equation for the temperature [115], in which the conductivity depends on temperature.

With more work, similar models in higher dimensions can be formulated.

13.4 Non-linear Heisenberg Model

We noted that in the Ising model we had to introduce the stochastic matrix T, and its parameter λ, by hand; the size of λ governs all irreversible effects, such as decay and diffusion. In the analogous quantum model, the Heisenberg model, there is a Hamiltonian, which determines the conservative dynamics. Krylov argued that the rate of dissipation should be related to the dynamics, that is, how mixing it is. We do not completely agree; the dissipation might be related to the mixing properties of the dynamics of the fast variables, exactly the part omitted from the description. But it would be nice to keep the arbitrary addition of dissipation to a minimum. We now try to follow the dictum, that dissipation is loss of information, never of energy. The Hamiltonian should, therefore, be the full microscopic Hamiltonian of the system. Another point is that, in a reduced description, in which the fast variables are eliminated from the algebra, the von Neumann entropy of the remaining, slow, variables does not include the entropy of the fast variables; in that case a fully probabilistic treatment is not possible. We now show how dissipation can be introduced by the technique of higher-order temperatures, in the spirit of Ingarden, without these disadvantages. There is still some freedom, in the selection of which constraints are chosen to be preserved by the map maximising the entropy. That is, we are free to choose the level of description. In this way, a given microscopic theory can have many dissipative versions. In none of them is the Hamiltonian altered, or any degrees of freedom lost. Let us illustrate this idea with the Heisenberg ferromagnet. The space Λ is taken to be finite in this section, and at each site $x \in \Lambda$ we have the spin algebra $\mathbf{M}_2$. Define $\mathcal{A} = \bigotimes_x \mathbf{M}_2(x)$, and choose the Hamiltonian $H = -\sum J\mathbf{S}_x.\mathbf{S}_y$, where the sum is over neighbours, and the vector $\mathbf{S} = \hbar\boldsymbol{\sigma}/2$, where $\boldsymbol{\sigma}$ is the triplet of Pauli matrices. The energy on each bond will be regarded as within the level of description; let us also include the three components of the spin on each site within the level of description, but no other spin correlations. Clearly, this is our choice, and, as emphasised by Ingarden, is part of the model. The fewer variables that we include, the more randomness we introduce in one time-step. We have already shown that the Hamiltonian defines a one-parameter group of automorphisms $\tau(t)$ of $\mathcal{A}$. Thus the reversible time-evolution is the map $A \mapsto e^{iHt}Ae^{-iHt} = A\tau(t)$; recall that automorphisms act on the right. We must decide on the size of the time-step t_0, the choice of which is part of the model; in fact, only the product Jt_0 enters, so we have not really introduced another parameter. Let us define the map Q on the set of states of $\mathcal{A}$ by the Jaynes-Ingarden method: $Q\rho$ is the (unique) state determined by ρ by the rule that its entropy is the largest among all states with the same values of all the chosen operators: thus

$$\begin{aligned} Q\rho(\mathbf{S_x}.\mathbf{S_y}) &= \rho(\mathbf{S_x}.\mathbf{S_y}) && \text{for all neighbours } x, y \in \Lambda \\ Q\rho(\mathbf{S_x}) &= \rho(\mathbf{S}_x) && \text{for all } x \in \Lambda. \end{aligned}$$

We call Q the thermalising map. Then we define one step in the dynamics by $\rho \mapsto \tau^*\rho$, where

$$\tau^*\rho(A) = Q\rho(A\tau(t_0)).$$

By construction, this obeys both laws of thermodynamics. The states ρ on any orbit are all defined by density matrices of the form

$$\rho = Z^{-1}\exp\{-\sum_{x,y}\beta(x,y)J\mathbf{S}_x.\mathbf{S}_y + \sum_x \boldsymbol{\mu}(x).\mathbf{S}_x\}$$

where the double sum is over neighbours. This has the form of a canonical state with space-dependent coupling and an external magnetic field, but it is not interpreted as an equilibrium state. The equations of motion of the local intensive variables are very implicit. We take the limit as $t_0 \to 0$ to simplify the formulae. Let X be an operator in the set defining the level of description;

for example, $X = S_x^{(i)}$ or $X = \mathbf{S}_x.\mathbf{S}_y$, where x, y are nearest neighbours. The time-evolution of the mean of X then obeys the equation of motion

$$\hbar \frac{d}{dt} \rho_{\beta,\boldsymbol{\mu}}(X) = i\rho_{\beta,\boldsymbol{\mu}}([H,X]). \tag{13.16}$$

In this equation, β and $\boldsymbol{\mu}$ stand for the collection of local intensive variables, which are functions of time. We get one such equation for each X, exactly the same as the number of variables β, $\boldsymbol{\mu}$. The left-hand side of the equation can be written in terms of the betas and mus and their time derivatives by writing

$$\rho_{\beta,\boldsymbol{\mu}}(X) = Tr\left(Z^{-1}\exp\{-J\sum_{x,y}\beta(x,y)\mathbf{S}_x.\mathbf{S}_y + \sum_x \boldsymbol{\mu}(x).\mathbf{S}_x\}X\right). \tag{13.17}$$

The right-hand side can similarly be written in terms of β and μ, and no time derivatives of these variables appears on the right. In this way, we get a family of first-order differential equations for the local intensive variables. The left-hand side is linear in the time-derivatives. This defines the model we shall call the dissipative Heisenberg spin model. We note that the means of the three components of the total spin are not changed by the dynamics, because of the rotation-invariance of the Hamiltonian. If we impose periodic conditions, then the total momentum is conserved in mean too. Thus the dynamics does not lead to the canonical state at large times, but to a grand canonical state; there is not enough mixing by the (very symmetric) Hamiltonian to reach the canonical state.

Let us look at the general features of the model. The thermalising map Q does not alter the mean values of any of the relevant variables, and the changes in one time-step, which constitute the visible movement towards equilibrium, are caused by the microscopic dynamics $\tau(t)$. Krylov would have liked that. The dynamics is given by a smooth but very non-linear flow through the states of the quantum system, and no energy or entropy of microscopic variables are omitted by the thermalising map. The von Neumann entropy can therefore be safely interpreted as the physical entropy, if we multiply by Boltzmann's constant. There is a reduced description in that the number of parameters needed to describe the dynamics is much less than that needed to describe a general density matrix. For example, if there are N sites, and N bonds, as in one dimension with periodic boundary conditions, then we need $3N$ components for μ and N components for β, compared with the number of parameters, $2^{2N} - 1$, needed to describe a density matrix on $\mathbf{M}_{2^N}$. Both laws of thermodynamics hold, and the final destination of the dynamics depends on the initial state and also on the mixing properties of the Hamiltonian. The rate of dissipation is partly determined by the Hamiltonian, but can be changed by altering the level of description. For example, if only the Hamiltonian H is chosen for the set defining the level of description, then the system reaches equilibrium in one time-step, however short. If every operator is chosen, then no dissipation can occur, and the dynamics is reversible, given by the Hamiltonian.

The theory is easily generalised to other choices of $\mathcal{A}$ and H. For example, in chemical kinetics with $n-1$ hard-core atoms, we choose $\mathcal{A}_x = \mathbf{M}_n$. We can cause reactions to occur by choosing a Hamiltonian to include polynomials in the creators and annihilators which allow the desired transitions between occupied and unoccupied neighbouring sites. The level of description should then be chosen to include the local energy of the bonds, as well as the occupation numbers of the atoms at each site.

We can now return to the question, what are the many-time correlations in a theory based on the use of Jaynes states? Suppose that one time-step is the map $\rho(t) \mapsto QT^{\dagger}\rho(t)$, where T is bistochastic, and Q replaces a state by the appropriate Jaynes state. We have a flow through the states, but it is non-linear and cannot be thrown onto the algebra simply by duality. Suppose that ρ_0 is the initial state, and let $A_1, \ldots, A_n$ be observables. We want to give some meaning to the expression

$$W(t_1, t_2, \ldots, t_n) = \rho_0 \left(A_1(t_1)A_2(t_2)\ldots A_n(t_n)\right) .$$

If the dissipation is zero, this is just the many-body Wightman function. If the dissipation is not zero, but is small, this should retain some meaning. Because of the non-linearity, we expect the meaning of the time-evolution $A(t)$ of the observables to depend on the initial state, and that W is not linear in ρ_0. If the dissipation is small then $A(t)$ should be nearly generated by the bistochastic map T, which is one step of a semigroup. The non-linear part will be time-dependent, and we cannot expect to get a semigroup τ_t; the best we can hope for is a two-parameter family $\tau(s,t)$ such that $A(s+t) = A(s)\tau(s,t)$. This is because the dynamics of the approach to equilibrium of the state $\rho(t)$ depends on the current state, $\rho(t)$ itself. We can hope that $\tau(s,t)$ is a linear map, but not an automorphism (even the linear case, $A \mapsto TA$, is not an automorphism). Finally, the change in a time-step must be related to the change in the expectation value given by the orbit through the states:

$$\langle \rho(t+dt), A(t)\rangle = \langle \rho(t), A(t+dt)\rangle \ . \tag{13.18}$$

If $TA(t)$ happens to be one of the priviledged observables whose mean value is not changed by the map Q, then we can choose $A(t+dt) = TA(t)$. It is not so clear what to do with a general operator. The states $\rho(t)$ obtained from the dynamics are all faithful states, being of the form of canonical states for some field of local betas and mus. There is then a scalar product on $\mathcal{A}$ given by $\langle A, B\rangle_t = Tr(\rho(t)A^*B)$. From this we get the norm on $\mathcal{A}$, say $\|\bullet\|_t$. We have seen that $\rho(t)$ itself is the state of smallest dual norm; in particular it has the smallest dual norm among all states with the given mean values of the observables that define the level of description. These are the "observables" in Ingarden's theory, whose mean value is not changed by Q. Let $\mathcal{A}_o$ be the linear span of these elements. The identity is always an element of $\mathcal{A}_o$, since the trace of a state is preserved to be 1 by Q. Define on $\mathcal{A}$ the family of orthogonal projection operators $Q_*(t)$, which project onto $\mathcal{A}_o$ in the scalar-product $\langle \bullet, \bullet \rangle_t$; then a possible choice for one time-step at time t is the linear map on the algebra given by

$$A \mapsto A\tau_* = TQ_*(t+1)A. \tag{13.19}$$

We now show that Eqn. (13.18) holds, where $\rho(t+1)$ is the state ρ' whose action on any A is

$$\langle \rho', A \rangle = \langle \rho(t), TQ_*(t+1)A \rangle.$$

First, $I \in \mathcal{A}_o$, so our map preserves I. Secondly, if $A \in \mathcal{A}_o$ then $Q_*(t+1)A = A$ and the state ρ' gives the same mean as $T^\dagger \rho$ for this operator, as required. Finally, denote the norm dual to $\| \bullet \|_t$ by $\| \bullet \|_{-t}$, in analogy with the notation $\| \bullet \|_{-\beta}$. Also, write $A = A_o + A^\perp$ where $A_o \in \mathcal{A}_o$ and $A^\perp$ is orthogonal to $\mathcal{A}_o$ in the scalar product $\langle \bullet, \bullet \rangle_{t+1}$. We note that $\langle T^\dagger \rho(t), A^\perp \rangle = 0$, since $I \in \mathcal{A}_o$. Let ρ be any state giving the same mean as $T^\dagger \rho(t)$ for all operators in $\mathcal{A}_o$. Then

$$\begin{aligned}
\|\rho'\|_{-t-1} &= \sup_A \{ |\langle \rho', A \rangle| : \|A\|_{t+1} = 1 \} \\
&= \sup_A \{ |\langle T^\dagger \rho(t), A_o \rangle| : \|A\|_{t+1} = 1 \} \\
&= \sup_A \{ |\langle \rho, A_o \rangle| : \|A\|_{t+1} = 1 \} \\
&\le \sup_A \{ \|\rho\|_{-t-1} \, \|A_o\|_{t+1} : \|A\|_{t+1} = 1 \} \\
&\le \|\rho\|_{-t-1} \,.
\end{aligned}$$

Therefore ρ' is the state of smallest $\| \bullet \|_{-t-1}$-norm with these mean values, and therefore coincides with $QT^\dagger \rho(t)$. To be satisfactory, the time-step should be a positive map; this is the case in many examples, though this fact is not obvious. Then define

$$A(t) = A\tau_*^t \,, \text{ for any } A \in \mathcal{A} \text{ and } t = 1, 2, \ldots \,.$$

If $\mathcal{A}_o$ is invariant under T then our dynamics reduces to a linear process on $\mathcal{A}_o$; if the non-invariance of $\mathcal{A}_o$ is small, then so is the non-linearity. Our proposed dynamical equation is a linear map on the algebra depending on $\rho(t)$. The Wightman functions $\rho_0 (A_1(t_1) \ldots A_n(t_n))$ can then be defined for any initial state ρ_0. The dissipation is determined by the choice of $\mathcal{A}_o$, the dynamics T^t and the size of one time-step.

So far, we have taken the size of the time-step in which the state relaxes to the Jaynes state to be 1 unit, the same as the time-step given by the dynamical operator $T^{\dagger}$. More generally, we can take these to be different. Thus, let κ be the time needed to relax to the Jaynes state and take $T_\lambda^{\dagger}$ to be the dynamical map in a time $\lambda < \kappa$. Let us suppose that in the time λ a fraction λ/κ of the state has relaxed. By this I mean that the change in the state ρ in the time-step λ is given by

$$\rho(t+\lambda) = (1-\lambda/\kappa)T_\lambda^{\dagger}\rho(t) + (\lambda/\kappa)QT_\lambda^{\dagger}\rho(t) . \tag{13.20}$$

Suppose that $X^{\dagger}$ is the infinitesimal generator of a one-parameter group $T_\lambda^{\dagger}$, so that

$$T_\lambda^{\dagger} = I + \lambda X^{\dagger} + O(\lambda^2) .$$

Then we can take the limit of Eqn. (13.20) to get

$$\frac{d\rho}{dt} = X^{\dagger}\rho - \kappa^{-1}(1-Q)\rho . \tag{13.21}$$

If we can solve this in continuous time to get $\rho(t)$ from an initial state ρ_0, then the Heisenberg dynamics is given by

$$\frac{dA(t)}{dt} = XA(t) - \kappa^{-1}\left(1 - Q_*(t)\right)A(t) . \tag{13.22}$$

This linear equation is "guided" by the solution to the non-linear equation Eqn. (13.21) for the state, inasmuch as $Q_*(t)$ is the projection onto $\mathcal{A}_o$ in the norm $\| \bullet \|_t$, which depends on $\rho(t)$. In the special case when $X = 0$ then any faithful state ρ_0 with $\Delta\rho \in \mathcal{A}_o$ is a fixed point and $Q_*(t) = Q_*$, independent of time. Then the solution to Eqn. (13.22) for any $A \in \mathcal{A}$ is

$$A(t) = Q_*A + exp\{-t/\kappa\}(1-Q_*)A$$

which shows an exponential relaxation with time-constant equal to κ.

In this book we have taken a practical view of model-building; we close with some remarks on the deeper foundations of the subject. In his appendix to [73] Sinai remarks that Krylov's call for a model with dynamical mixing led Kolmogorov (and later Sinai himself) to ask for tests that distinguish the concept of random as opposed to deterministic sequences. There are deterministic sequences that for practical purposes cannot be shown not to be random. The sequence of digits in the decimal expansion of π might be such a sequence. This led eventually to the theory of chaos, which studies the case in which the dependence of the state $\rho(t)$ at time t on the initial state $\rho(0)$ at time 0 is so sensitive to the detail of $\rho(0)$ that we can do just as well by replacing $\rho(t)$

with a mixed state, such as that of the Jaynes-Ingarden theory. The ergodic programme becomes justified if there are two time-scales in the problem; the variables very sensitive to the initial conditions must be treated statistically, and no detectable error is introduced thereby if the dynamics were at this scale not distinguishable from random. Further, when we have a theory with local structure, as in the Heisenberg model, then the reversible time-evolution of the correlations between remote parts becomes a very intricate problem, whose *complexity* increases rapidly as the distance between the points increases. These correlations cease to lie within the possible level of description. We are therefore led to Ingarden's formulation, because these correlations cannot be *calculated.* They can be *measured* and, according to statistical dynamics, will be found to be well described by the state $Q\rho$, at least after a long enough time and if they are far enough apart. To justify these claims takes us into the traditional (hard) subjects of ergodic theory and the thermodynamic limit; but this is enough for a book.

13.5 Exercises

Exercise 13.23 Show that the state of maximum entropy, subject to having mean energy $\overline{E}$ and a given variance V, is of the form

$$\rho = Z^{-1}\exp\{-\beta H - \beta'(H-\overline{E})^2\} \tag{13.24}$$

where β' and Z are determined by $\operatorname{Tr}\rho_{\beta,\beta'} = 1$ and $\rho(H-\overline{E})^2 = V$.

Exercise 13.25 Find the general form of the state of three spin 1/2 atoms which maximises the entropy subject to fixed values of the three mean spins S_1, S_2, S_3, and of the three Ising bond energies $-JS_1S_2$, $-JS_2S_3$ and $-JS_3S_1$.

Exercise 13.26 Derive Eqn. (13.10) from Eqn. (13.9) in the approximation where μ_x and $\beta - \beta_x$ are small.

Exercise 13.27 Show that

$$Z = 2e^{3\beta J} + 6e^{-\beta J}$$

for the dynamic Ising model at equilibrium.

Exercise 13.28 Solve Eqn. (13.11), (13.12) and (13.13) for μ_x to obtain Eqn. (13.14), and write Eqn. (13.10) in the form of Eqn. (13.15).

Bibliography

[1] J. von Neumann, *The Mathematical Foundations of Quantum Mechanics*, Princeton, 1955.

[2] C.H. Collie, *Kinetic Theory and Entropy*, Longman, 1982.

[3] H. Hasagawa and T. Nakagomi, *On the Characterization of the Stationary State of a Class of Dynamical Semigroups*, Journ. of Statistical Physics, **23** 639-652 (1980).

[4] R. S. Ingarden, *Towards Mesoscopic Thermodynamics: Small Systems in Higher-order States*, Open Systems and Information Dynamics, **1**, 75-102 (1992).

[5] B. Simon, *The Statistical Mechanics of Lattice Gases I*, Princeton University Press, 1993.

[6] R.J. Glauber, *Time-dependent Statistics for the Ising Model*, Journal of Mathematical Physics, **4**, 294 (1963).

[7] E. T. Jaynes, *Information Theory and Statistical Mechanics*, Physical Review, **106**, 620-630; and II, **108**, 171-190 (1957).

[8] R.H. Fowler, *Statistical Mechanics*, 2nd Ed., Cambridge, 1936.

[9] R.C. Tolman, *Foundations of Statistical Mechanics*, Oxford, 1938.

[10] W. T. Grandy *The Origins of Entropy and Irreversibility*, Open Systems and Information Dynamics, **1**, 183-196 (1992).

[11] For a modern introduction, see J. Neveu, *Mathematical Foundations of the Calculus of Probability*, Holden-Day, San Francisco, (1965).

[12] F. Riesz and B. St.-Nagy, *Functional Analysis*, Frederick Ungar Publishing, New York, 1955.

[13] J. Marcinkiewicz, Mathematisches Zeitschrift, **44**, 612-618 (1939).

[14] R. A. Kunze and I. E. Segal, *Integrals and Operators*, Springer, 1978.

[15] A. Renyi, *The Theory of Probability*, North-Holland, 1970.

[16] P. Alberti and A. Uhlmann, *Stochasticity and Partial Order*, VEG Deutscher Verlag, Berlin, 1981.

[17] S. Koseki, *Isothermal Statistical Dynamics*, Ph. D. Thesis, King's College, 1993.

[18] G.D. Birkhoff, Univ. Nac. Tucuman Rev. Ser. **A5**, 147 (1946). See: T. Ando, *Majorization, Doubly Stochastic Matrices and Comparison of Eigenvalues*, Sopporo, 1982; Linear Algebra and its Applications, **118**, 163-248 (1983).

[19] O. Penrose, *Foundations of Statistical Mechanics*, Pergamon, 1970.

[20] W. Thirring, *Quantum Mechanics of Large Systems*, Springer-Verlag, 1980.

[21] D. Revuz, *Markov Chains*, North-Holland, 1975.

[22] H. P. McKean, *Propagation of Chaos for a Class of Nonlinear Parabolic Equations*, in: Lecture Series in Differential Equations II, A. K. Aziz (Ed.), 177, Van Nostrand Reinhold, 1969.

[23] R. Monaco and L. Preziosi, *Fluid Dynamic Applications of the Discrete Boltzmann Equation*, World Scientific, 1991. See also N. Bellomo, A. Palczewski and G. Toscani, *Mathematical Topics in Nonlinear Kinetic Theory*, World Scientific, 1988.

[24] R. F. Streater, *Stochastic Models of Cotransport*, Transport Theory and Statistical Physics, **22**, 1-37 (1993).

[25] J. E. Broadwell, *Study of Rarified Shear Flow by Discrete Velocity Method*, Journal of Fluid Mechanics, **7**, 401-414 (1964).

[26] R. F. Streater, *Convection in a Gravitational Field,* Journal of Statistical Physics, **77**, 441-448 (1994).

[27] J. Keizer, *Statistical Thermodynamics of Nonequilibrium Processes*, Springer, New York, 1987.

[28] H. Eyring, S. H. Lin, and S. M. Lin, *Basic Chemical Kinetics*, Wiley, New York, 1980.

[29] S. Koseki and R. F. Streater, *The Statistical Dynamics of Activity-led Reactions*, Open Systems and Information Dynamics, **2**, 77-94 (1993).

[30] A. S. Wightman, Appendix, in R. Israel, *Convexity in the Theory of Lattice Gases*, Princeton University Press, 1979.

[31] R. F. Streater, *The F-Theorem for Stochastic Models,* Annals of Physics, **218**, 255-278, (1992).

[32] R. F. Streater, *The Free-energy Theorem*, pp. 137-147 of "Quantum and Non-commutative Analysis", Ed. H. Araki, K. R. Ito, A. Kishimoto, and I. Ojima, Kluwer, 1993.

[33] D. B. Shear, *Stability and uniqueness of the equilibrium point in chemical reaction systems*, J. Chemical Physics, **48**, 4144-4147 (1968).

[34] F. Horn and R. Jackson, *General Mass Action Kinetics*, Archive for Rational Mechanics and Analysis, **49**, 81 (1972).

[35] F. Horn, *Necessary and Sufficient Conditions for Complex Balancing in Chemical Kinetics*, Archive for Rational Mechanics and Analysis **49**, 172-186 (1972-73).

[36] B. L. Clarke, *Stability of Complex Reaction Networks*, Advances in Chemical Physics, **43**, 1-216 (1980).

[37] L. Rondoni, *Long Time Behaviour of "X-led" Reactions*, Journal of Mathematical Physics, **35**, 1778-1795 (1994).

[38] L. Rondoni, *A Stochastic Treatment of Reaction and Diffusion*, Ph.D. thesis, Virginia Tech., (1991). L. Rondoni, *Complex Chemical Reactions: a Probabilistic Approach*, Nuclear Science and Engineering, **112**, 392-405, 1992. See also: D. A. McQuarrie, *A stochastic Approach to Chemical Kinetics*, Supplement Review Series in Applied Probability, **8**, Methuen, 1967; I. Oppenheim, K. E. Shuler, and G. H. Weiss, *Stochastic Processes in Chemical Physics: the Master Equation*, MIT Press, 1977.

[39] R. F. Streater, *Convergence of the Iterated Boltzmann Map*, Publications of the Research Institute in Mathematical Sciences, Kyoto, **20**, 913-927 (1984).

[40] R. A. Fisher, Proceedings of the Cambridge Philosophical Society, **22**, 700 (1925).

[41] S-I. Amari, *Differential Geometric Methods in Statistics*, Lecture Notes in Statistics, **28**, Springer-Verlag, New York, 1985.

[42] R. Haag, N. M. Hugenholtz and M. Winnink, *On the Equilibrium States in Quantum Statistical Mechanics*, Communications in Mathematical Physics **5**, 215-236 (1967).

[43] R. F. Streater, *Detailed Balance and Free Energy*, Proceedings of the Conference on Quantum Information Theory and Open Systems, Kyoto, 1993, (Ed. M. Ohya et al.), RIMS Kokyuroku Series, Kyoto.

[44] R. F . Streater, *Entropy and KMS Norms*, pp. 405-413, in "Stochastic Analysis and Applications in Physics", Edited by A. I. Cardoso, M. de Faria, J. Potthoff, R. Sénéor, and L. Streit, Kluwer, 1994.

[45] R. F. Streater, *Statistical Dynamics with Thermal Noise*, Proc. Edinburgh Conference on Stochastic Processes, Edited by A. Etheridge, Cambridge University Press, 1995.

[46] R. F. Streater, *Statistical Dynamics with Sources and Sinks*, pp 15-20 in: Quantum Chemistry and Technology in the Mesoscopic Level, Ed. H. Hasagawa, Physical Society of Japan, 1994.

[47] R. F. Streater, *Heat Conduction in a Fermionic Crystal*, Journal of Physics A: Math. Gen. **26**, 1553-1557 (1993).

[48] J. F. C. Kingman, Journal of Applied Probability, **6**, 1 (1969).

[49] E. B. Davies, *A Model of Heat Conduction*, Journal of Statistical Physics, **18**, 161-170 (1978).

[50] L. Rondoni and R. F. Streater, *The Statistical Dynamics of the Brussellator*, Open Systems and Information Dynamics, **2**, 175-194, Torun (1994).

[51] L. Rondoni and R. F. Streater, *The Brussellator with Thermal Noise*, Proceedings of the XXVI Symposium on Mathematical Physics, Torun, 1993. pp 59-64, Ed. by A. Jamiolkowski, W.A. Majewski, M. Michalski, and R. Mrugala, Nicolas Copernicus University Press, 1994.

[52] O. Platt, *Ordinary Differential Equations*, Holden-Day, San Francisco, 1971.

[53] M. J. Freidlin and A. D. Wentzell, *Random Perturbations of a Dynamical System*, Springer-Verlag, 1984.

[54] Y. Kifer, *Random Perturbations of Dynamical Systems*, Birkhaüser, Boston, 1988.

[55] D. Amit, *Modelling Brain Function*, Cambridge University Press, 1989.

[56] E. B. Davies, *Quantum Theory of Open Systems*, Academic Press, 1976.

[57] I. E. Segal, *Mathematical Problems of Relativistic Physics*, American Mathematical Society, Providence, 1963.

[58] O. Bratteli and D. W. Robinson, *Operator Algebras and Quantum Statistical Mechanics*, Vols. I and II, Springer-Verlag, (1979, 1981).

[59] A. Einstein, P. Podolsky, and N. Rosen, *Can quantum-mechanical description of reality be considered complete?* Physical Review, **47**, 777-780 (1935).

[60] R. F. Streater, *Convergence of the Quantum Boltzmann Map*, Communications in Mathematical Physics, **98**, 177-185 (1985).

[61] D. Ruelle, *Statistical Mechanics: Rigorous Results*, W. A. Benjamin, New York, 1969.

[62] M. Ohya and D. Petz, *Quantum Entropy and its Uses*, Springer-Verlag, 1994.

[63] R. F. Streater, *Spontaneous Breakdown of Symmetry in Axiomatic Theory*, Proc. Royal Society, **278 A**, 510-518, (1965).

[64] R. F. Streater, *Symmetry Groups and Non-abelian Cohomology*, Communications in Mathematical Physics, **132**, 201-215 (1990).

[65] E. Størmer, *Positive Linear Maps of Operator Algebras*, Acta Mathematica, **110**, 233-278 (1963). See also: E. Størmer, *Positive Linear Maps of C^*-algebras*, Lecture Notes in Physics, **29**, Springer-Verlag, 1974.

[66] S. L. Tregub, Soviet Math. **30**, 105 (1986).

[67] L. J. Landau and R. F. Streater, *On Birkhoff's Theorem for Doubly Stochastic Completely Positive Maps of Matrix Algebras*, Linear Algebra and its Applications, **193**, 107-127 (1993).

[68] W. F. Stinespring, *Positive Functions on C^*-algebras*, Proceedings of the American Mathematical Society, **6**, 211-216 (1955). If you can find it, see also the beautiful pamphlet, *Dilations of Irreversible Evolutions in Algebraic Quantum Theory*, Communications of the Dublin Institute for Advanced Studies, Series A, No. 24 (1977).

[69] K. Kraus, *General State Changes in Quantum Theory*, Annals of Physics, **64**, 311-335 (1971). See also: K. Kraus, *States, Effects and Operations*, Lecture Notes in Physics, **190**, Springer-Verlag, Berlin, 1983.

[70] B. Kümmerer and H. Maassen, *The Essentially Commutative Dilations of Dynamical Semigroups on* $\mathbf{M}_n$, Communications in Mathematical Physics, **109**, 1, (1987).

[71] M. D. Choi, *Positive Linear Maps on C^*-algebras*, Canadian Journal of Mathematics, **24**, 520-529 (1972). M. D. Choi, *Completely Positive Linear Maps on Complex Matrices*, Linear Algebra and Applications, **10**, 285-290, (1975).

[72] R. Alicki and J. Messer, *Nonlinear Quantum Dynamical Semigroups for Many-body Open Systems*, Journal of Statistical Physics, **32**, 299-312, (1983). See also R. Alicki and K. Lendl, *Quantum Dynamical Semigroups and Applications*, Lecture Notes in Physics, **286**, Springer, 1987.

[73] N. S. Krylov, *Works on the Foundation of Statistical Physics*, Princeton University Press, 1979.

[74] I. E. Segal, *Foundations of the Theory of Infinite-dimensional Systems, II*, Canadian Journal Of Mathematics, **13**, 1-18 (1961).

[75] H. Araki, *Hamiltonian Formalism in Quantum Field Theory*, Journal of Mathematical Physics, **1**, 492 (1960).

[76] D. W. Robinson, *The Ground State of the Bose Gas*, Communications in Mathematical Physics, **1**, 159-174 (1965).

[77] D. W. Robinson, *A Theorem Concerning the Positive Metric*, Communications in Mathematical Physics, **1**, 89-94 (1965).

[78] H. Araki and E. J. Woods, *Representations of the Canonical Commutation Relations Describing a Nonrelativistic Free Bose Gas*, Journal of Mathematical Physics, **4**, 637-662 (1963).

[79] J. Manuceau and A. Verbeure, *Quasifree states of the CCR Algebra and Bogoliubov Transformations*, Communications in Mathematical Physics, **9**, 293-302 (1968).

[80] M. Moshinsky and P. Winternitz, *Quadratic Hamiltonians in Phase Space and their Eigenstates*, Journal of Mathematical Physics, **21**, 1667-1681 (1980).

[81] E. H. Wichmann, *Density Matrices arising from Incomplete Measurements*, Journal of Mathematical Physics, **4**, 884-896 (1963).

[82] R. F. Streater, *Entropy and the Central Limit Theorem in Quantum Mechanics*, Journal of Physics **A 20**, 4321-4330 (1987).

[83] O. E. Lanford and D. W. Robinson, *Approach to Equilibrium of Free Quantum Systems*, Communications in Mathematical Physics, **24**, 193-219 (1972).

[84] A. Katz, *Principles of Statistical Mechanics; The Information Theory Approach*, W. H. Freeman and Company, 1967.

[85] P. Jordan and E. P. Wigner, *Über das Paulische Äquivalenzverbot*, Zur Physik, **47**, 631-651 (1928)

[86] E. Balslev and A. Verbeure, *States on Clifford Algebras*, Communications in Mathematical Physics, **7**, 55-76 (1968).

[87] D. Shale and W. Stinespring, *States on the Clifford Algebra*, Annals of Mathematics, **80**, 365 (1964).

[88] R. F. Streater and A. S. Wightman, *PCT, Spin and Statistics and All That*, Benjamin-Cummings, 1964. Student edition, Addison Wesley, 1989.

[89] E. Balslev, J. Manuceau and A. Verbeure, *Representations of Anticommutation Relations and Bogoliubov Transformations*, Communications in Mathematical Physics, **8**, 313-326 (1968).

[90] H. Araki and W. Wyss, *Representations of Canonical Anticommutation Relations*, Helvetica Physica Acta, **37**, 136-159 (1964).

[91] E. A. Uehling and G. E. Uhlenbeck, *Transport Phenomena in Bose-Einstein and Fermi-Dirac Gas*, Physical Review, **43**, 552-561 (1933). For a more modern version, see: R. Balescu, *Equilibrium and Nonequilibrium Statistical Mechanics*, John Wiley, 1975.

[92] A. Einstein, *Zur Quantentheorie der Strahlung*, Physikalische Zeitschrift, **18**, 121 (1917). Translated into English in: B. L. van der Waerden, *Sources of Quantum Mechanics*, Dover, New York, 1969, pp. 28-32.

[93] L. van Hove, *Quantum Mechanical Perturbations giving rise to a Statistical Transport Equation*, Physica, **21**, 517-540 (1955).

[94] T. G. Ho, L. J. Landau and A. J. Wilkins, *On the Weak Coupling Limit of a Fermi Gas in a Random Potential*, Reviews in Mathematical Physics, **5**, 209-298 (1993).

[95] L. J. Landau, *Observation of Quantum Particles on a Large Space-time Scale*, Journal of Statistical Physics, **77**, 259-310 (1994).

[96] G. Kiegerl and F. Schürrer, *Energy Conservation and H-Theorem in the Scalar Non-linear Equation and its Multigroup Representation*, Physics Letters A, **148**, 158-163 (1990).

[97] T. Kato, *Perturbation Theory for Linear Operators*, Springer-Verlag, 1966.

[98] R. Haag, *On Quantum Field Theory*, Mat.-Fys. Medd. Danske Videns. Selskapet, **29**, No.12, 1-37 (1955).

[99] V. Gorini, A. Kossakowski and E. C. G. Sudarshan, *Completely Positive Dynamical Semigroups of N-level Systems*, Journal of Mathematical Physics, **17**, 821-825 (1976).

[100] G. Lindblad, *Completely Positive Maps and Entropy Inequalities*, Communications in Mathematical Physics, **40**, 147-151 (1975).

[101] W. A. Majewski and R. F. Streater, *Detailed Balance and Quantum Dynamical Semigroups*, Preprint, G'dansk University, 1994.

[102] W. A. Majewski, *The Detailed Balance Condition in Quantum Statistical Mechanics*, Journal of Mathematical Physics, **25**, 614- (1984). See also: W. A. Majewski, *Dynamical Semigroups in the Algebraic Formulation of Statistical Mechanics*, Fortschritte der Physik, **32**, 89 (1984).

[103] H. Spohn and J. L. Lebowitz, *Irreversible Thermodynamics for Quantum Systems Weakly Coupled to Thermal Reservoirs*, Advances in Chemical Physics, **38**, 127 (1977).

[104] H. Fröhlich, *Long-range Correlations and Energy Storage in Biological Systems*, International Journal of Quantum Chemistry, **2**, 641-649 (1968).

[105] G. Sewell, *Quantum Theory of Collective Phenomena*, Oxford University Press, 1986.

[106] N. G. Duffield, *The Continuum Limit of Dissipative Dynamics in H. Fröhlich's Pumped Phonon System*, Helvetica Physica Acta, **61**, 363-378 (1988).

[107] R. E. Peierls, *On Ising's Model of Ferromagnetism*, Proceedings of the Cambridge Philosophical Society, **32**, 477-481 (1936).

[108] R. B. Griffiths, *Peierls' Proof of Spontaneous Magnetization in a Two-dimensional Ising Ferromagnet*, Physical Review A, **136**, 437-439 (1964). See also: R. B. Griffiths, *Rigorous Results and Theorems*, in: Phase Transitions and Critical Phenomena, Editors: C. Domb and M. S. Green, Vol. 1, Academic Press, 1972.

[109] R. F. Streater, *The Heisenberg Ferromagnet as a Quantum Field Theory*, Communications in Mathematical Physics, **6**, 233-247 (1967).

[110] R. Kubo, *Statistical Mechanical Theory of Irreversible Processes*, Journal of the Physical Society of Japan **12**, 570-586 (1957).

[111] P. C. Martin and J. Schwinger, *Theory of many-particle systems: I*, Physical Review **115**, 1352-1373 (1959).

[112] D. W. Robinson, *The Statistical Mechanics of Quantum Spin Systems*, Communications in Mathematical Physics **7**, 337-348 (1968).

[113] G. Lindblad, On the Generators of Quantum Dynamical Semigroups, Communications in Mathematical Physics, **48**, 119-130 (1976).

[114] W. Pusz and S. L. Woronowicz, *Passive States and KMS States for General Quantum Systems*, Communications in Mathematical Physics, **58**, 273-290 (1978).

[115] C. Connaughton and R. F. Streater, *Dissipative Dynamics of the Ising Model*, to appear.

Index